BASIC MECHANICAL ENGINEERING

By the Same Author :

- *Thermal Engineering*
- *Engineering Thermodynamics*
- *I.C. Engines*
- *Automobile Engineering*
- *Power Plant Engineering*
- *Manufacturing Technology*
  (Manufacturing Processes)

BASIC MECHANICAL ENGINEERING

For
B.E. 1st Year (CMELRPTA 107)

Mahatma Gandhi University, Kerala

By
R.K. RAJPUT
M.E. (Heat Power Engg.) Hons.—Gold Medalist ; Grad. (Mech. Engg. & Elect. Engg.) ;
M.I.E. (India) ; M.S.E.S.I. ; M.I.S.T.E. ; C.E. (India)
PATIALA (Punjab)

LAXMI PUBLICATIONS (P) LTD

BANGALORE • CHENNAI • COCHIN • GUWAHATI • HYDERABAD
JALANDHAR • KOLKATA • LUCKNOW • MUMBAI • RANCHI
NEW DELHI • BOSTON, USA

Published by :
LAXMI PUBLICATIONS (P) LTD
113, Golden House, Daryaganj,
New Delhi-110002

Phone : 011-43 53 25 00
Fax : 011-43 53 25 28

www.laxmipublications.com
info@laxmipublications.com

© *All rights reserved with the Publishers. No part of this publication may be reproduced, stored in a retrieval system, or transmitted in any form or by any means, electronic, mechanical, photocopying, recording or otherwise without the prior written permission of the publisher.*

Price : Rs. 125.00 Only. First Edition : 2008

OFFICES

India		**USA**
ⓒ Bangalore	080-26 61 15 61	**Boston**
ⓒ Chennai	044-24 34 47 26	11, Leavitt Street, Hingham,
ⓒ Cochin	0484-239 70 04	MA 02043, USA
ⓒ Guwahati	0361-254 36 69, 251 38 81	
ⓒ Hyderabad	040-24 65 23 33	
ⓒ Jalandhar	0181-222 12 72	
ⓒ Kolkata	033-22 27 43 84	
ⓒ Lucknow	0522-220 95 78	
ⓒ Mumbai	022-24 91 54 15, 24 92 78 69	
ⓒ Ranchi	0651-221 47 64	

EME-0751-125-BASIC MECHANICAL ENGG (KE) C—15979/08/06
Typeset at : Goswami Associates, Delhi. *Printed at* : Mehra Offset Printer, Delhi.

CONTENTS

Chapters *Pages*

Syllabus (xv)

Module–1

1. **BASIC CONCEPTS OF THERMODYNAMICS** 3—31
   1.1. Definition of Thermodynamics 3
   1.2. Thermodynamic Systems 3
      1.2.1. System, boundary and surroundings 3
      1.2.2. Closed system 4
      1.2.3. Open system 4
      1.2.4. Isolated system 4
      1.2.5. Adiabatic system 5
      1.2.6. Homogeneous system 5
      1.2.7. Heterogeneous system 5
   1.3. Macroscopic and Microscopic Points of View 5
   1.4. Pure Substance 6
   1.5. Thermodynamic Equilibrium 6
   1.6. Properties of Systems 6
   1.7. State 6
   1.8. Process 7
   1.9. Cycle 7
   1.10. Point Function 7
   1.11. Path Function 8
   1.12. Temperature 8
   1.13. Zeroth Law of Thermodynamics 8
   1.14. Ideal Gas 9
   1.15. Pressure 9
      1.15.1. Definition of pressure 9
      1.15.2. Unit for pressure 10
      1.15.3. Types of pressure measurement devices 10
      1.15.4. Mechanical-type instruments 10
   1.16. Specific Volume 11
   1.17. Reversible and Irreversible Processes 14
   1.18. Energy, Work and Heat 15
      1.18.1. Energy 15
      1.18.2. Work and heat 16
   1.19. Reversible Work 17
   Highlights 27
   Objective Type Questions 28

(v)

Chapters		Pages

| | *Theoretical Questions* | 30 |
| | *Unsolved Examples* | 30 |

2. FIRST LAW OF THERMODYNAMICS — 32—80

2.1.	Internal Energy	32
2.2.	Law of Conservation of Energy	32
2.3.	First Law of Thermodynamics	32
2.4.	Application of First Law to a Process	34
2.5.	Energy—a Property of System	34
2.6.	Perpetual Motion Machine of the First Kind—PMM1	35
2.7.	Energy of an Isolated System	36
2.8.	The Perfect Gas	36
2.8.1.	The characteristic equation of state	36
2.8.2.	Specific heats	37
2.8.3.	Joule's law	38
2.8.4.	Relationship between two specific heats	38
2.8.5.	Enthalpy	39
2.8.6.	Ratio of specific heats	40
2.9.	Application of First Law of Thermodynamics to Non-flow or Closed System	40
	Additional Typical Worked Examples	70
	Highlights	74
	Objective Type Questions	75
	Theoretical Questions	77
	Unsolved Examples	78

3. GAS POWER CYCLES — 81—106

3.1.	Definition of a Cycle	81
3.2.	Air Standard Efficiency	81
3.3.	The Carnot Cycle	82
3.4.	Constant Volume or Otto Cycle	87
3.5.	Constant Pressure or Diesel Cycle	95
	Highlights	103
	Objective Type Questions	104
	Theoretical Questions	105
	Unsolved Examples	105

Module–2

4. INTERNAL COMBUSTION ENGINES — 109—167

4.1.	Heat Engines	109
4.2.	Development of I.C. Engines	110
4.3.	Classification of I.C. Engines	110
4.4.	Applications of I.C. Engines	111
4.5.	Basic Idea of I.C. Engines	111

Chapters		Pages
4.6.	Different Parts of I.C. Engines	112
4.7.	Terms Connected with I.C. Engines	136
4.8.	Working Cycles	137
4.9.	Indicator Diagram	138
4.10.	Four-Stroke Cycle Engines	138
4.11.	Two-Stroke Cycle Engines	144
4.12.	Comparison of Four-Stroke and Two-Stroke Cycle Engines	146
4.13.	Comparison of Spark Ignition (S.I.) and Combustion Ignition (C.I.) Engines	147
4.14.	Comparison Between a Petrol Engine and a Diesel Engine	148
4.15.	How to Tell a Two-Stroke Cycle Engine from a Four-Stroke Cycle Engine ?	149
4.16.	Ignition System (Petrol Engines)	149
4.17.	Cooling Systems	153
4.18.	Lubrication Systems	159
	Highlights	164
	Objective Type Questions	164
	Theoretical Questions	167

5. REFRIGERATION AND AIR-CONDITIONING — 168—193

			Pages
5.1.	Fundamentals of Refrigeration		168
	5.1.1.	Introduction	168
	5.1.2.	Elements of refrigeration systems	169
	5.1.3.	Refrigeration systems	169
	5.1.4.	Co-efficient of performance (C.O.P.)	169
	5.1.5.	Standard rating of a refrigeration machine	170
5.2.	Simple Vapour Compression System		170
	5.2.1.	Introduction	170
	5.2.2.	Simple vapour compression cycle	170
	5.2.3.	Functions of parts of a simple vapour compression system	171
5.3.	Domestic Refrigerator		171
5.4.	Vapour Absorption System		173
	5.4.1.	Introduction	173
	5.4.2.	Simple vapour absorption system	173
	5.4.3.	Practical vapour absorption system	174
	5.4.4.	Comparison between vapour compression and vapour absorption systems	175
5.5.	Refrigerants		175
	5.5.1.	Classification of refrigerants	176
	5.5.2.	Desirable properties of an ideal refrigerant	177
	5.5.3.	Properties and uses of commonly used refrigerants	179
5.6.	Air-Conditioning		182
	5.6.1.	Introduction	182
	5.6.2.	Air-conditioning systems	182
	5.6.3.	Applications of air-conditioning	189
	Highlights		189
	Objective Type Questions		191
	Theoretical Questions		192

Module–3

6. POWER TRANSMISSION — 197—229

- 6.1. Introduction ... 197
- 6.2. Belts and Belt Drives ... 197
  - 6.2.1. Flat belts ... 197
  - 6.2.2. V-belts ... 198
  - 6.2.3. Round belts ... 199
  - 6.2.4. Belt drive ... 199
  - 6.2.5. Applications of belt drives ... 199
  - 6.2.6. Velocity ratio of belt drive ... 200
  - 6.2.7. Length of belt ... 202
  - 6.2.8. Power transmitted by a belt ... 205
  - 6.2.9. Ratio of tensions ... 205
  - 6.2.10. Centrifugal tension ... 208
  - 6.2.11. Condition for transmission of maximum (absolute) power ... 211
  - 6.2.12. Initial tension ... 212
  - 6.2.13. V-belt and rope drive ... 214
- 6.3. Chains and Chain Drives ... 216
  - 6.3.1. Roller chain drive ... 216
  - 6.3.2. Silent chain drive ... 217
  - 6.3.3. Advantages and disadvantages of chain drive ... 217
- 6.4 Gear Drive ... 218
  - 6.4.1. Introduction ... 218
  - 6.4.2. Advantages and disadvantages of toothed gearing ... 218
  - 6.4.3. Definitions ... 218
  - 6.4.4. Types of gears ... 220
  - 6.4.5. Types of gear trains ... 221
  - 6.4.6. Simple gear train ... 221
  - 6.4.7. Compound gear train ... 222
  - 6.4.8. Epicyclic (or planetary) gear train ... 224
- Highlights ... 225
- Objective Type Questions ... 226
- Theoretical Questions ... 228
- Unsolved Examples ... 228

Module–4

7. POWER PLANTS, NON-CONVENTIONAL ENERGY SOURCES, HYDRAULIC AND STEAM TURBINES — 233—280

- 7.1. Power Plants ... 233
  - 7.1.1. Sources of energy ... 233
  - 7.1.2. Types of power plants ... 233
  - 7.1.3. Steam power plant ... 233
  - 7.1.4. Hydro-electric power plant ... 235

Chapters			Pages
	7.1.5. Nuclear power plant	...	239
	7.1.6. Simple gas turbine plant	...	239
	7.1.7. Diesel engine power plant	...	240
7.2.	Non-Conventional Energy Sources	...	241
	7.2.1. Introduction	...	241
	7.2.2. Description of non-conventional sources of energy	...	242
7.3.	Hydraulic Turbines	...	253
	7.3.1. Introduction to hydraulic turbines	...	253
	7.3.2. Classification of hydraulic turbines	...	253
	7.3.3. Impulse turbines—Pelton wheel	...	255
	7.3.4. Reaction turbines	...	257
	7.3.5. Tubular or bulb turbines	...	263
	7.3.6. Run away speed	...	264
	7.3.7. Draft tube	...	265
	7.3.8. Specific speed	...	266
	7.3.9. Cavitation	...	267
	7.3.10. Selection of hydraulic turbines	...	267
7.4.	Steam Turbines	...	269
	7.4.1. Introduction to steam turbines	...	269
	7.4.2. Classification of steam turbines	...	269
	7.4.3. Advantages of steam turbine over the steam engines	...	270
	7.4.4. Description of common types of steam turbines	...	271
	7.4.5. Methods of reducing wheel or rotor speed (compounding methods)	...	273
	7.4.6. Differences between impulse and reaction turbines	...	277
	Highlights	...	277
	Objective Type Questions	...	278
	Theoretical Questions	...	279

Module-5

8. MACHINE TOOLS   283—328

8.1.	Classification of Cutting Tools	...	283
8.2.	Types of Chips	...	284
8.3.	Cutting Tool Materials	...	284
8.4.	Cutting Conditions	...	284
8.5.	Machinability	...	285
8.6.	Tool Life	...	285
8.7.	Forces of a Single Point Tool	...	286
8.8.	Machining Processes	...	286
8.9.	Machine Tools	...	287
8.10.	Lathe	...	288
	8.10.1. Introduction	...	288
	8.10.2. Parts of lathe	...	289
	8.10.3. Size and specifications of lathe	...	289
	8.10.4 Types of lathe	...	290

Chapters	Pages
8.10.5. Lathe tools	290
8.10.6. Lathe operation	291
8.10.7. Lathe accessories	299
8.10.8. Eccentric turning	301
8.10.9. Thread rolling	301
8.10.10. Cutting speed, feed and depth of cut	302
8.10.11. Testing of lathes	302
8.10.12. Exercise	303
8.11. Drilling Machines	306
8.11.1. Introduction	306
8.11.2. Specifications of a drilling machine	306
8.11.3. Operations performed	307
8.11.4. Classification of drilling machines	308
8.11.5. Cutting speeds and feeds	310
8.11.6. Work holding devices	310
8.11.7. Drill holding devices	310
8.11.8. Drilling machine tools	311
8.12. Shaping Machine (Shaper)	311
8.12.1. Introduction	311
8.12.2. Classification of shapers	311
8.12.3. Principal parts	311
8.12.4. Specifications of a shaper	312
8.12.5. Operations performed	313
8.12.6. Tools used	313
8.13. Planing Machine (Planer)	313
8.13.1. Introduction	313
8.13.2. Comparison between planer and shaper	314
8.13.3. Types of planer	314
8.13.4. Principal parts of a planer	314
8.13.5. Size of a planer	315
8.13.6. Standard clamping devices	315
8.13.7. Planer operations	316
8.14. Milling Machine	316
8.14.1. General aspects	316
8.14.2. Specifications of a milling machine	317
8.14.3. Types of milling machines	317
8.14.4. Main parts of a horizontal milling machine	318
8.14.5. Types of milling cutters	318
8.14.6. Milling operations	319
8.14.7. Cutting speed, feed and depth of cut	321
8.15. Grinding Machines	322
8.15.1. Introduction	322
8.15.2. Types of grinding machines	323
8.15.3. The grinding wheel	323
8.15.4. Abrasives	324
8.15.5. Selection of grinding wheels	324
8.15.6. Wheel shapes	324
8.15.7. Mounting of wheels	325

Chapters	Pages
8.15.8. Wheel truing	326
Highlights	326
Objective Type Questions	326
Theoretical Questions	327

9. MANUFACTURING PROCESSES 329—394

- 9.1. Moulding and Casting ... 329
  - 9.1.1. Introduction ... 329
  - 9.1.2. Mould making ... 334
  - 9.1.3. Core ... 338
  - 9.1.4. Moulding sand ... 340
  - 9.1.5. Foundry hand tools ... 344
  - 9.1.6. Melting equipment ... 349
  - 9.1.7. Casting ... 350
  - 9.1.8. Advantages of casting process ... 350
  - 9.1.9. Preparation of a casting ... 350
  - 9.1.10. Design of a casting ... 351
  - 9.1.11. Casting processes ... 355
  - 9.1.12. Defects in castings ... 359
  - 9.1.13. Cleaning of castings ... 360
  - 9.1.14. Inspection of castings ... 361
  - 9.1.15. Exercise ... 363
- 9.2. Forging ... 363
  - 9.2.1. Introduction ... 363
  - 9.2.2. Advantages and disadvantages of forging ... 364
  - 9.2.3. Applications of forging ... 364
  - 9.2.4. Classification of forging ... 364
  - 9.2.5. Hand forging ... 370
  - 9.2.6. Machine forging ... 372
  - 9.2.7. Defects in forging ... 373
  - 9.2.8. Heat treatment of forgings ... 373
  - 9.2.9. Exercise ... 375
- 9.3. Rolling ... 377
- 9.4. Welding ... 377
  - 9.4.1. Pressure welding ... 380
  - 9.4.2. Fusion welding

1. Moulding and Casting

Highlights ... 389
Objective Type Questions ... 390
Theoretical Questions ... 391

2. Forging

Highlights ... 391
Objective Type Questions ... 392
Theoretical Questions ... 392

3. Welding

Highlights ... 393
Objective Type Questions ... 393
Theoretical Questions ... 394

Chapters	Pages
8.18. Wheel truing	326
Highlights	326
Objective Type Questions	326
Theoretical Questions	327
9. MANUFACTURING PROCESSES	328—394
9.1. Moulding and Casting	329
9.1.1. Introduction	329
9.1.2. Mould making	329
9.1.3. Core	337
9.1.4. Moulding sand	338
9.1.5. Foundry hand tools	340
9.1.6. Melting equipment	341
9.1.7. Casting	349
9.1.8. Advantages of casting process	350
9.1.9. Preparation of a casting	350
9.1.10. Design of a casting	350
9.1.11. Casting processes	351
9.1.12. Defects in castings	352
9.1.13. Cleaning of casting	359
9.1.14. Inspection of castings	360
9.1.15. Exercise	361
9.2. Forging	362
9.2.1. Introduction	362
9.2.2. Advantages and disadvantages of forging	363
9.2.3. Applications of forging	364
9.2.4. Classification of forging	364
9.2.5. Hand forging	367
9.2.6. Machine forging	370
9.2.7. Defects in forging	372
9.2.8. Heat treatment of forgings	373
9.2.9. Exercise	375
9.3. Rolling	377
9.4. Welding	377
9.4.1. Pressure welding	380
9.4.2. Fusion welding	380

I. Moulding and Casting

Highlights	389
Objective Type Questions	390
Theoretical Questions	391

2. Forging

Highlights	391
Objective Type Questions	392
Theoretical Questions	392

3. Welding

Highlights	393
Objective Type Questions	393
Theoretical Questions	394

PREFACE TO THE FIRST EDITION

This book has been written specifically to meet exhaustively the requirements of the subject of *"Basic Mechanical Engineering"* (strictly according to the syllabus) for 1st year B.E. examination (CMELRPTA 107) of *Mahatma Gandhi University, Kerala.*

This book has been divided into **5 units** containing **9 chapters** in all. The book includes comprehensive treatment of the subject matter under wide range of topics mentioned in the syllabus, with a large number of solved examples to support the text. Besides this, *Highlights, Objective Type Questions, Theoretical Questions* and *Unsolved Examples* have been added at the end of each chapter to make the book a comprehensive unit in all respects.

The author's thanks are due to his wife Ramesh Rajput for extending all co-operation during preparation of the manuscript and proof reading of the book.

Although every care has been taken to remove all the mistakes and misprints yet some of them might have been remained unnoticed. The author shall feel grateful if these errors/misprints are pointed out by the readers. Constructive criticism of the book will be warmly received.

—**Author**

SYLLABUS

MAHATMA GANDHI UNIVERSITY, KERALA

BASIC MECHANICAL ENGINEERING

CMELRPTA 107 1-1-0

Module-1
Thermodynamics : Basic concepts and definitions, Gas laws, specific heat—Universal gas constant—Isothermal, adiabatic and polytropic processes, work done and heat transferred : Carnot, Otto and Diesel Cycles—air standard efficiency.

Module-2
I.C. Engines : Working of two stroke and four stroke engines—petrol and diesel engines—fuel systems, injector and carburetor—ignition systems—lubrication and cooling systems.

Refrigeration and air-conditioning : methods of refrigeration—vapour compression and vapour absorption systems—block diagrams and general descriptions—winter and summer air-conditioning systems—general description.

Module-3
Power transmission : Methods of transmission—belt, rope, chain and gear drives. Fields of application, calculation of length of belt—expression for ratio of belt tension. Velocity ratio and slip—simple problems—velocity ratio and choice of gear wheels—simple problems.

Module-4
Power plants : General layout of hydraulic, diesel, thermal and nuclear power plants, non-conventional energy sources, general description only.

Types of hydraulic turbines—selection of turbines depending upon head, discharge and specific speed—steam turbines—reaction and impulse turbines—compounding methods.

Module-5
Simple description of general purpose machines like lathe, shaping machines, drilling machine, milling machine and grinding machine.

Manufacturing process : moulding and casting, forging, rolling, welding—arc welding—gas welding (simple descriptions only).

MODULE – 1

Chapters :

1. **Basic Concepts of Thermodynamics**
2. **First Law of Thermodynamics**
3. **Gas Power Cycles**

MODULE - 1

Chapter
1. Basic Concepts of Thermodynamics
2. First Law of Thermodynamics
3. Gas Power Cycles

1

Basic Concepts of Thermodynamics

1.1. Definition of thermodynamics. 1.2. Thermodynamic systems—System, boundary and surroundings—Closed system—open system—Isolated system—Adiabatic system—Homogeneous system—Heterogeneous system. 1.3. Macroscopic and microscopic points of view. 1.4. Pure substance. 1.5. Thermodynamic equilibrium. 1.6. Properties of systems. 1.7. State. 1.8. Process. 1.9. Cycle. 1.10. Point function. 1.11. Path function. 1.12. Temperature. 1.13. Zeroth law of thermodynamics. 1.14. Ideal gas. 1.15. Pressure—Definition of pressure—Unit for pressure—Types of pressure measurement devices—Mechanical-type instruments. 1.16. Specific volume. 1.17. Reversible and irreversible processes. 1.18. Energy, work and heat—Energy—Work and heat. 1.19. Reversible work—Highlights—Objective Type Questions—Theoretical Questions— Unsolved Examples.

1.1. DEFINITION OF THERMODYNAMICS

Thermodynamics may be *defined* as follows :

- *Thermodynamics is an axiomatic science which deals with the relations among heat, work and properties of system which are in equilibrium. It describes state and changes in state of physical systems.*

Or

Thermodynamics is *the science of the regularities governing processes of energy conversion.*

Or

Thermodynamics is *the science that deals with the interaction between energy and material systems.*

Thermodynamics, basically entails *four laws* or axioms known as Zeroth, First, Second and Third law of thermodynamics.

- The *First law* throws light on *concept of internal energy.*
- The *Zeroth law* deals with *thermal equilibrium* and establishes a *concept of temperature.*
- The *Second law* indicates the limit of *converting heat into work* and introduces the *principle of increase of entropy.*
- The *Third law* defines the *absolute zero of entropy.*

These laws are based on experimental observations and have no *mathematical proof.* Like all physical laws, these laws are based on *logical reasoning.*

1.2. THERMODYNAMIC SYSTEMS

1.2.1. System, Boundary and Surroundings

System. A system is *a finite quantity of matter or a prescribed region of space* (Refer to Fig. 1.1)

Boundary. The *actual or hypothetical envelope enclosing the system* is the boundary of the system. The boundary may be fixed or it may move, as and when a system containing a gas is compressed or expanded. The boundary may be *real* or *imaginary*. It is not difficult to envisage a real boundary but an example of imaginary boundary would be one drawn around a system consisting of the fresh mixture about to enter the cylinder of an I.C. engine together with the remanants of the last cylinder charge after the exhaust process (Refer to Fig. 1.2).

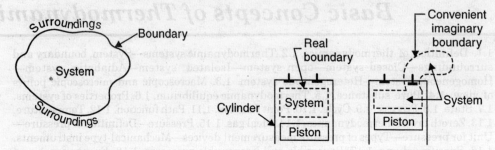

Fig. 1.1. The system.   Fig. 1.2. The real and imaginary boundaries.

1.2.2. Closed System

Refer to Fig. 1.3. *If the boundary of the system is impervious to the flow of matter, it is called a **closed system**.* An *example* of this system is mass of gas or vapour contained in an engine cylinder, the boundary of which is drawn by the cylinder walls, the cylinder head and piston crown. Here the *boundary is continuous and no matter may enter or leave*.

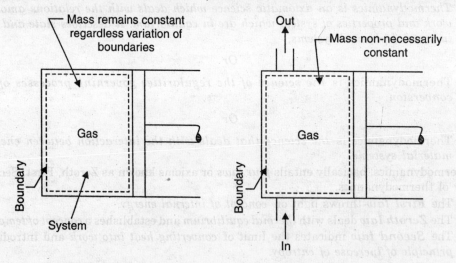

Fig. 1.3. Closed system.   Fig. 1.4. Open system.

1.2.3. Open System

Refer to Fig. 1.4. An open system is one in which *matter flows into or out of the system*. Most of the engineering systems are open.

1.2.4. Isolated System

An isolated system is that system *which exchanges neither energy nor matter with any other system or with environment*.

BASIC CONCEPTS OF THERMODYNAMICS

1.2.5. Adiabatic System

An adiabatic system is one *which is thermally insulated from its surroundings*. It can, however, *exchange work with its surroundings*. If it does not, it becomes an isolated system.

Phase. A phase is a quantity of matter which is homogeneous throughout in chemical composition and physical structure.

1.2.6. Homogeneous System

A system which consists of a single phase is termed as *homogeneous system*. Examples : Mixture of air and water vapour, water plus nitric acid and octane plus heptane.

1.2.7. Heterogeneous System

A system which consists of two or more phases is called a *heterogeneous system*. Examples : Water plus steam, ice plus water and water plus oil.

1.3. MACROSCOPIC AND MICROSCOPIC POINTS OF VIEW

Thermodynamic studies are undertaken by the following two different approaches.
1. Macroscopic approach—(*Macro* mean *big* or *total*)
2. Microscopic approach—(*Micro* means *small*)

These approaches are discussed (in a comparative way) below :

S. No.	Macroscopic approach	Microscopic approach
1.	In this approach a certain quantity of matter is considered *without* taking into account the events occurring at molecular level. In other words this approach to thermodynamics is concerned with *gross or overall behaviour*. This is known as *classical thermodynamics*.	The approach considers that the system is made up of a very large number of discrete particles known as *molecules*. These molecules have different velocities and energies. The values of these energies are constantly changing with time. This approach to thermodynamics which is concerned directly with the *structure of the matter* is known as *statistical thermodynamics*.
2.	The analysis of macroscopic system requires simple mathematical formulae.	The behaviour of the system is found by using statistical methods as the number of molecules is very large. So advanced statistical and mathematical methods are needed to explain the changes in the system.
3.	The values of the properties of the system are their average values. For example, consider a sample of a gas in a closed container. The *pressure* of the gas is the average value of the pressure exerted by millions of individual molecules. Similarly the *temperature* of this gas is the average value of translational kinetic energies of millions of individual molecules. These properties like *pressure* and *temperature* can be measured very easily. *The changes in properties can be felt by our senses.*	The properties like *velocity, momentum*, impulse, kinetic energy, force of impact etc. which describe the molecule *cannot be easily measured by instruments. Our senses cannot feel them.*
4.	In order to describe a system only a few properties are needed.	Large number of variables are needed to describe a system. So the approach is complicated.

Note. Although the macroscopic approach seems to be different from microscopic one, there exists a relation between them. Hence when both the methods are applied to a particular system, they give the same result.

1.4. PURE SUBSTANCE

A pure substance is one that has a homogeneous and invariable chemical composition even though there is a change of phase. In other words, it is a system which is (a) homogeneous in composition, (b) homogeneous in chemical aggregation. Examples : Liquid, water, mixture of liquid water and steam, mixture of ice and water. The mixture of liquid air and gaseous air is not a pure substance.

1.5. THERMODYNAMIC EQUILIBRIUM

A system is in *thermodynamic equilibrium* if the temperature and pressure at all points are same ; there should be no velocity gradient ; the chemical equilibrium is also necessary. Systems under temperature and pressure equilibrium but not under chemical equilibrium are sometimes said to be in metastable equilibrium conditions. *It is only under thermodynamic equilibrium conditions that the properties of a system can be fixed.*

Thus for attaining a state of *thermodynamic equilibrium* the following *three* types of equilibrium states must be achieved :

1. **Thermal equilibrium.** The temperature of the system does not change with time and has same value at all points of the system.

2. **Mechanical equilibrium.** There are no unbalanced forces within the system or between the surroundings. The pressure in the system is same at all points and does not change with respect to time.

3. **Chemical equilibrium.** No chemical reaction takes place in the system and the chemical composition which is same throughout the system does not vary with time.

1.6. PROPERTIES OF SYSTEMS

A property of a system is a characteristic of the system which depends upon its state, but not upon how the state is reached. There are two sorts of property :

1. **Intensive properties.** These properties *do not depend on the mass of the system.* Examples : Temperature and pressure.

2. **Extensive properties.** These properties *depend on the mass of the system. Example* : Volume. Extensive properties are often divided by mass associated with them to obtain the intensive properties. For example, if the volume of a system of mass m is V, then the specific volume of matter within the system is $\frac{V}{m} = v$ which is an intensive property.

1.7. STATE

State is the condition of the system at an instant of time as described or measured by its *properties. Or each unique condition of a system is called a* **state**.

It follows from the definition of state that each property has a single value at each state. Stated differently, all properties are *state* or *point functions*. Therefore, all properties are identical for identical states.

On the basis of the above discussion, we can determine if a given variable is *property* or not by applying the following *tests* :

— *A variable is a property, if and only if, it has a single value at each equilibrium state.*

— A variable is a property, if and only if, the change in its value between any two prescribed equilibrium states is single-valued.

Therefore, *any variable whose change is fixed by the end states is a property.*

1.8. PROCESS

A process occurs when the system undergoes a change in a state or an energy transfer at a steady state. A process may be *non-flow* in which a fixed mass within the defined boundary is undergoing a change of state. *Example* : A substance which is being heated in a closed cylinder undergoes a *non-flow process* (Fig. 1.3). *Closed systems undergo non-flow processes.* A process may be a *flow process in which mass is entering and leaving through the boundary of an open system.* In a steady flow process (Fig. 1.4) mass is crossing the boundary from surroundings at entry, and an equal mass is crossing the boundary at the exit so that the total mass of the system remains constant. In an open system it is necessary to take account of the work delivered from the surroundings to the system at entry to cause the mass to enter, and also of the work delivered from the system at surroundings to cause the mass to leave, as well as any heat or work crossing the boundary of the system.

Quasi-static process. Quasi means '*almost*'. A quasi-static process is also called a *reversible process. This process is a succession of equilibrium states and infinite slowness is its characteristic feature.*

1.9. CYCLE

Any process or series of processes whose end states are identical is termed a **cycle**. The processes through which the system has passed can be shown on a state diagram, but a complete section of the path requires in addition a statement of the heat and work crossing the boundary of the system. Fig. 1.5 shows such a cycle in which a system commencing at condition '1' changes in pressure and volume through a path 123 and returns to its initial condition '1'.

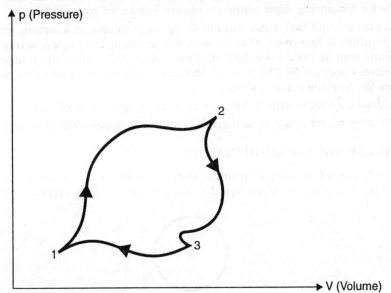

Fig. 1.5. Cycle of operations.

1.10. POINT FUNCTION

When two properties locate a point on the graph (co-ordinate axes) then those properties are called as **point function**.

Examples. Pressure, temperature, volume etc.

$$\int_1^2 dV = V_2 - V_1 \text{ (an } exact\ differential\text{)}.$$

1.11. PATH FUNCTION

There are certain quantities which cannot be located on a graph by a *point* but are given by the *area* or so, on that graph. In that case, the area on the graph, pertaining to the particular process, *is a function of the path of the process*. Such quantities are called **path functions**.

Examples. Heat, work etc.

Heat and work are *inexact differentials*. Their change cannot be written as difference between their end states.

Thus $\int_1^2 \delta Q \neq Q_2 - Q_1$, and is shown as $_1Q_2$ or Q_{1-2}

Similarly $\int_1^2 \delta W \neq W_2 - W_1$, and is shown as $_1W_2$ or W_{1-2}

Note. The operator δ is used to denote inexact differentials and operator d is used to denote exact differentials.

1.12. TEMPERATURE

- *The temperature is a thermal state of a body which distinguishes a hot body from a cold body.* The temperature of a body is *proportional to the stored molecular energy i.e.,* the average molecular kinetic energy of the molecules in a system. (A particular molecule does not have a temperature, it has energy. The gas as a system has temperature).
- Instruments for measuring *ordinary temperatures* are known as **thermometers** and those for measuring *high temperatures* are known as **pyrometers**.
- It has been found that a gas will not occupy any volume at a certain temperature. This temperature is known as *absolute zero temperature*. The temperatures measured with absolute zero as basis are called *absolute temperatures*. Absolute temperature is stated in degrees centigrade. The point of absolute temperature is found to occur at 273.15°C below the freezing point of water.

Then : Absolute temperature = Thermometer reading in °C + 273.15.

Absolute temperature is degree centigrade is known as degrees kelvin, denoted by K (SI unit).

1.13. ZEROTH LAW OF THERMODYNAMICS

- *'Zeroth law of thermodynamics'* states that if two systems are each equal in temperature to a third, they are equal in temperature to each other.

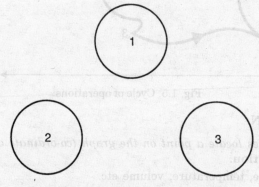

Fig. 1.6. Zeroth law of thermodynamics.

BASIC CONCEPTS OF THERMODYNAMICS

Example. Refer to Fig. 1.6. System '1' may consist of a mass of gas enclosed in a rigid vessel fitted with a pressure gauge. If there is no change of pressure when this system is brought into contact with system '2' a block of iron, then the two systems are equal in temperature (assuming that the systems 1 and 2 do not react each other chemically or electrically). Experiment reveals that if system '1' is brought into contact with a third system '3' again with no change of properties then systems '2' and '3' will show no change in their properties when brought into contact provided they do not react with each other chemically or electrically. Therefore, '2' and '3' must be in equilibrium.

- This law was enunciated by R.H. Fowler in the year 1931. However, since the first and second laws already existed at that time, it was designated as *zeroth law* so that it *precedes* the first and second laws *to form a logical sequence.*

1.14. IDEAL GAS

From experimental observations it has been established that an ideal gas (to a good approximation) behaves according to the simple equation

$$pV = mRT \qquad \ldots(1.1)$$

where p, V and T are the pressure, volume and temperature of gas having mass m and R is a constant for the gas known as its **gas constant**.

Eqn. (1.1) can be written as ;

$$pv = RT \qquad \ldots(1.2)$$

(where $v = V/m$)

In reality there is no gas which can be qualified as an ideal or perfect gas. However *all gases tend to ideal or perfect gas behaviour at all temperatures as their pressure approaches **zero pressure.***

For two states of the gas, Eqn. (1.1) can be written as ;

$$\frac{p_1 V_1}{T_1} = \frac{p_2 V_2}{T_2}$$

or

$$\frac{T_2}{T_1} = \frac{p_2}{p_1} \times \frac{V_2}{V_1} \qquad \ldots(1.3)$$

With the help of this eqn. (1.3), the temperatures can be measured or compared.

1.15. PRESSURE

1.15.1. Definition of Pressure

Pressure is defined as a *force per unit area*. Pressures are exerted by gases, vapours and liquids. The instruments that we generally use, however, record pressure as the difference between two pressures. Thus, it is the *difference between the pressure exerted by a fluid of interest and the ambient atmospheric pressure.* Such devices indicate the pressure either above or below that of the atmosphere. When it is *above the atmospheric pressure*, it is termed *gauge pressure* and is *positive*. When it is *below* atmospheric, it is *negative* and is known as *vacuum*. Vacuum readings are given in millimetres of mercury or millimetres of water below the atmosphere.

It is necessary to establish an absolute pressure scale which is independent of the changes in atmospheric pressure. A pressure of absolute zero can exist only in complete vacuum. *Any pressure measured above the absolute zero of pressure* is termed an '*absolute pressure*'.

A schematic diagram showing the *gauge pressure, vacuum pressure* and the *absolute pressure* is given in Fig. 1.7.

Mathematically :

(i) Absolute pressure = Atmospheric pressure + Gauge pressure

$$p_{abs.} = p_{atm.} + p_{gauge}.$$

(ii) Vacuum pressure = Atmospheric pressure – Absolute pressure.

Vacuum is defined as the *absence of pressure*. A *perfect vacuum* is obtained when *absolute pressure is zero*, at this instant *molecular momentum is zero*.

Atmospheric pressure is measured with the help of barometer.

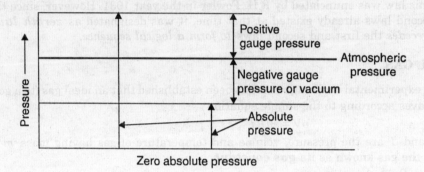

Fig. 1.7. Schematic diagram showing gauge, vacuum and absolute pressures.

1.15.2. Unit for Pressure

The fundamental SI unit of pressure is N/m^2 (sometimes called *pascal*, Pa) or bar. 1 bar = 10^5 N/m^2 = 10^5 Pa. Standard atmospheric pressure = 1.01325 bar = 0.76 m (or 760 mm) Hg.

Low pressures are often expressed in terms of mm of water or mm of mercury. This is an abbreviated way of saying that the pressure is such that which will support a liquid column of stated height.

1.15.3. Types of Pressure Measurement Devices

The pressure may be measured by means of indicating gauges or recorders. These instruments may be mechanical, electro-mechanical, electrical or electronic in operation.

1. **Mechanical instruments.** These instruments may be classified into following two groups :

— The *first group* includes those instruments in which the *pressure* measurement is made by *balancing an unknown force with a known force*.

— The *second group* includes those employing *quantitative deformation of an elastic member for pressure measurement*.

2. **Electro-mechanical instruments.** These instruments usually *employ a mechanical means for detecting the pressure and electrical means for indicating or recording the detected pressure.*

3. **Electronic instruments.** Electronic pressure measuring instruments normally depend on some physical change that can be detected and indicated or recorded electronically.

1.15.4. Mechanical-type Instruments

The mechanical-type instruments are classified as follows :

1. **Manometer gauges**

   (i) U-tube manometer   (ii) Cistern manometer

   (iii) Micro-manometer etc.

BASIC CONCEPTS OF THERMODYNAMICS

2. **Pressure gauges**
   (i) Bourdon tube pressure gauge (ii) Diaphragm gauge
   (iii) Vacuum gauge.

1.16. SPECIFIC VOLUME

The *specific volume* of a system is the volume occupied by the unit mass of the system. The symbol used is v and units are ; for example, m^3/kg. The symbol V will be used for volume. (Note that specific volume is *reciprocal of density*).

Example 1.1. *Convert the following readings of pressure to kPa assuming that barometer reads 760 mm of Hg.*

(i) 80 cm of Hg (ii) 30 cm Hg vacuum
(iii) 1.35 m H_2O gauge (iv) 4.2 bar.

Solution. Assuming density of Hg, ρ_{Hg} = 13.596 × 1000 kg/m³

Pressure of 760 mm of Hg will be

$$= \rho \times g \times h = 13.596 \times 1000 \times 9.806 \times \frac{760}{1000}$$

$$= 101325 \text{ Pa} = 101.325 \text{ kPa}.$$

(i) **Pressure of 80 cm of Hg**

$$= \frac{800}{760} \times 101.325 = \mathbf{106.65 \text{ kPa}.} \quad \text{(Ans.)}$$

(ii) **30 cm Hg vacuum**

$$= 76 - 30 = 46 \text{ cm of Hg absolute.}$$

Pressure due to 46 cm of Hg

$$= \frac{460}{760} \times 101.325 = \mathbf{61.328 \text{ kPa}.} \quad \text{(Ans.)}$$

(iii) **Pressure due to 1.35 m H_2O gauge**

$$= 1000 \times 9.806 \times 1.35 = 13238 \text{ Pa} = \mathbf{13.238 \text{ kPa}.} \quad \text{(Ans.)}$$

(iv) **4.2 bar**

$$= 4.2 \times 10^2 \text{ kPa} = \mathbf{420 \text{ kPa}.} \quad \text{(Ans.)}$$

Note. Pressure of 1 atmosphere
$$= 760 \text{ mm of Hg}$$
$$= 101325 \text{ N/m}^2.$$

The above values are standard. To get this value we have to use ρ_{Hg} = 13596 kg/m³ and g = 9.806 m/s². When we use ρ_{Hg} = 13600 kg/m³ and g = 9.81 m/s², we get p_{atm} = 101396 N/m² which is slightly different from 101325 N/m². It is recommended that for pressure of 1 atm. the value 101325 N/m² should be used.

Example 1.2. *On a piston of 10 cm diameter a force of 1000 N is uniformly applied. Find the pressure on the piston.*

Solution. Diameter of the piston d = 10 cm (= 0.1 m)
Force applied on the piston, F = 1000 N

∴ Pressure on the piston, $p = \dfrac{\text{Force}}{\text{Area}} = \dfrac{F}{A} = \dfrac{1000}{\pi/4 \times (0.1)^2}$

$$= 127307 \text{ N/m}^2 = \mathbf{127.307 \text{ kN/m}^2.} \quad \text{(Ans.)}$$

Example 1.3. *A tube contains an oil of specific gravity 0.9 to a depth of 120 cm. Find the gauge pressure at this depth (in kN/m^2).*

Solution. Specific gravity of oil = 0.9

Depth of oil in the tube, h = 120 cm = (1.2 m)

We know that
$$p = wh$$
$$= \rho.g.h, \rho \text{ being the mass density}$$
$$= (0.9\, \rho_w) \times g \times h, \rho_w \text{ being mass density of water}$$

$$\left[\text{Specific gravity} = \frac{\rho}{\rho_w}\right]$$

$$= 0.9 \times 1000 \times 9.81 \times 1.2 \text{ N/m}^2$$
$$= 10594.8 \text{ N/m}^2 = \textbf{10.595 kN/m}^2. \quad \textbf{(Ans.)}$$

Example 1.4. *A vacuum recorded in the condenser of a steam power plant is 740 mm of Hg. Find the absolute pressure in the condenser in Pa. The barometric reading is 760 mm of Hg.*

Solution. Vacuum recorded in the condenser = 740 mm of Hg

Barometric reading = 760 mm of Hg

We know that,

Absolute pressure in the condenser

$$= \text{Barometric reading} - \text{vacuum in the condenser}$$
$$= 760 - 740 = 20 \text{ mm of Hg}$$
$$= 20 \times 133.4 \text{ N/m}^2 \qquad (\because 1 \text{ mm of Hg} = 133.4 \text{ N}/\text{m}^2)$$
$$= 2668 \text{ N/m}^2 = \textbf{2668 Pa.} \quad \textbf{(Ans.)}$$

Example 1.5. *A vessel of cylindrical shape is 50 cm in diameter and 75 cm high. It contains 4 kg of a gas. The pressure measured with manometer indicates 620 mm of Hg above atmosphere when barometer reads 760 mm of Hg. Determine :*

(i) The absolute pressure of the gas in the vessel in bar.

(ii) Specific volume and density of the gas.

Solution. Diameter of the vessel, d = 50 cm (= 0.5 m)

Height of the vessel, h = 75 cm (= 0.75 m)

Mass of gas in the vessel, m = 4 kg

Manometer reading = 620 mm of Hg above atmosphere

Barometer reading = 760 mm of Hg

Now, volume of the vessel $= \frac{\pi}{4}d^2 \times h = \frac{\pi}{4} \times (0.5)^2 \times (0.75) = 0.147 \text{ m}^3.$

(i) **Total pressure in the vessel**

$$= 760 + 620 = 1380 \text{ mm of Hg}$$
$$= 1380 \times 133.4 \text{ N/m}^2 \qquad [\because 1 \text{ mm of Hg} = 133.4 \text{ N}/\text{m}^2]$$
$$= 1.841 \times 10^5 \text{ N/m}^2 = \textbf{1.841 bar.} \quad \textbf{(Ans.)} \qquad [\because 1 \text{ bar} = 10^5 \text{ N/m}^2]$$

(ii) **Specific volume** $= \dfrac{0.147}{4} = \textbf{0.03675 m}^3\textbf{/kg.} \quad \textbf{(Ans.)}$

BASIC CONCEPTS OF THERMODYNAMICS

Density $= \dfrac{4}{0.147} = 27.21$ kg/m³. **(Ans.)**

Example 1.6. *In a pipe line the pressure of gas is measured with a mercury manometer having one limb open to the atmosphere (Fig. 1.8). If the difference in the height of mercury in the two limbs is 550 mm, calculate the gas pressure.*

Given : Barometric reading = 761 mm of Hg
Acceleration due to gravity = 9.79 m/s²
Density of mercury = 13640 kg/m³.

Solution. At the plane LM, we have
$$p = p_0 + \rho g h$$
Now, $\qquad p_0 = \rho g h_0$
where h_0 = barometric height ; ρ = density of mercury ; p_0 = atmospheric pressure
$\therefore \qquad p = \rho g h_0 + \rho g h = \rho g (h_0 + h)$

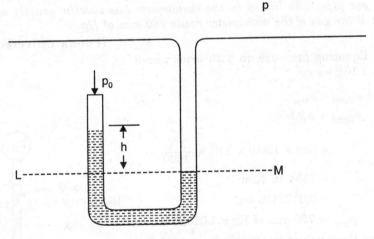

Fig. 1.8

$$= 13640 \times 9.79 \left(\dfrac{761}{1000} + \dfrac{550}{1000}\right) = 13640 \times 9.79\, (0.761 + 0.55)$$

$$= 175.065 \times 10^3 \text{ N/m}^2 = 175.065 \text{ kPa} = \mathbf{1.75 \text{ bar.}} \quad \textbf{(Ans.)}$$

Example 1.7. *A U-tube mercury manometer with one arm open to atmosphere is used to measure pressure in a steam pipe. The level of mercury in open arm is 97.5 mm greater than that in the arm connected to the pipe. Some of steam in the pipe condenses in the manometer arm connected to the pipe. The height of this column is 34 mm. The atmospheric pressure is 760 mm of Hg. Find the absolute pressure of steam.* **(Poona University, Nov. 1995)**

Solution. Equating the pressure in mm of Hg on both arms above the line X-X (Fig. 1.9), we get

$$p_{abs.} + p_{water} = p_{Hg} + p_{atm.}$$

Now, $p_{water} = \dfrac{34}{13.6} = 2.5$ mm of Hg.

$\therefore \quad p_{abs.} + 2.5 = 97.5 + 760$

or $\quad p_{abs.} = 97.5 + 760 - 2.5$

$= 855$ mm of Hg.

$= 855 \times p_{Hg} \times g \times 10^{-5}$ bar

$= \dfrac{855}{1000}$ (m) $\times (13.6 \times 1000)$ (kg/m^3)

$\times 9.81 \times 10^{-5}$

$= \mathbf{1.1407}$ **bar.** (Ans.)

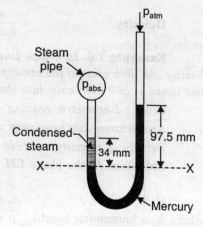

Fig. 1.9

Example 1.8. *A U-tube manometer is connected to a gas pipe. The level of the liquid in the manometer arm open to the atmosphere is 170 mm lower than the level of the liquid in the arm connected to the gas pipe. The liquid in the manometer has specific gravity of 0.8. Find the absolute pressure of the gas if the manometer reads 760 mm of Hg.*

(Poona University, Dec. 1996)

Solution. Equating pressure on both arms above the line X-X (Fig. 1.10), we get

$$p_{gas} + p_{liquid} = p_{atm.} \qquad ...(i)$$

Now, $\quad p_{liquid} = \rho.g.h$

$= (0.8 \times 1000) \times 9.81 \times \dfrac{170}{1000}$

$= 1334.16$ N/m^2

$= 0.0133416$ bar

$p_{atm.} = 760$ mm of Hg $= 1.01325$ bar

Substituting these values in eqn. (*i*) above, we have

$p_{gas} + 0.0133416 = 1.01325$

$\therefore \quad p_{gas} = \mathbf{0.9999}$ **bar.** (Ans.)

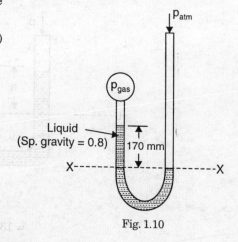

Fig. 1.10

1.17. REVERSIBLE AND IRREVERSIBLE PROCESSES

Reversible process. *A reversible process (also sometimes known as quasi-static process) is one which can be stopped at any stage and reversed so that the system and surroundings are exactly restored to their initial states* (Fig. 1.11).

This process has the following *characteristics* :

1. It must pass through the same states on the reversed path as were initially visited on the forward path.

2. This process when undone will leave no history of events in the surroundings.

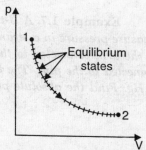

Fig. 1.11. Reversible process.

3. It must pass through a continuous series of equilibrium states.

No real process is truely reversible but some processes may approach reversibility, to close approximation.

Examples. Some examples of nearly reversible processes are :

(*i*) Frictionless relative motion.

(*ii*) Expansion and compression of spring.

(*iii*) Frictionless adiabatic expansion or compression of fluid.

(*iv*) Polytropic expansion or compression of fluid.

(*v*) Isothermal expansion or compression.

(*vi*) Electrolysis.

Irreversible process. An *irreversible process* is one *in which heat is transferred through a finite temperature.*

Examples.

(*i*) Relative motion with friction (*ii*) Combustion

(*iii*) Diffusion (*iv*) Free expansion

(*v*) Throttling (*vi*) Electricity flow through a resistance

(*vii*) Heat transfer (*viii*) Plastic deformation.

An irreversible process is usually represented by a dotted (or discontinuous) line joining the end states to indicate that the intermediate states are indeterminate (Fig. 1.12).

Irreversibilities are of *two types* :

1. **External irreversibilities.** *These are associated with dissipating effects outside the working fluid.*

Example. *Mechanical friction occurring during a process due to some external source.*

2. **Internal irreversibilities.** *These are associated with dissipating effects within the working fluid.*

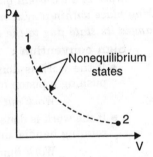

Fig. 1.12. Irreversible process.

Example. *Unrestricted expansion of gas, viscosity and inertia of the gas.*

1.18. ENERGY, WORK AND HEAT

1.18.1. Energy

Energy is a general term embracing *energy in transition* and *stored energy*. The stored energy of a substance may be in the forms of *mechanical energy* and *internal energy* (other forms of stored energy may be chemical energy and electrical energy). Part of the stored energy may take the form of either potential energy (which is the gravitational energy due to height above a chosen datum line) or kinetic energy due to velocity. The balance part of the energy is known as *internal energy*. In a **non-flow process** usually there is no change of potential or kinetic energy and hence change of mechanical energy will not enter the calculations. In a **flow process**, however, *there may be changes in both potential and kinetic energy and these must be taken into account while considering the changes of stored energy.* **Heat and work** are *the forms of energy in transition*. These are the only forms in which energy can cross the boundaries of a system. *Neither heat nor work can exist as stored energy.*

1.18.2. Work and Heat
Work

Work is said to be done when a *force moves through a distance*. If a part of the boundary of a system undergoes a displacement under the action of a pressure, the work done W is the product of the force (pressure × area), and the distance it moves in the direction of the force. Fig. 1.13 (a) illustrates this with the conventional piston and cylinder arrangement, the heavy line defining the boundary of the system. Fig. 1.13 (b) illustrates another way in which work might be applied to a system. A force is exerted by the paddle as it changes the momentum of the fluid, and since this force moves during rotation of the paddle room work is done.

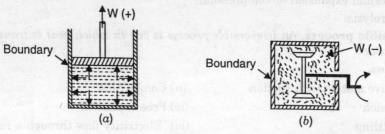

Fig. 1.13

Work is a transient quantity which only appears at the boundary while a change of state is taking place within a system. Work is 'something' which appears at the boundary when a system changes its state due to the movement of a part of the boundary under the action of a force.

Sign convention :

- If the work is done *by* the system *on* the surroundings, *e.g.*, when a fluid expands pushing a piston outwards, the work is said to be *positive*.

i.e.,   Work output of the system = + W

- If the work is done *on* the system *by* the surroundings, *e.g.*, when a force is applied to a rotating handle, or to a piston to compress a fluid, the work is said to be *negative*.

i.e.,   Work input to system = − W

Heat

Heat (denoted by the symbol Q), may be, defined in an analogous way to work as follows :

"Heat is 'something' which appears at the boundary when a system changes its state due to a difference in temperature between the system and its surroundings".

Heat, like work, is a transient quantity which only appears at the boundary while a change is taking place within the system.

It is apparent that neither δW or δQ are exact differentials and therefore any integration of the elemental quantities of work or heat which appear during a change from state 1 to state 2 must be written as

$$\int_1^2 \delta W = W_{1-2} \text{ or } {}_1W_2 \text{ (or } W\text{), and}$$

$$\int_1^2 \delta Q = Q_{1-2} \text{ or } {}_1Q_2 \text{ (or } Q\text{)}$$

Sign convention :

If the heat flows *into* a system *from* the surroundings, the quantity is said to be *positive* and, conversely, if heat flows *from* the system to the surroundings it is said to be *negative*.

BASIC CONCEPTS OF THERMODYNAMICS

In other words :
Heat received by the system = + Q
Heat rejected or given up by the system = – Q.

Comparison of Work and Heat

Similarities :

(i) Both are *path functions and inexact differentials*.

(ii) Both are boundary phenomenon *i.e.*, both are recognized at the boundaries of the system as they cross them.

(iii) Both are associated with a process, not a state. Unlike properties, work or heat has no meaning at a state.

(iv) Systems possess energy, but not work or heat.

Dissimilarities :

(i) In heat transfer temperature difference is required.

(ii) In a stable system there cannot be work transfer, however, there is no restriction for the transfer of heat.

(iii) The sole effect external to the system could be reduced to rise of a weight but in the case of a heat transfer other effects are also observed.

1.19. REVERSIBLE WORK

Let us consider an ideal frictionless fluid contained in a cylinder above a piston as shown in Fig. 1.14. *Assume that the pressure and temperature of the fluid are uniform and that there is no friction between the piston and the cylinder walls.*

Let A = Cross-sectional area of the piston,

p = Pressure of the fluid at any instant,

$(p - dp) A$ = Restraining *force exerted* by the surroundings on the piston, and

dl = The distance moved by the piston under the action of the *force exerted*.

Then work done by the fluid on the piston is given by force times the distance moved,

i.e.,   Work done by the fluid

$= (pA) \times dl = pdV$

(where dV = a small increase in volume)

Or considering unit mass

Work done = pdv (where v = specific volume)

This is only *true* when (*a*) the *process is frictionless* and (*b*) *the difference in pressure between the fluid and its surroundings during the process is infinitely small*. Hence when a reversible process takes place between states 1 and 2, we have

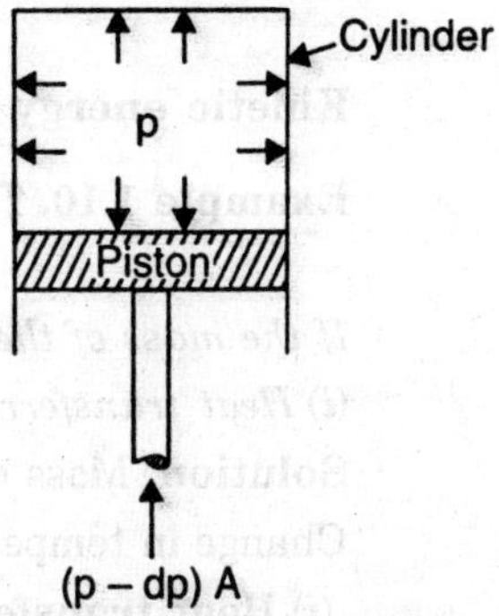

Work done by the unit mass of fluid = $\int_1^2 pdv$ ...(1.4)

Fig. 1.14

When a fluid undergoes a reversible process a series of state points can be joined up to form a line on a diagram of properties. The work done by the fluid during any reversible process is therefore given by the area under the line of process plotted on a *p-v* diagram (Fig. 1.15).

i.e., Work done = Shaded area in Fig. 1.15

$$= \int_1^2 pdv.$$

When p can be expressed in terms of v then the integral, $\int_1^2 pdv$, can be evaluated.

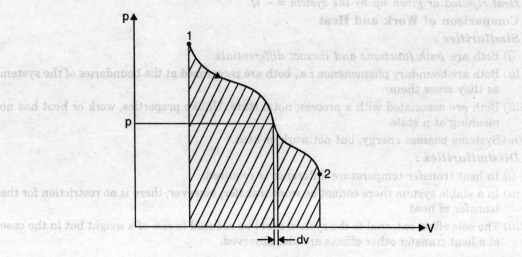

Fig. 1.15

Example 1.9. *An artificial satellite revolves round the earth with a relative velocity of 800 m/s. If acceleration due to gravity is 9 m/s² and gravitational force is 3600 N, calculate its kinetic energy.*

Solution. Relatively velocity of satellite, $v = 800$ m/s

Acceleration due to gravity, $g = 9$ m/s²

Gravitational force, $m.g = 3600$ N

$\therefore$ Mass, $m = \dfrac{3600}{g} = \dfrac{3600}{9} = 400$ kg.

Kinetic energy $= \dfrac{1}{2} mv^2 = \dfrac{1}{2} \times 400 \times (800)^2$ J $= 128 \times 10^6$ J or **128 MJ.** **(Ans.)**

Example 1.10. *The specific heat capacity of the system during a certain process is given by*
$$c_n = (0.4 + 0.004\, T)\ kJ/kg°C.$$
If the mass of the gas is 6 kg and its temperature changes from 25°C to 125°C find :
(i) Heat transferred ; *(ii) Mean specific heat of the gas.*

Solution. Mass of the gas, $m = 6$ kg

Change in temperature of the gas $= 25°C$ to $125°C$

(i) Heat transferred, Q :

We know that heat transferred is given by,

$$Q = \int m\, c_n\, dT = 6 \int_{25}^{125} (0.4 + 0.004\, T)\, dT$$

$$= 6 \left[0.4\, T + 0.04 \left(\dfrac{T^2}{2}\right) \right]_{25}^{125}$$

BASIC CONCEPTS OF THERMODYNAMICS

$$= 6[0.4\,(125 - 25) + 0.002\,(125^2 - 25^2)]$$
$$= 6(40 + 30) = \mathbf{420\ kJ.}\ \mathbf{(Ans.)}$$

(ii) **Mean specific heat of the gas, c_n :**
$$Q = m.c_n.dT$$
i.e., $\qquad 420 = 6 \times c_n \times (125 - 25)$

$$\therefore\qquad c_n = \frac{420}{6 \times 100} = \mathbf{0.7\ kJ/kg°C.}\ \mathbf{(Ans.)}$$

Example 1.11. *A temperature scale of certain thermometer is given by the relation*
$$t = a \ln p + b$$
where a and b are constants and p is the thermometric property of the fluid in the thermometer. If at the ice point and steam point the thermometric properties are found to be 1.5 and 7.5 respectively what will be the temperature corresponding to the thermometric property of 3.5 on Celsius scale.
(Poona University, Nov. 1997)

Solution. $\qquad t = a \ln p + b \qquad\qquad\qquad\qquad$ ...(given)
On Celsius scale :
$$\text{Ice point} = 0°C, \text{ and}$$
$$\text{Steam point} = 100°C$$
∴ From given conditions, we have
$$0 = a \ln 1.5 + b \qquad\qquad\qquad ...(i)$$
and $\qquad 100 = a \ln 7.5 + b \qquad\qquad\qquad ...(ii)$
i.e., $\qquad 0 = a \times 0.4054 + b \qquad\qquad\qquad ...(iii)$
and $\qquad 100 = a \times 2.015 + b \qquad\qquad\qquad ...(iv)$

Subtracting (iii) from (iv), we get
$$100 = 1.61a$$
or $\qquad a = 62.112$

Substituting this value in eqn. (iii), we get
$$b = -0.4054 \times 62.112 = -25.18$$
∴ When $p = 3.5$ the value of temperature is given by,
$$t = 62.112 \ln (3.5) - 25.18 = \mathbf{52.63°C.}\ \mathbf{(Ans.)}$$

Example 1.12. *A thermocouple with test junction at t°C on gas thermometer scale and reference junction at ice point gives the e.m.f. as*
$$e = 0.20\,t - 5 \times 10^{-4} t^2\ mV.$$
The millivoltmeter is calibrated at ice and steam points. What will be the reading on this thermometer where the gas thermometer reads 70°C ?

Solution. $\qquad e = 0.20\,t - 5 \times 10^{-4} t^2\ mV \qquad\qquad$ ...(given)
At ice point : When $t = 0°C$, $e = 0$
At steam point : When $t = 100°C$,
$$e = 0.20 \times 100 - 5 \times 10^{-4} \times (100)^2 = 15\ mV$$
Now, $\qquad$ when $t = 70°C$,
$$e = 0.20 \times 70 - 5 \times 10^{-4} \times (70)^2 = 11.55\ mV$$
∴ When the gas thermometer reads 70°C the thermocouple will read
$$t = \frac{100 \times 11.55}{15} = \mathbf{77°C.}\ \mathbf{(Ans.)}$$

☞ **Example 1.13.** *Comment whether the following quantities can be called as properties or not* :

(i) $\int p dV$, (ii) $\int V dp$, and (iii) $\int p dV + \int V dp$.

Solution. (i) $\int p dV$:

p is a function of V and integral can only be evaluated if relation between p and V is known. It is thus an *inexact differential* and hence **not a property.** **(Ans.)**

(ii) $\int V dp$:

It is **not a property** for the *same reason* as mentioned in (i). **(Ans.)**

(iii) $\int p dV + \int V dp$:

$$\int p dV + \int V dp = \int p dV + V dp = \int d(pV) = pV.$$

Thus the integral can be evaluated without knowing the relation between p and V. It is an *exact differential* and hence **it is a property.** **(Ans.)**

Example 1.14. *Gas from a cylinder of compressed helium is used to inflate an inelastic flexible balloon, originally folded completely flat, to a volume 0.6 m³. If the barometer reads 760 mm Hg, what is the amount of work done upon the atmosphere by the balloon ? Sketch the system before and after the process.*

Solution. Refer Fig. 1.16. The firm line B_1 shows the boundary of the system before the process, and dotted line B_2 shows the boundary after the process.

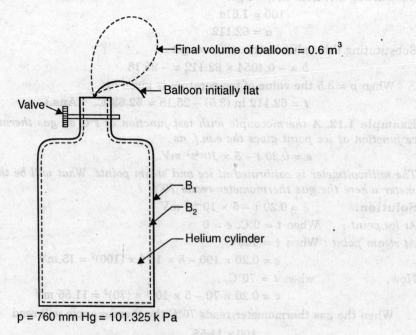

Fig. 1.16

BASIC CONCEPTS OF THERMODYNAMICS

The displacement work,

$$W_d = \int_{cylinder} pdV + \int_{balloon} pdV = 0 + \int_{balloon} pdV$$

$$= 101.325 \times 0.6 \qquad [\because dV = 0.6 \text{ m}^3]$$

$$= \mathbf{60.795 \text{ kJ.}} \quad \mathbf{(Ans.)}$$

This is a positive work, because the *work is done by the system*. Work done by the atmosphere is – 60.795 kJ. Since the wall of the cylinder is rigid there is no *pdV*-work involved in it.

It is assumed that the pressure in the balloon is atmospheric at all times, since the balloon fabric is light, inelastic and unstressed. If the balloon were elastic and stressed during the filling process, the work done by the gas would be *greater* than 60.795 kJ by an amount equal to the work done in stretching the balloon, although the displacement work done by atmosphere is still – 60.795 kJ. However, if the system includes both the gas and the balloon, the displacement work should be 60.795 kJ, as estimated above.

Example 1.15. *Determine the work done by the air which enters into an evacuated vessel from atmosphere when the valve is opened. The atmospheric pressure is 1.013 bar and 1.5 m³ of air at atmospheric condition enters into the vessel.*

Solution. Fig. 1.17 shows the initial and final condition of the system.

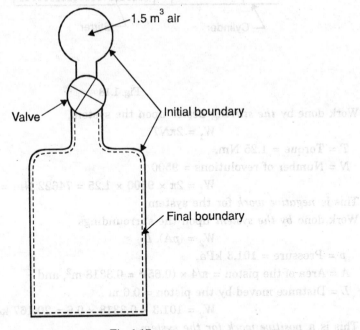

Fig. 1.17

No work is done by the boundary in contact with the vessel as the boundary does not move. Work is done by the external boundary at constant pressure.

$$\therefore W = \int_{V_1}^{V_2} p \, dV = \int_{1.5}^{0} p \, dV \qquad [\because V_1 = 1.5 \text{ m}^3 \text{ and } V_2 = 0]$$

$$= p(0 - 1.5) = 1.013 \times 10^5 \times (-1.5)$$

$$= -1.5195 \times 10^5 \text{ J} = \mathbf{-151.95 \text{ kJ.}} \quad \mathbf{(Ans.)}$$

Since the free air boundary is contracting, the work done by the system is negative, and the surroundings do positive work upon the system.

☞ **Example 1.16.** *A piston and cylinder machine containing a fluid system has a stirring device as shown in Fig. 1.18. The piston is frictionless, and it is held down against the fluid due to atmospheric pressure of 101.3 kPa. The stirring device is turned 9500 revolutions with an average torque against the fluid of 1.25 Nm. Meanwhile the piston of 0.65 m diameter moves out 0.6 m. Find the net work transfer for the system.*

Solution. Refer Fig. 1.18.

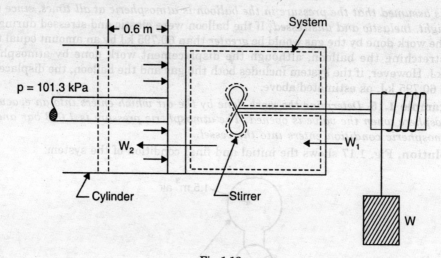

Fig. 1.18

Work done by *the stirring device* upon the system,
$$W_1 = 2\pi NT$$
where T = Torque = 1.25 Nm,
N = Number of revolutions = 9500
$$W_1 = 2\pi \times 9500 \times 1.25 = 74622 \text{ Nm} = 74.622 \text{ kJ}$$
This is *negative work* for the system.

Work done *by the system* upon the surroundings
$$W_2 = (pA) \cdot L$$
where p = Pressure = 101.3 kPa,
A = Area of the piston = $\pi/4 \times (0.65)^2 = 0.3318 \text{ m}^2$, and
L = Distance moved by the piston = 0.6 m
$$W_2 = 101.3 \times 0.3318 \times 0.6 = 20.167 \text{ kJ}$$
This is a *positive work for the system.*

Hence, the net work transfer for the system
$$W_{net} = W_1 + W_2 = -74.622 + 20.167 = -54.455 \text{ kJ.} \quad \textbf{(Ans.)}$$

Example 1.17. *A diesel engine piston which has an area of 45 cm^2 moves 5 cm during part of suction stroke. 300 cm^3 of fresh air is drawn in from the atmosphere. The pressure in the cylinder during suction stroke is 0.9×10^5 N/m^2 and the atmospheric pressure is 1.013×10^5 N/m^2. The difference between the suction and atmospheric pressure is accounted for flow resistance in the suction pipe and inlet valve. Find the net work done during the process.*

BASIC CONCEPTS OF THERMODYNAMICS

Solution. Area of diesel engine piston
$$= 45 \text{ cm}^2 = 45 \times 10^{-4} \text{ m}^2$$

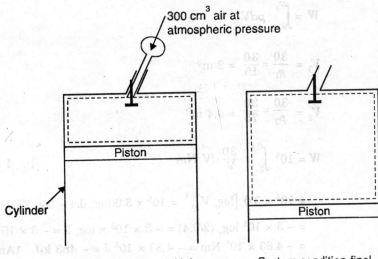

Fig. 1.19

Amount of fresh air drawn in from atmosphere
$$= 300 \text{ cm}^3 = 300 \times 10^{-6} \text{ m}^3$$

The pressure inside the cylinder during suction stroke
$$= 0.9 \times 10^5 \text{ N/m}^2$$

Atmospheric pressure $\quad = 1.013 \times 10^5 \text{ N/m}^2$

Initial and final conditions of the system are shown in Fig. 1.19.

Net work done = Work done by free air boundary + work done on the piston

The work done by the free air = –ve because *boundary contracts*

The work done by the cylinder on the piston = +ve because the *boundary expands*

$\therefore \quad$ **Net work done** $= \int_{\text{Piston}} p dV + \int_{\substack{\text{Free air} \\ \text{boundary}}} p dV$

$$= \left[0.9 \times 10^5 \times 45 \times 10^{-4} \times \frac{5}{100} - 1.013 \times 10^5 \times 300 \times 10^{-6} \right]$$

$$= [20.25 - 30.39] = -\ 10.14 \text{ Nm or J. (Ans.)}$$

Example 1.18. *The properties of a closed system change following the relation between pressure and volume as pV = 3.0 where p is in bar V is in m^3. Calculate the work done when the pressure increases from 1.5 bar to 7.5 bar.*

Solution. Initial pressure, $\quad p_1 = 1.5$ bar

Final pressure, $\quad\quad\quad\quad\quad p_2 = 7.5$ bar

Relation between p and V, $\quad pV = 3.0$

Work done, W :

The work done during the process is given by

$$W = \int_{V_1}^{V_2} p\,dV$$

$$V_1 = \frac{3.0}{p_1} = \frac{3.0}{1.5} = 2 \text{ m}^3$$

$$V_2 = \frac{3.0}{p_2} = \frac{3.0}{7.5} = 0.4 \text{ m}^3$$

$$\therefore \quad W = 10^5 \int_2^{0.4} \frac{3.0}{V} dV \text{ Nm} \qquad [\because 1 \text{ bar} = 10^5 \text{ N/m}^2]$$

$$= 10^5 \times 3.0 \left[\log_e V\right]_2^{0.4} = 10^5 \times 3.0 (\log_e 0.4 - \log_e 2)$$

$$= -3 \times 10^5 \log_e (2/0.4) = -3 \times 10^5 \times \log_e 5 = -3 \times 10^5 \times 1.61 \text{ Nm}$$

$$= -4.83 \times 10^5 \text{ Nm} = -4.83 \times 10^5 \text{ J} = \mathbf{-483 \text{ kJ}}. \quad \textbf{(Ans.)}$$

Example 1.19. *To a closed system 150 kJ of work is supplied. If the initial volume is 0.6 m^3 and pressure of the system changes as $p = 8 - 4V$, where p is in bar and V is in m^3, determine the final volume and pressure of the system.*

Solution. Amount of work supplied to a closed system = 150 kJ

Initial volume = 0.6 m^3

Pressure-volume relationship, $p = 8 - 4V$

The work done during the process is given by

$$W = \int_{V_1}^{V_2} p\,dV$$

$$= 10^5 \int_{0.6}^{V_2} (8 - 4V) dV = 10^5 \left[8V - 4 \times \frac{V^2}{2}\right]_{0.6}^{V}$$

$$= 10^5 [8(V_2 - 0.6) - 2(V_2^2 - 0.6^2)]$$

$$= 10^5 [8V_2 - 4.8 - 2V_2^2 + 0.72]$$

$$= 10^5 [8V_2 - 2V_2^2 - 4.08] \text{ Nm or J}$$

But this work is equal to -150×10^3 J as this work is supplied to the system.

$$\therefore \quad -150 \times 10^3 = 10^5 [8V_2 - 2V_2^2 - 4.08]$$

or $\quad 2V_2^2 - 8V_2 + 2.58 = 0$

$$V_2 = \frac{8 \pm \sqrt{64 - 4 \times 2 \times 2.58}}{4} = \frac{8 \pm 6.585}{4} = 0.354 \text{ m}^3$$

Positive sign is incompatible with the present problem, therefore it is not considered.

$\therefore \quad$ Final volume, $V_2 = \mathbf{0.354 \text{ m}^3}$. **(Ans.)**

and $\quad$ final pressure, $p_2 = 8 - 4V = 8 - 4 \times 0.354$

$$= 6.584 \text{ bar} = \mathbf{6.584 \times 10^5 \text{ N/m}^2 \text{ or Pa}}. \quad \textbf{(Ans.)}$$

BASIC CONCEPTS OF THERMODYNAMICS

REVERSIBLE WORK

Example 1.20. *A fluid at a pressure of 3 bar, and with specific volume of 0.18 m^3/kg, contained in a cylinder behind a piston expands reversibly to a pressure of 0.6 bar according to a law, $p = \dfrac{C}{v^2}$ where C is a constant. Calculate the work done by the fluid on the piston.*

Solution. Refer Fig. 1.20.

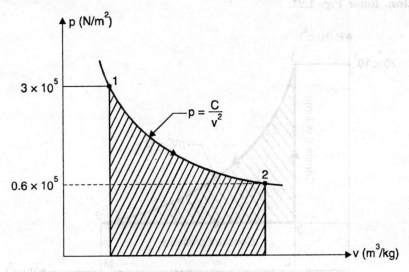

Fig. 1.20

$$p_1 = 3 \text{ bar} = 3 \times 10^5 \text{ N/m}^2$$
$$v_1 = 0.18 \text{ m}^3/\text{kg}$$

$$\text{Work done} = \text{Shaded area} = \int_1^2 p\, dv$$

i.e.,

$$\text{Work done, } W = \int_1^2 \frac{C}{v^2}\, dv = C\int_1^2 \frac{dv}{v^2} = C\left|\frac{v^{-2+1}}{-2+1}\right|_{v_1}^{v_2}$$

$$= C\left[-v^{-1}\right]_{v_1}^{v_2} = C\left[-\frac{1}{v}\right]_{v_1}^{v_2} = C\left[\frac{1}{v_1} - \frac{1}{v_2}\right] \quad \ldots(i)$$

Also

$$C = pv^2 = p_1 v_1^2 = 3 \times 0.18^2 = 0.0972 \text{ bar } (m^3/kg)^2$$

and

$$v_2 = \sqrt{\frac{C}{p_2}} = \sqrt{\frac{0.0972}{0.6}} = 0.402 \text{ m}^3/\text{kg}$$

Substituting the values of C, v_1 and v_2 in eqn. (i), we get

Work done,

$$W = 0.0972 \times 10^5 \left[\frac{1}{0.18} - \frac{1}{0.402}\right] \text{ Nm/kg}$$

$$= \textbf{29840 Nm/kg.} \quad \textbf{(Ans.)}$$

☞ **Example 1.21.** *A cylinder contains 1 kg of a certain fluid at an initial pressure of 20 bar. The fluid is allowed to expand reversibly behind a piston according to a law pV^2 = constant until the volume is doubled. The fluid is then cooled reversibly at constant pressure until the piston regains its original position ; heat is then supplied reversibly with the piston firmly locked in position until the pressure rises to the original value of 20 bar. Calculate the net work done by the fluid, for an initial volume of 0.05 m^3.*

Solution. Refer Fig. 1.21.

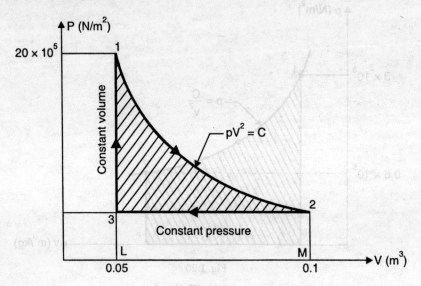

Fig. 1.21

Mass of fluid, m = 1 kg

p_1 = 20 bar = 20 × 10^5 N/m^2

V_1 = 0.05 m^3

Considering the process 1-2

$$p_1 V_1^2 = p_2 V_2^2$$

∴ $p_2 = p_1 \left(\dfrac{V_1}{V_2}\right)^2 = 20 \left(\dfrac{V_1}{2V_1}\right)^2$ [∵ $V_2 = 2V_1$ (given)]

$$= \dfrac{20}{4} = 5 \text{ bar}$$

Work done by the fluid from 1 to 2 = Area 12 ML1 = $\displaystyle\int_1^2 p\, dV$

i.e., $W_{1-2} = \displaystyle\int_{v_1}^{v_2} \dfrac{C}{V^2} dV$, where $C = p_1 V_1^2 = 20 \times 0.05^2$ bar m^6

∴ $W_{1-2} = 10^5 \times 20 \times 0.0025 \left[-\dfrac{1}{V}\right]_{0.05}^{0.1}$

$$= 10^5 \times 20 \times 0.0025 \left(\frac{1}{0.05} - \frac{1}{0.1}\right) = 50000 \text{ Nm}$$

Work done on fluid from 2 to 3
$$= \text{Area } 32ML3 = p_2(V_2 - V_3) = 10^5 \times 5 \times (0.1 - 0.05) = 25000 \text{ Nm}$$

Work done during the process 3-1
$$= 0, \text{ because piston is locked in position} \quad (i.e., \textit{Volume remains constant})$$

∴ **Net work done by the fluid**
$$= \text{Enclosed area } 1231 = 50000 - 25000$$
$$= \textbf{25000 Nm. (Ans.)}$$

HIGHLIGHTS

1. *Thermodynamics* is an axiomatic science which deals with the relations among heat, work and properties of systems which are in equilibrium. It basically entails four laws or axioms known as *Zeroth, First, Second* and *Third* law of thermodynamics.
2. A *system* is a finite quantity of matter or a prescribed region of space.
   A system may be a *closed, open* or *isolated* system.
3. A *phase* is a quantity of matter which is homogeneous throughout in chemical composition and physical structure.
4. A *homogeneous system* is one which consists of a *single phase*.
5. A *heterogeneous system* is one which consists of *two or more phases*.
6. A *pure substance* is one that has a homogeneous and invariable chemical composition even though there is a change of phase.
7. A system is in *thermodynamic equilibrium* if temperature and pressure at all points are same ; there should be no *velocity gradient*.
8. A *property of a system* is a characteristic of the system which depends upon its state, but not upon how the state is reached.
   Intensive properties do not depend on the mass of the system.
   Extensive properties depend on the mass of the system.
9. *State* is the condition of the system at an instant of time as described or measured by its properties. Or each unique condition of a system is called a state.
10. A *process* occurs when the system undergoes a change in state or an energy transfer takes place at a steady state.
11. Any process or series of processes whose end states are identical is termed a *cycle*.
12. The *pressure* of a system is the force exerted by the system on unit area of boundaries. Vacuum is defined as the absence of pressure.
13. A *reversible process* is one which can be stopped at any stage and reversed so that the system and surroundings are exactly restored to their initial states.
    An *irreversible process* is one in which heat is transferred through a finite temperature.
14. Zeroth law of thermodynamics states that if two systems are each equal in temperature to a third, they are equal in temperature to each other.
15. Infinite slowness is the characteristic feature of a quasi-static process. A quasi-static process is a succession of equilibrium states. It is also called a reversible process.

OBJECTIVE TYPE QUESTIONS

Choose the Correct Answer :

1. A definite area or space where some thermodynamic process takes place is known as
   (a) thermodynamic system
   (b) thermodynamic cycle
   (c) thermodynamic process
   (d) thermodynamic law.

2. An open system is one in which
   (a) heat and work cross the boundary of the system, but the mass of the working substance does not
   (b) mass of working substance crosses the boundary of the system but the heat and work do not
   (c) both the heat and work as well as mass of the working substances cross the boundary of the system
   (d) neither the heat and work nor the mass of the working substances cross the boundary of the system.

3. An isolated system
   (a) is a specified region where transfer of energy and/or mass take place
   (b) is a region of constant mass and only energy is allowed to cross the boundaries
   (c) cannot transfer either energy or mass to or from the surroundings
   (d) is one in which mass within the system is not necessarily constant
   (e) none of the above.

4. In an extensive property of a thermodynamic system
   (a) extensive heat is transferred
   (b) extensive work is done
   (c) extensive energy is utilised
   (d) all of the above
   (e) none of the above.

5. Which of the following is an intensive property of a thermodynamic system ?
   (a) Volume
   (b) Temperature
   (c) Mass
   (d) Energy.

6. Which of the following is the extensive property of a thermodynamic system ?
   (a) Pressure
   (b) Volume
   (c) Temperature
   (d) Density.

7. When two bodies are in thermal equilibrium with a third body they are also in thermal equilibrium with each other. This statement is called
   (a) Zeroth law of thermodyamics
   (b) First law of thermodynamics
   (c) Second law of thermodynamics
   (d) Kelvin Planck's law.

8. The temperature at which the volume of a gas becomes zero is called
   (a) absolute scale of temperature
   (b) absolute zero temperature
   (c) absolute temperature
   (d) none of the above.

9. The value of one bar (in SI units) is equal to
   (a) 100 N/m^2
   (b) 1000 N/m^2
   (c) 1×10^4 N/m^2
   (d) 1×10^5 N/m^2
   (e) 1×10^6 N/m^2.

10. The absolute zero pressure will be
    (a) when molecular momentum of the system becomes zero
    (b) at sea level
    (c) at the temperature of -273 K
    (d) under vacuum conditions
    (e) at the centre of the earth.

11. Absolute zero temperature is taken as
    (a) $-273°$C
    (b) $273°$C
    (c) $237°$C
    (d) $-373°$C.

12. Which of the following is *correct* ?
    (a) Absolute pressure = gauge pressure + atmospheric pressure
    (b) Gauge pressure = absolute pressure + atmospheric pressure
    (c) Atmospheric pressure = absolute pressure + gauge pressure
    (d) Absolute pressure = gauge pressure − atmospheric pressure.

BASIC CONCEPTS OF THERMODYNAMICS

13. The unit of energy in SI units is
    (a) Joule (J)
    (b) Joule metre (Jm)
    (c) Watt (W)
    (d) Joule/metre (J/m).

14. One watt is equal to
    (a) 1 Nm/s
    (b) 1 N/min
    (c) 10 N/s
    (d) 100 Nm/s
    (e) 100 Nm/m.

15. One joule (J) is equal to
    (a) 1 Nm
    (b) kNm
    (c) 10 Nm/s
    (d) 10 kNm/s.

16. The amount of heat required to raise the temperature of 1 kg of water through 1°C is called
    (a) specific heat at constant volume
    (b) specific heat at constant pressure
    (c) kilo calorie
    (d) none of the above.

17. The heating and expanding of a gas is called
    (a) thermodynamic system
    (b) thermodynamic cycle
    (c) thermodynamic process
    (d) thermodynamic law.

18. A series of operations, which take place in a certain order and restore the initial condition is known as
    (a) reversible cycle
    (b) irreversible cycle
    (c) thermodynamic cycle
    (d) none of the above.

19. The condition for the reversibility of a cycle is
    (a) the pressure and temperature of the working substance must not differ, appreciably, from those of the surroundings at any stage in the process
    (b) all the processes, taking place in the cycle of operation, must be extremely slow
    (c) the working parts of the engine must be friction free
    (d) there should be no loss of energy during the cycle of operation
    (e) all of the above
    (f) none of the above.

20. In an irreversible process, there is a
    (a) loss of heat
    (b) no loss of heat
    (c) gain of heat
    (d) no gain of heat.

21. The main cause of the irreversibility is
    (a) mechanical and fluid friction
    (b) unrestricted expansion
    (c) heat transfer with a finite temperature difference
    (d) all of the above
    (e) none of the above.

22. According to kinetic theory of heat
    (a) temperature should rise during boiling
    (b) temperature should fall during freezing
    (c) at low temperature all bodies are in solid state
    (d) at absolute zero there is absolutely no vibration of molecules
    (e) none of the above.

23. A system comprising a single phase is called a
    (a) closed system
    (b) open system
    (c) isolated system
    (d) homogeneous system
    (e) heterogeneous system.

Answers

1. (a) 2. (c) 3. (c) 4. (e) 5. (b) 6. (b) 7. (a)
8. (b) 9. (d) 10. (a) 11. (a) 12. (a) 13. (a) 14. (a)
15. (a) 16. (c) 17. (b) 18. (c) 19. (e) 20. (a) 21. (d)
22. (d) 23. (d).

THEORETICAL QUESTIONS

1. Define a thermodynamic system. Differentiate between open system, closed system and an isolated system.
2. How does a homogeneous system differ from a heterogeneous system?
3. What do you mean by a pure substance?
4. Explain the following terms:
   (i) State,   (ii) Process, and   (iii) Cycle.
5. Explain briefly zeroth law of thermodynamics.
6. What is a quasi-static process?
7. What do you mean by 'reversible work'?

UNSOLVED EXAMPLES

1. A cylindrical vessel of 60 cm diameter and 80 cm height contains 3.2 kg of a gas. The pressure measured with manometer indicates 60 cm of Hg above atmosphere when barometer reads 760 mm of Hg. Find:
   (i) The absolute pressure of the gas in the vessel in bar, and (ii) Specific volume and density of gas.
   [**Ans.** 1.81 bar ; 0.09 m^3/kg ; 11.11 kg/m^3]

2. A force of 1600 N is applied uniformly on a piston of 8 cm diameter. Determine the pressure on the piston.
   [**Ans.** 319 kN/m^2]

3. An oil of specific gravity 0.8 is contained in a tube to a depth of 80 cm. Determine the gauge pressure at this depth in kN/m^2.
   [**Ans.** 6.275 kN/m^2]

4. A vacuum recorded in the condenser of a steam power plant is 745 mm of Hg. Find the absolute pressure in the condenser in Pa. The barometer reading is 760 mm of Hg.
   [**Ans.** 2262 Pa]

5. The pressure of gas in a pipe line is measured with a mercury manometer having one limb open to the atmosphere. If the difference in the height of mercury in the two limbs is 562 mm, calculate the gas pressure. The barometer reads 761 mm Hg, the acceleration due to gravity is 9.79 m/s^2, and the density of mercury is 13640 kg/m^3.
   [**Ans.** 1.77 bar]

6. A turbine is supplied with steam at a gauge pressure of 1.4 MPa. After expansion in the turbine the steam flows into a condenser which is maintained at a vacuum of 710 mm Hg. The barometric pressure is 772 mm Hg. Express the inlet and exhaust steam pressures in pascals (absolute). Take the density of mercury as 13600 kg/m^3.
   [**Ans.** 1.503 MPa ; 8.27 kPa]

7. Gas from a bottle of compressed helium is used to inflate an inelastic flexible balloon, originally folded completely flat to a volume of 0.5 m^3. If the barometer reads 760 mm Hg, what is the amount of work done upon the atmosphere by the balloon?
   [**Ans.** 50.66 kJ]

8. A rubber balloon (flexible) is to be filled with hydrogen from a storage cylinder which contains hydrogen at 5 bar pressure until it has a volume of 1 m^3. The atmospheric pressure may be taken as 1.013×10^5 N/m^2. Determine the work done by the system comprising the hydrogen initially in the bottle.
   [**Ans.** 101.3 kJ]

9. When the valve of the evacuated bottle is opened, atmospheric air rushes into it. If the atmospheric pressure is 101.325 kPa, and 1.2 m^3 of air (measured at atmospheric conditions) enters into the bottle, calculate the work done by the air.
   [**Ans.** – 60.8 kJ]

10. A pump forces 1.2 m^3/min of water horizontally from an open well to a closed tank where the pressure is 0.9 MPa. Compute the work the pump must do upon the water in an hour just to force the water into the tank against the pressure.
    [**Ans.** 15972 kJ]

11. 14.5 litres of gas at 172 MN/m^2 is expanded at constant pressure until its volume becomes 130.5 litres. Determine the work done by the gas.
    [**Ans.** 199.5 kJ]

12. 1 kg of a fluid is compressed reversibly according to a law $pv = 0.25$ where p is in bar and v is in m^3/kg. The final volume is 1/4 of the initial volume. Calculate the work done on the fluid and sketch the process on a p-v diagram.
    [**Ans.** 34660 Nm]

BASIC CONCEPTS OF THERMODYNAMICS

13. A gas having a volume of 0.05 m³ and pressure of 6.9 bar expands reversibly in a cylinder behind a piston according to law $pv^{1.2}$ = constant until the volume is 0.08 m³. Calculate the work done by the gas. Also sketch the process on a p-v diagram. **[Ans. 15300 Nm]**

14. A certain fluid at 10 bar is contained in a cylinder behind a piston, the initial volume being 0.05 m³. Calculate the work done by the fluid when it expands reversibly, (a) At constant pressure to final volume of 0.2 m³ ; (b) According to linear law to final volume of 0.2 m³ and a final pressure of 2 bar ; (c) According to a law pV = constant to a final volume of 0.1 m³ ; (d) According to law pV^3 = constant to a final volume of 0.06 m³. Sketch all processes on p-V diagram. **[Ans. 150000 ; 90000 ; 34700 ; 7640 Nm]**

15. A fluid undergoes the following processes :
    (i) Heated reversibly at a constant pressure of 1.05 bar until it has a specific volume of 0.1 m³/kg.
    (ii) It is then compressed reversibly according to a law pv = constant to a pressure of 4.2 bar.
    (iii) It is then allowed to expand reversibly according to a law $pv^{1.3}$ = constant.
    (iv) Finally it is heated at constant volume back to initial conditions.
    The work done in the constant pressure process is 515 Nm and the mass of fluid present is 0.2 kg. Calculate the net work done on or by the fluid in the cycle and sketch the cycle on a p-v diagram. **[Ans.– 422 Nm]**

2

First Law of Thermodynamics

2.1. Internal energy. 2.2. Law of conservation of energy. 2.3. First law of thermodynamics. 2.4. Application of first law to a process. 2.5. Energy—A property of system. 2.6. Perpetual motion machine of the first kind—PMM 1. 2.7. Energy of an isolated system. 2.8. The perfect gas—The characteristic equation of state—Specific heats—Joule's law—Relationship between two specific heats—Enthalpy—Ratio of specific heats. 2.9. Application of first law of thermodynamics to non-flow or closed system—Highlights—Objective Type Questions—Theoretical Questions—Unsolved Examples.

2.1. INTERNAL ENERGY

It is the heat energy stored in a gas. If a certain amount of heat is supplied to a gas the result is that temperature of gas may increase or volume of gas may increase thereby doing some external work or both temperature and volume may increase ; but it will be decided by the conditions under which the gas is supplied heat. *If during heating of the gas the temperature increases its internal energy will also increase.*

Joule's law of internal energy states that internal energy of a perfect gas is a *function of temperature only*. In other words, internal energy of a gas is dependent on the temperature change only and is not affected by the change in pressure and volume.

We do not know how to find the absolute quantity of internal energy in any substance ; however, what is needed in engineering is the change of internal energy (ΔU).

2.2. LAW OF CONSERVATION OF ENERGY

In the early part of nineteenth century the scientists developed the concept of energy and hypothesis that it can be neither created nor destroyed ; this came to be known as the *law of the conservation of energy*. The first law of thermodynamics is merely one statement of this general law/principle with particular reference to heat energy and mechanical energy *i.e.,* work.

2.3. FIRST LAW OF THERMODYNAMICS

It is observed that when a system is made to undergo a complete cycle then net work is done *on* or *by* the system. Consider a cycle in which net work is done by the system. Since energy cannot be created, this mechanical energy must have been supplied from some source of energy. Now the system has been returned to its initial state : Therefore, its *intrinsic* energy is unchanged, and hence the mechanical energy has not been provided by the system itself. The only other energy involved in the cycle is the heat which was supplied and rejected in various processes. Hence, by the law of conservation of energy, the net work done by the system is equal to the net heat supplied to the system. The First Law of Thermodynamics can, therefore, be stated as follows :

FIRST LAW OF THERMODYNAMICS

"**When a system undergoes a thermodynamic cycle then the net heat supplied to the system from the surroundings is equal to net work done by the system on its surroundings**".

or
$$\oint dQ = \oint dW$$

where $\oint$ represents the sum for a complete cycle.

The First Law of Thermodynamics *cannot be proved analytically, but experimental evidence has repeatedly confirmed its validity*, and since no phenomenon has been shown to contradict it, the first law is accepted as a *law of nature*. It may be remarked that no restriction was imposed which limited the application of first law to reversible energy transformation. Hence the first law applies to reversible as well as irreversible transformations : For non-cyclic process, a more general formulation of first law of thermodynamics is required. A new concept which involves a term called *internal energy* fulfils this need.

— The First Law of Thermodynamics may also be stated as follows :

"**Heat and work are mutually convertible but since energy can neither be created nor destroyed, the total energy associated with an energy conversion remains constant**".

Or

— "**No machine can produce energy without corresponding expenditure of energy, *i.e.*, it is impossible to construct a perpetual motion machine of first kind**".

Fig. 2.1 shows the experiment for checking first law of thermodynamics.

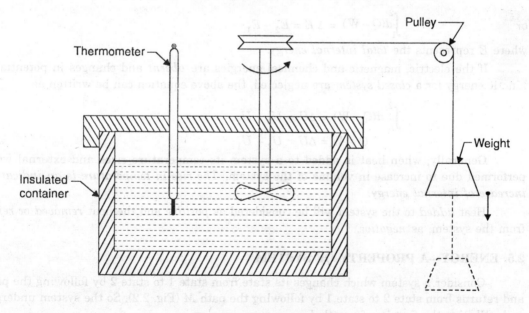

Fig. 2.1. Heat and work.

The work input to the paddle wheel is measured by the fall of weight, while the corresponding temperature rise of liquid in the insulated container is measured by the thermometer. It is already known to us from experiments on heat transfer that temperature rise can also be produced by heat transfer. The experiments show : (i) A definite quantity of work is always required to accomplish the same temperature rise obtained with a unit amount of heat. (ii) Regardless of whether the temperature of liquid is raised by work transfer or heat transfer, the liquid can be returned by heat transfer in opposite direction to the identical state from which it started. The above results lead to the inference that *work and heat* are different forms of something more general, which is called *energy*.

— It can be stated as an invariable experience that whenever a physical system passes through a complete cycle the algebraic sum of the work transfers during the cycle $\oint dW$ bears a definite ratio to the algebraic sum of the heat transfers during the cycle, $\oint dQ$. This may be expressed by the equation,

$$\oint dW = J \oint dQ \qquad \ldots(2.1)$$

where J is the proportionality constant and is known as *Mechanical Equivalent of heat*. In S.I. units its value is unity, *i.e.*, 1 Nm/J.

2.4. APPLICATION OF FIRST LAW TO A PROCESS

When a process is executed by a system, the *change in stored energy of the system is numerically equal to the net heat interactions minus the net work interaction during the process*.

$$\therefore \qquad E_2 - E_1 = Q - W$$

$$\therefore \qquad \Delta E = Q - W \qquad [\text{or } Q = \Delta E + W]$$

or

$$\int_1^2 d(Q - W) = \Delta E = E_2 - E_1 \qquad \ldots(2.2)$$

where E represents the *total internal energy*.

If the electric, magnetic and chemical energies are *absent* and changes in potential and kinetic energy for a *closed system* are neglected, the above equation can be written as

$$\int_1^2 d(Q - W) = \Delta U = U_2 - U_1 \qquad \ldots(2.3)$$

$$\therefore \qquad Q - W = \Delta U = U_2 - U_1 \qquad \ldots(2.4)$$

Generally, when heat is added to a system its temperature rises and external work is performed due to increase in volume of the system. *The rise in temperature is an indication of increase of internal energy*.

Heat *added* to the system will be considered as *positive* and the heat *removed* or *rejected*, from the system, as *negative*.

2.5. ENERGY—A PROPERTY OF SYSTEM

Consider a system which changes its state from state 1 to state 2 by following the path L, and returns from state 2 to state 1 by following the path M (Fig. 2.2). So the system undergoes a cycle. Writing the first law for path L

$$Q_L = \Delta E_L + W_L \qquad \ldots(2.5)$$

and for path M

$$Q_M = \Delta E_M + W_M \qquad \ldots(2.6)$$

FIRST LAW OF THERMODYNAMICS

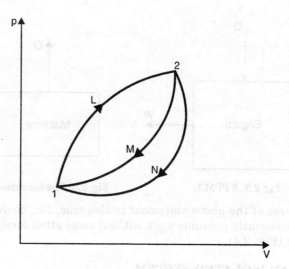

Fig. 2.2. Energy—a property of system.

The processes L and M together constitute a cycle, for which

$$\oint dW = \oint dQ$$

$$W_L + W_M = Q_L + Q_M$$

or
$$Q_L - W_L = W_M - Q_M \qquad \ldots(2.7)$$

From equations (2.5), (2.6) and (2.7), it yields

$$\Delta E_L = -\Delta E_M \qquad \ldots(2.8)$$

Similarly, had the system returned from state 2 to state 1 by following the path N instead of path M

$$\Delta E_L = -\Delta E_N \qquad \ldots(2.9)$$

From equations (2.8) and (2.9),

$$\Delta E_M = \Delta E_N \qquad \ldots(2.10)$$

Thus, it is seen that the *change in energy between two states of a system is the same, whatever path the system may follow in undergoing that change of state.* If some arbitrary value of energy is assigned to state 2, the value of energy at state 1 is fixed independent of the path the system follows. Therefore, *energy has a definite value for every state of the system.* Hence, it is *a point function and a property of the system.*

2.6. PERPETUAL MOTION MACHINE OF THE FIRST KIND—PMM 1

The first law of thermodynamics states the general principle of the conservation of energy. Energy is neither created nor destroyed, but only gets transformed from one form to another. *There can be no machine which would continuously supply mechanical work without some form of energy disappearing simultaneously* (Fig. 2.3). Such a *fictitious machine* is called a **perpetual motion machine of the first kind,** or in brief, PMM 1. A PMM 1 is thus **impossible.**

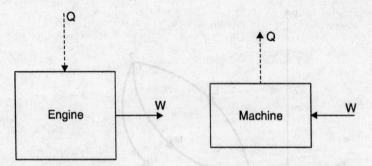

Fig. 2.3. A PPM 1.   Fig. 2.4. The converse of PMM 1.

— The converse of the above statement is also true, *i.e.*, there can be no machine which would continuously consume work without some other form of energy appearing simultaneously (Fig. 2.4).

2.7. ENERGY OF AN ISOLATED SYSTEM

An isolated system is one in which there is no interaction of the system with the surroundings.

For an isolated system,
$$dQ = 0, dW = 0$$
The first law of thermodynamics gives
$$dE = 0$$
or
$$E = \text{constant}$$

The energy of an isolated system is always constant.

2.8. THE PERFECT GAS

2.8.1. The Characteristic Equation of State

— At temperatures that are considerably in excess of critical temperature of a fluid, and also at very low pressure, the vapour of fluid tends to obey the equation

$$\frac{pv}{T} = \text{constant} = R$$

In practice, no gas obeys this law rigidly, but many gases tend towards it.

An imaginary ideal gas which obeys this law is called a *perfect gas*, and the equation $\frac{pv}{T} = R$, is called the *characteristic equation of a state of a perfect gas*. The constant R is called the *gas constant*. Each perfect gas has a different gas constant.

Units of R are Nm/kg K or kJ/kg K.

Usually, the characteristic equation is written as
$$pv = RT \qquad \ldots(2.11)$$
or for m kg, occupying V m^3
$$pV = mRT \qquad \ldots(2.12)$$

— The characteristic equation in *another form,* can be derived by using kilogram-mole as a unit.

FIRST LAW OF THERMODYNAMICS

The *kilogram-mole* is defined as a quantity of a gas equivalent to M kg of the gas, where M is the molecular weight of the gas (*e.g.*, since the molecular weight of oxygen is 32, then 1 kg mole of oxygen is equivalent to 32 kg of oxygen).

As per definition of the kilogram-mole, for m kg of a gas, we have

$$m = nM \qquad ...(2.13)$$

where n = number of moles.

Note. Since the standard of mass is the kg, kilogram-mole will be written simply as mole.

Substituting for m from Eqn. (2.13) in Eqn. (2.12) gives

$$pV = nMRT$$

or

$$MR = \frac{pV}{nT}$$

According to *Avogadro's hypothesis* the volume of 1 mole of any gas is the same as the volume of 1 mole of any other gas, when the gases are at the same temperature and pressure. Therefore, $\frac{V}{n}$ is the same for all gases at the same value of p and T. That is the quantity $\frac{pV}{nT}$ is a *constant* for all gases. This constant is called *universal gas constant*, and is given the symbol, R_0.

i.e.,

$$MR = R_0 = \frac{pV}{nT}$$

or

$$pV = nR_0 T \qquad ...(2.14)$$

Since $MR = R_0$, then

$$R = \frac{R_0}{M} \qquad ...(2.15)$$

It has been found experimentally that the volume of 1 mole of any perfect gas at 1 bar and 0°C is approximately 22.71 m³.

Therefore from Eqn. (2.14),

$$R_0 = \frac{pV}{nT} = \frac{1 \times 10^5 \times 22.71}{1 \times 273.15}$$

$$= 8314.3 \text{ Nm/mole K}$$

Using Eqn. (2.15), the gas constant for any gas can be found when the molecular weight is known.

Example. For oxygen which has a molecular weight of 32, the gas constant

$$R = \frac{R_0}{M} = \frac{8314}{32} = 259.8 \text{ Nm/kg K.}$$

2.8.2. Specific Heats

— The specific heat of a solid or liquid is usually defined as the *heat required to raise unit mass through one degree temperature rise.*

— For small quantities, we have

$$dQ = mcdT$$

where m = Mass,
  c = Specific heat, and
  dT = Temperature rise.

For a gas there are an infinite number of ways in which heat may be added between any two temperatures, and hence a *gas could have an infinite number of specific heats*. However, only two specific heats for gases are defined.

and   Specific heat at constant volume, c_v
          Specific heat at constant pressure, c_p.
We have
$$dQ = m\, c_p\, dT \text{ For a reversible non-flow process at } constant\ pressure \quad ...(2.16)$$
and
$$dQ = m\, c_v\, dT \text{ For a reversible non-flow process at } constant\ volume \quad ...(2.17)$$

The values of c_p and c_v, for a perfect gas, are constant for any one gas at all pressures and temperatures. Hence, integrating Eqns. (2.16) and (2.17), we have

Flow of heat in a reversible constant pressure process
$$= mc_p\,(T_2 - T_1) \quad ...(2.18)$$
Flow of heat in a reversible constant volume process
$$= mc_v\,(T_2 - T_1) \quad ...(2.19)$$

In case of *real gases*, c_p and c_v *vary with temperature,* but a suitable *average value* may be used for most practical purposes.

2.8.3. Joule's Law

Joule's law states as follows :

"The internal energy of a perfect gas is a function of the absolute temperature only."

i.e., $u = f(T)$

To evaluate this function let 1 kg of a perfect gas be heated at constant volume.

According to non-flow energy equation,
$$dQ = du + dW$$
$dW = 0$, since volume remains constant
$$\therefore \quad dQ = du$$
At constant volume for a perfect gas, from Eqn. (2.17), for 1 kg
$$dQ = c_v dT$$
$\therefore$
$$dQ = du = c_v dT$$
and integrating $u = c_v\, T + K$, K being constant.

According to Joule's law $u = f(T)$, which means that internal energy varies linearly with absolute temperature. Internal energy can be made zero at any arbitrary reference temperature. For a perfect gas it can be assumed that $u = 0$ when $T = 0$, hence constant K is zero.

i.e.,   Internal energy, $u = c_v\, T$ for a perfect gas ...(2.20)

or   For mass m, of a perfect gas

Internal energy, $U = mc_v\, T$ ...(2.21)

For a perfect gas, in any process between states 1 and 2, we have from Eqn. (2.21)

Gain in internal energy,
$$U_2 - U_1 = mc_v\,(T_2 - T_1) \quad ...(2.22)$$

Eqn. (2.22) gives the gains of internal energy for a perfect gas between two states *for any process, reversible* or *irreversible*.

2.8.4. Relationship Between Two Specific Heats

Consider a perfect gas being heated at constant pressure from T_1 to T_2.

According to non-flow equation,
$$Q = (U_2 - U_1) + W$$
Also for a perfect gas,
$$U_2 - U_1 = mc_v\,(T_2 - T_1)$$
$$Q = mc_v\,(T_2 - T_1) + W$$

FIRST LAW OF THERMODYNAMICS

In a constant pressure process, the work done by the fluid,
$$W = p(V_2 - V_1)$$
$$= mR(T_2 - T_1)$$

$$\left[\begin{array}{l} \because \ p_1 V_1 = mRT_1 \\ p_2 V_2 = mRT_2 \\ p_1 = p_2 = p \text{ in this case} \end{array} \right]$$

On substituting
$$Q = mc_v(T_2 - T_1) + mR(T_2 - T_1) = m(c_v + R)(T_2 - T_1)$$

But for a constant pressure process,
$$Q = mc_p(T_2 - T_1)$$

By equating the two expressions, we have
$$m(c_v + R)(T_2 - T_1) = mc_p(T_2 - T_1)$$
$$\therefore \quad c_v + R = c_p$$

or
$$c_p - c_v = R \qquad \ldots(2.23)$$

Dividing both sides by c_v, we get
$$\frac{c_p}{c_v} - 1 = \frac{R}{c_v}$$

$$\therefore \quad c_v = \frac{R}{\gamma - 1} \qquad \ldots[2.23\,(a)]$$

(where $\gamma = c_p/c_v$)

Similarly, dividing both sides by c_p, we get
$$c_p = \frac{\gamma R}{\gamma - 1} \qquad \ldots[2.23\,(b)]$$

$$\left[\begin{array}{l} \text{In M.K.S. units}: c_p - c_v = \dfrac{R}{J}\,;\, c_v = \dfrac{R}{J(\gamma - 1)},\, c_p = \dfrac{\gamma R}{(\gamma - 1)J} \\ \text{In SI units the value of } J \text{ is unity.} \end{array} \right]$$

2.8.5. Enthalpy

— One of the fundamental quantities which occur invariably in thermodynamics is the sum of internal energy (u) and pressure volume product (pv). This sum is called **Enthalpy** (h).

i.e.,
$$h = u + pv \qquad \ldots(2.24)$$

— The enthalpy of a fluid is the property of the fluid, since it consists of the sum of a property and the product of the two properties. Since enthalpy is a property like internal energy, pressure, specific volume and temperature, it can be introduced into any problem whether the process is a flow or a non-flow process.

The total enthalpy of mass, m, of a fluid can be
$$H = U + pV, \text{ where } H = mh.$$

For a **perfect gas**,
Referring equation (2.24),
$$h = u + pv$$

$$= c_v T + RT \qquad [\because pv = RT]$$
$$= (c_v + R)T$$
$$= c_p T \qquad [\because c_p = c_v + R]$$

i.e., $\qquad h = c_p T$

and $\qquad H = mc_p T.$

(Note that, since it has been assumed that $u = 0$ at $T = 0$, then $h = 0$ at $T = 0$).

2.8.6. Ratio of Specific Heats

The ratio of specific heat at constant pressure to the specific heat at constant volume is given the symbol γ (gamma).

i.e., $\qquad \gamma = \dfrac{c_p}{c_v}$ ...(2.25)

Since $c_p = c_v + R$, it is clear that c_p must be *greater* than c_v for any perfect gas. It follows, therefore, that the ratio, $\dfrac{c_p}{c_v} = \gamma$ is *always greater than unity*.

In general, the approximate values of γ are as follows :

For *monoatomic* gases such as *argon, helium* = 1.6.

For *diatomic* gases such as *carbon monoxide, hydrogen, nitrogen* and *oxygen* = 1.4.

For triatomic gases such as *carbon dioxide* and *sulphur dioxide* = 1.3.

For some hydro-carbons the value of γ is quite low.

[*e.g.*, for ethane $\gamma = 1.22$, and for isobutane $\gamma = 1.11$]

2.9. APPLICATION OF FIRST LAW OF THERMODYNAMICS TO NON-FLOW OR CLOSED SYSTEM

1. Reversible Constant Volume Process (v = constant)

In a constant volume process the working substance is contained in a rigid vessel, hence the boundaries of the system are immovable and no work can be done on or by the system, other than paddle-wheel work input. It will be assumed that *'constant volume'* implies zero work unless stated otherwise.

Fig. 2.5 shows the system and states before and after the heat addition at constant volume.

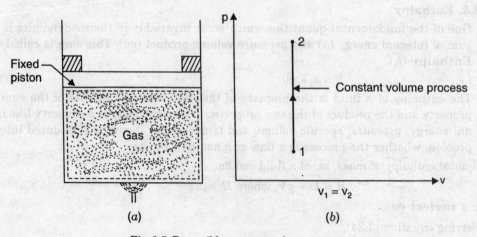

Fig. 2.5. Reversible constant volume process.

FIRST LAW OF THERMODYNAMICS

Considering mass of the working substance *unity* and applying first law of thermodynamics to the process

$$Q = (u_2 - u_1) + W \qquad \qquad ...(2.26)$$

The work done $\quad W = \int_1^2 p dv = 0$ as $dv = 0$.

$$\therefore \quad Q = (u_2 - u_1) = c_v(T_2 - T_1) \qquad \qquad ...[2.27\ (a)]$$

where c_v = Specific heat at constant volume.

For mass, m, of working substance

$$Q = U_2 - U_1 = mc_v(T_2 - T_1) \qquad \qquad ...[2.27\ (b)]$$
$$[\because \quad mu = U]$$

2. Reversible Constant Pressure Process (p = constant)

It can be seen from Fig. 2.5 (b) that when the boundary of the system is *inflexible* as in a constant volume process, then the pressure rises when heat is supplied. Hence for a constant pressure process, the boundary must move against an external resistance as heat is supplied ; for instance a gas [Fig. 2.6 (a)] in a cylinder behind a piston can be made to undergo a constant pressure process. Since the *piston is pushed through a certain distance* by the force exerted by the gas, then the work is done by the gas on its surroundings.

Fig. 2.6 shows the system and states before and after the heat addition at constant pressure.

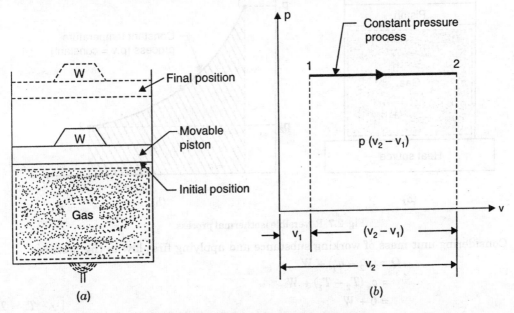

Fig. 2.6. Reversible constant pressure process.

Considering unit mass of working substance and applying first law of thermodynamics to the process

$$Q = (u_2 - u_1) + W$$

The work done, $\quad W = \int_1^2 p dv = p(v_2 - v_1)$

$$\therefore \quad Q = (u_2 - u_1) + p(v_2 - v_1) = u_2 - u_1 + pv_2 - pv_1$$
$$= (u_2 + pv_2) - (u_1 + pv_1) = h_2 - h_1 \qquad [\because \ h = u + pv]$$
or
$$Q = h_2 - h_1 = c_p (T_2 - T_1) \qquad \qquad ...(2.28)$$

where h = Enthalpy (specific), and

c_p = Specific heat at constant pressure.

For mass, m, of working substance

$$Q = H_2 - H_1 = mc_p (T_2 - T_1) \qquad \qquad ...[2.28\ (a)]$$
$$[\because \ mh = H]$$

3. Reversible Temperature or Isothermal Process (pv = constant, T = constant)

A process at a constant temperature is called an isothermal process. When a working substance in a cylinder behind a piston expands from a high pressure to a low pressure there is a tendency for the temperature to fall. In an isothermal expansion heat must be added continuously in order to keep the temperature at the initial value. Similarly in an isothermal compression heat must be removed from the working substance continuously during the process.

Fig. 2.7 shows the system and states before and after the heat addition at constant temperature.

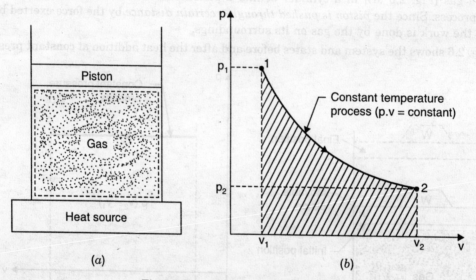

Fig. 2.7. Reversible isothermal process.

Considering unit mass of working substance and applying first law to the process

$$Q = (u_2 - u_1) + W$$
$$= c_v (T_2 - T_1) + W$$
$$= 0 + W \qquad \qquad [\because \ T_2 = T_1]$$

The work done, $W = \int_1^2 p\,dv$

In this case pv = constant or $p = \dfrac{C}{v}$ (where C = constant)

$$\therefore \qquad W = \int_{v_1}^{v_2} C\frac{dv}{v} = C[\log_e v]_{v_1}^{v_2} = C \log_e \frac{v_2}{v_1}$$

FIRST LAW OF THERMODYNAMICS

The constant C can either be written as p_1v_1 or as p_2v_2, since
$$p_1v_1 = p_2v_2 = \text{constant}, C$$

i.e.,
$$W = p_1v_1 \log_e \frac{v_2}{v_1} \text{ per unit mass of working substance}$$

or
$$W = p_2v_2 \log_e \frac{v_2}{v_1} \text{ per unit mass of working substance}$$

$\therefore$
$$Q = W = p_1v_1 \log_e \frac{v_2}{v_1} \qquad \qquad ...(2.29)$$

For mass, m, of the working substance
$$Q = p_1V_1 \log_e \frac{V_2}{V_1} \qquad \qquad ...[2.29\,(a)]$$

or
$$Q = p_1V_1 \log_e \frac{p_1}{p_2} \qquad \left[\because \frac{V_2}{V_1} = \frac{p_1}{p_2}\right] \qquad ...[2.29\,(b)]$$

4. Reversible Adiabatic Process (pv^γ = constant)

An **adiabatic process** *is one in which no heat is transferred to or from the fluid during the process.* Such a process can be reversible or irreversible. The reversible adiabatic non-flow process will be considered in this section.

Considering unit mass of working substance and applying first law to the process
$$Q = (u_2 - u_1) + W$$
$$0 = (u_2 - u_1) + W$$

or
$$W = (u_1 - u_2) \text{ for any adiabatic process} \qquad \qquad ...(2.30)$$

Eqn. (2.30) is true for an adiabatic process whether the process is reversible or not. In an adiabatic expansion, the work done W by the fluid is at the expense of a reduction in the internal energy of the fluid. Similarly in an adiabatic compression process all the work done on the fluid goes to increase the internal energy of the fluid.

For an adiabatic process to take place, perfect thermal insulation for the system must be available.

To derive the law pv^γ = constant :

To obtain a law relating p and v for a reversible adiabatic process let us consider the non-flow energy equation in differential form,
$$dQ = du + dW$$

For a *reversible process*
$$dW = pdv$$

$\therefore$
$$dQ = du + pdv = 0$$

(Since for an adiabatic process $Q = 0$)
Also for a perfect gas
$$pv = RT \text{ or } p = \frac{RT}{v}$$

Hence substituting,
$$du + \frac{RTdv}{v} = 0$$

Also
$$u = c_vT \text{ or } du = c_vdT$$

$\therefore$
$$c_vdT + \frac{RTdv}{v} = 0$$

Dividing both sides by T, we get

$$c_v \frac{dT}{T} + \frac{Rdv}{v} = 0$$

Integrating

$$c_v \log_e T + R \log_e v = \text{constant}$$

Substituting $T = \frac{pv}{R}$

$$c_v \log_e \frac{pv}{R} + R \log_e v = \text{constant}$$

Dividing throughout both sides by c_v

$$\log_e \frac{pv}{R} + \frac{R}{c_v} \cdot \log_e v = \text{constant}$$

Again

$$c_v = \frac{R}{(\gamma - 1)} \quad \text{or} \quad \frac{R}{c_v} = \gamma - 1$$

Hence substituting

$$\log_e \frac{pv}{R} + (\gamma - 1) \log_e v = \text{constant}$$

$$\therefore \quad \log_e \frac{pv}{R} + \log_e v^{\gamma - 1} = \text{constant}$$

$$\log_e \frac{pv \times v^{\gamma - 1}}{R} = \text{constant}$$

i.e.,

$$\log_e \frac{pv^\gamma}{R} = \text{constant}$$

i.e.,

$$\frac{pv^\gamma}{R} = e^{\text{constant}} = \text{constant}$$

or

$$pv^\gamma = \text{constant} \qquad \ldots(2.31)$$

Expression for work W :

A reversible adiabatic process for a perfect gas is shown on a p-v diagram in Fig. 2.8 (b).

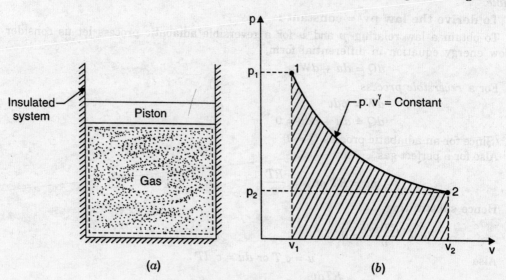

Fig. 2.8. Reversible adiabatic process.

FIRST LAW OF THERMODYNAMICS

The work done is given by the shaded area, and this area can be evaluated by integration.

i.e.,
$$W = \int_{v_1}^{v_2} p \, dv$$

Therefore, since $pv^\gamma = $ constant, C, then

$$W = \int_{v_1}^{v_2} C \frac{dv}{v^\gamma} \qquad \left[\because p = \frac{C}{v^\gamma}\right]$$

i.e.,
$$W = C \int_{v_1}^{v_2} \frac{dv}{v^\gamma} = C \left| \frac{v^{-\gamma+1}}{-\gamma+1} \right|_{v_1}^{v_2}$$

$$= C \left(\frac{v_2^{-\gamma+1} - v_1^{-\gamma+1}}{1-\gamma} \right) = C \left(\frac{v_1^{-\gamma+1} - v_2^{-\gamma+1}}{\gamma-1} \right)$$

The constant in this equation can be written as $p_1 v_1^\gamma$ or as $p_2 v_2^\gamma$. Hence,

$$W = \frac{p_1 v_1^\gamma v_1^{-\gamma+1} - p_2 v_2^\gamma v_2^{-\gamma+1}}{\gamma-1} = \frac{p_1 v_1 - p_2 v_2}{\gamma-1}$$

i.e.,
$$W = \frac{p_1 v_1 - p_2 v_2}{\gamma-1} \qquad \ldots(2.32)$$

or
$$W = \frac{R(T_1 - T_2)}{\gamma-1} \qquad \ldots(2.33)$$

Relationship between T and v, and T and p :

By using equation $pv = RT$, the relationship between T and v, and T and p, may be derived as follows :

i.e.,
$$pv = RT$$

$$\therefore \quad p = \frac{RT}{v}$$

Putting this value in the equation $pv^\gamma = $ constant

$$\frac{RT}{v} \cdot v^\gamma = \text{constant}$$

i.e.,
$$Tv^{\gamma-1} = \text{constant} \qquad \ldots(2.34)$$

Also $v = \dfrac{RT}{p}$; hence substituting in equation $pv^\gamma = $ constant

$$p \left(\frac{RT}{p} \right)^\gamma = \text{constant}$$

$$\therefore \quad \frac{T^\gamma}{p^{\gamma-1}} = \text{constant}$$

or
$$\frac{T}{(p)^{\frac{\gamma-1}{\gamma}}} = \text{constant} \qquad \ldots(2.35)$$

Therefore, for a reversible adiabatic process for a perfect gas between states 1 and 2, we can write :

From Eqn. (2.31),

$$p_1 v_1^\gamma = p_2 v_2^\gamma \quad \text{or} \quad \frac{p_2}{p_1} = \left(\frac{v_1}{v_2}\right)^\gamma \qquad \ldots(2.36)$$

From Eqn. (2.34),

$$T_1 v_1^{\gamma-1} = T_2 v_2^{\gamma-1} \quad \text{or} \quad \frac{T_2}{T_1} = \left(\frac{v_1}{v_2}\right)^{\gamma-1} \qquad \ldots(2.37)$$

From Eqn. (2.35),

$$\frac{T_1}{(p_1)^{\frac{\gamma-1}{\gamma}}} = \frac{T_2}{(p_2)^{\frac{\gamma-1}{\gamma}}} \quad \text{or} \quad \frac{T_2}{T_1} = \left(\frac{p_2}{p_1}\right)^{\frac{\gamma-1}{\gamma}} \qquad \ldots(2.38)$$

From Eqn. (2.30), the work done in an adiabatic process per kg of gas is given by $W = (u_1 - u_2)$. The gain in internal energy of a perfect gas is given by equation :

$$u_2 - u_1 = c_v (T_2 - T_1) \qquad \text{(for 1 kg)}$$

$$\therefore \quad W = c_v (T_1 - T_2)$$

Also, we know that

$$c_v = \frac{R}{\gamma - 1}$$

Hence substituting, we get

$$W = \frac{R(T_1 - T_2)}{\gamma - 1}$$

Using equation, $pv = RT$

$$W = \frac{p_1 v_1 - p_2 v_2}{\gamma - 1}$$

This is the same expression obtained before as Eqn. (2.32).

5. Polytropic Reversible Process (pv^n = constant)

It is found that many processes in practice approximate to a reversible law of form pv^n = constant, where n is a constant. Both vapours and perfect gases obey this type of law closely in many non-flow processes. Such processes are *internally reversible*.

We know that for any reversible process,

$$W = \int p\, dv$$

For a process in pv^n = constant, we have

$$p = \frac{C}{v^n}, \text{ where } C \text{ is a constant}$$

$$\therefore \quad W = C \int_{v_1}^{v_2} \frac{dv}{v^n} = C \left| \frac{v^{-n+1}}{-n+1} \right| = C \left(\frac{v_2^{-n+1} - v_1^{-n+1}}{-n+1} \right)$$

i.e.,

$$W = C \left(\frac{v_1^{-n+1} - v_2^{-n+1}}{n-1} \right) = \frac{p_1 v_1^n v_1^{-n+1} - p_2 v_2^n v_2^{-n+1}}{n-1}$$

(since the constant C, can be written as $p_1 v_1^n$ or as $p_2 v_2^n$)

FIRST LAW OF THERMODYNAMICS

i.e., $\quad\quad\quad$ Work done, $W = \dfrac{p_1 v_1 - p_2 v_2}{n - 1}$ $\quad\quad$ ...(2.39)

or $\quad\quad\quad\quad\quad\quad W = \dfrac{R(T_1 - T_2)}{n - 1}$ $\quad\quad$ ...(2.40)

Eqn. (2.39) is true for any working substance undergoing a reversible polytropic process. It follows also that for any polytropic process, we can write

$$\dfrac{p_2}{p_1} = \left(\dfrac{v_1}{v_2}\right)^n \quad\quad ...(2.41)$$

The following relations can be derived (following the same procedure as was done under reversible adiabatic process)

$$\dfrac{T_2}{T_1} = \left(\dfrac{v_1}{v_2}\right)^{n-1} \quad\quad ...(2.42)$$

$$\dfrac{T_2}{T_1} = \left(\dfrac{p_2}{p_1}\right)^{\frac{n-1}{n}} \quad\quad ...(2.43)$$

Heat transfer during polytropic process (for perfect gas pv = RT) :

Using non-flow energy equation, the heat flow/transfer during the process can be found,

i.e., $\quad\quad\quad\quad Q = (u_2 - u_1) + W$

$\quad\quad\quad\quad\quad\quad = c_v(T_2 - T_1) + \dfrac{R(T_1 - T_2)}{n - 1}$

i.e., $\quad\quad\quad\quad Q = \dfrac{R(T_1 - T_2)}{n - 1} - c_v(T_1 - T_2)$

Also $\quad\quad\quad\quad c_v = \dfrac{R}{(\gamma - 1)}$

On substituting,

$$Q = \dfrac{R}{n - 1}(T_1 - T_2) - \dfrac{R}{(\gamma - 1)}(T_1 - T_2)$$

i.e., $\quad\quad\quad\quad Q = R(T_1 - T_2)\left(\dfrac{1}{n - 1} - \dfrac{1}{\gamma - 1}\right)$

$\quad\quad\quad\quad\quad = \dfrac{R(T_1 - T_2)(\gamma - 1 - n + 1)}{(\gamma - 1)(n - 1)} = \dfrac{R(T_1 - T_2)(\gamma - n)}{(\gamma - 1)(n - 1)}$

$\therefore \quad\quad\quad\quad Q = \dfrac{(\gamma - n)}{(\gamma - 1)} \dfrac{R(T_1 - T_2)}{(n - 1)}$

or $\quad\quad\quad\quad Q = \left(\dfrac{\gamma - n}{\gamma - 1}\right) W \quad\quad \left[\because W = \dfrac{R(T_1 - T_2)}{(n - 1)}\right]$ ...(2.44)

In a polytropic process, the *index n depends only on the heat and work quantities* during the process. The various processes considered earlier are special cases of polytropic process for a perfect gas. For example,

$\quad\quad$ When $\quad n = 0$ $\quad\quad\quad\quad pv^0$ = constant *i.e.*, p = constant

$\quad\quad$ When $\quad n = \infty$ $\quad\quad\quad\quad pv^\infty$ = constant

or $\quad\quad p^{1/\infty} v$ = constant, $\quad\quad$ *i.e.*, v = constant

$\quad\quad$ When $\quad n = 1$ $\quad\quad\quad\quad pv$ = constant, *i.e.*, T = constant

$\quad\quad\quad\quad\quad\quad\quad\quad\quad\quad\quad\quad$ [since $(pv)/T$ = constant for a perfect gas]

$\quad\quad$ When $\quad n = \gamma$ $\quad\quad\quad\quad pv^\gamma$ = constant, *i.e.*, reversible adiabatic

This is illustrated on a *p-v* diagram in Fig. 2.9.

(*i*) State 1 to state *A* is *constant pressure cooling* ($n = 0$).
(*ii*) State 1 to state *B* is *isothermal compression* ($n = 1$).
(*iii*) State 1 to state *C* is *reversible adiabatic compression* ($n = \gamma$).
(*iv*) State 1 to state *D* is *constant volume heating* ($n = \infty$).

Similarly,

(*i*) State 1 to state *A'* is *constant pressure heating* ($n = 0$).
(*ii*) State 1 to state *B'* is *isothermal expansion* ($n = 1$).
(*iii*) State 1 to state *C'* is *reversible adiabatic expansion* ($n = \gamma$).
(*iv*) State 1 to state *D'* is *constant volume cooling* ($n = \infty$).

It may be noted that, since γ is always greater than unity, than process 1 to *C* must lie between processes 1 to *B* and 1 to *D* ; similarly, process 1 to *C'* must lie between processes 1 to *B'* and 1 to *D'*.

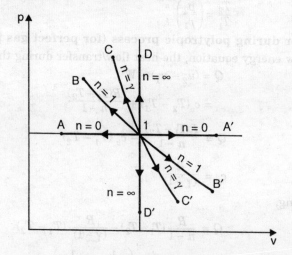

Fig. 2.9. Polytropic process.

6. Free Expansion

Consider two vessels 1 and 2 interconnected by a short pipe with a valve *A*, and perfectly thermally insulated [Fig. 2.10]. Initially let the vessel 1 be filled with a fluid at a certain pressure, and let 2 be completely evacuated. When the valve *A* is opened the fluid in 1 will expand rapidly to fill both vessels 1 and 2. The pressure finally will be lower than the initial pressure in vessel 1. This is known as *free or unresisted expansion*. The process is *highly irreversible* ; since the fluid is eddying continuously during the process. Now applying first law of thermodynamics (or non-flow energy equation) between the initial and final states,

$$Q = (u_2 - u_1) + W$$

In this process, no work is done on or by the fluid, since the boundary of the system does not move. No heat flows to or from the fluid since the system is well lagged. The process is therefore, *adiabatic but irreversible*.

i.e., $$u_2 - u_1 = 0 \quad \text{or} \quad u_2 = u_1$$

In a free expansion, therefore, the internal energy initially equals the initial energy finally.

For a perfect gas,
$$u = c_v T$$

FIRST LAW OF THERMODYNAMICS

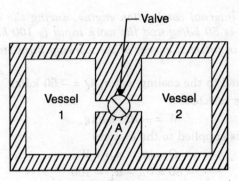

Fig. 2.10. Free expansion.

∴ For a free expansion of a perfect gas,
$$c_v T_1 = c_v T_2 \quad i.e., \quad T_1 = T_2$$

That is, for a perfect gas undergoing a free expansion, the initial temperature is equal to the final temperature.

Table 2.1. Summary of Processes for Perfect Gas (Unit mass)

Process	Index n	Heat added	$\int_1^2 p\,dv$	p, v, T relations	Specific heat, c
Constant pressure	$n = 0$	$c_p(T_2 - T_1)$	$p(v_2 - v_1)$	$\dfrac{T_2}{T_1} = \dfrac{v_2}{v_1}$	c_p
Constant volume	$n = \infty$	$c_v(T_2 - T_1)$	0	$\dfrac{T_1}{T_2} = \dfrac{p_1}{p_2}$	c_v
Constant temperature	$n = 1$	$p_1 v_1 \log_e \dfrac{v_2}{v_1}$	$p_1 v_1 \log_e \dfrac{v_2}{v_1}$	$p_1 v_1 = p_2 v_2$	∞
Reversible adiabatic	$n = \gamma$	0	$\dfrac{p_1 v_1 - p_2 v_2}{\gamma - 1}$	$p_1 v_1^\gamma = p_2 v_2^\gamma$ $\dfrac{T_2}{T_1} = \left(\dfrac{v_1}{v_2}\right)^{\gamma - 1}$ $= \left(\dfrac{p_2}{p_1}\right)^{\frac{\gamma - 1}{\gamma}}$	0
Polytropic	$n = n$	$c_n(T_2 - T_1)$ $= c_v\left(\dfrac{\gamma - n}{1 - n}\right) \times (T_2 - T_1)$ $= \dfrac{\gamma - n}{\gamma - 1} \times$ work done (non–flow)	$\dfrac{p_1 v_1 - p_2 v_2}{n - 1}$	$p_1 v_1^n = p_2 v_2^n$ $\dfrac{T_2}{T_1} = \left(\dfrac{v_1}{v_2}\right)^{n - 1}$ $= \left(\dfrac{p_2}{p_1}\right)^{\frac{n - 1}{n}}$	$c_n = c_v\left(\dfrac{\gamma - n}{1 - n}\right)$

Note. Equations must be used keeping dimensional consistence.

Example 2.1. *In an internal combustion engine, during the compression stroke the heat rejected to the cooling water is 50 kJ/kg and the work input is 100 kJ/kg.*

Calculate the change in internal energy of the working fluid stating whether it is a gain or loss.

Solution. Heat rejected to the cooling water, $Q = -50$ kJ/kg

(–ve sign since heat is rejected)

Work input, $\qquad W = -100$ kJ/kg

(–ve sign since work is supplied to the system)

Using the relation, $\qquad Q = (u_2 - u_1) + W$

$\qquad\qquad -50 = (u_2 - u_1) - 100$

or $\qquad\qquad u_2 - u_1 = -50 + 100 = 50$ kJ/kg

Hence, **gain in internal energy = 50 kJ/kg. (Ans.)**

Example 2.2. *In an air motor cylinder the compressed air has an internal energy of 450 kJ/kg at the beginning of the expansion and an internal energy of 220 kJ/kg after expansion. If the work done by the air during the expansion is 120 kJ/kg, calculate the heat flow to and from the cylinder.*

Solution. Internal energy at beginning of the expansion,

$$u_1 = 450 \text{ kJ/kg}$$

Internal energy after expansion,

$$u_2 = 220 \text{ kJ/kg}$$

Work done by the air during expansion,

$$W = 120 \text{ kJ/kg}$$

Heat flow, Q :

Using the relation, $\qquad Q = (u_2 - u_1) + W$

$\therefore \qquad\qquad Q = (220 - 450) + 120$

$\qquad\qquad\qquad = -230 + 120 = -110$ kJ/kg

Hence, **heat rejected by air = 110 kJ/kg. (Ans.)**

Example 2.3. *0.3 kg of nitrogen gas at 100 kPa and 40°C is contained in a cylinder. The piston is moved compressing nitrogen until the pressure becomes 1 MPa and temperature becomes 160°C. The work done during the process is 30 kJ.*

Calculate the heat transferred from the nitrogen to the surroundings.

c_v *for nitrogen* = 0.75 kJ/kg K.

Solution. Mass of nitrogen, $m = 0.3$ kg

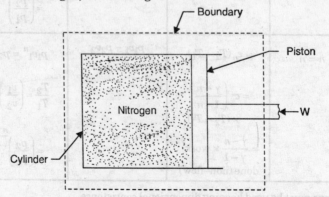

Fig. 2.11

FIRST LAW OF THERMODYNAMICS

Temperature before compression = 40°C or 313 K
Temperature after compression = 160°C or 433 K
The work done during the compression process, $W = -30$ kJ
According to first law of thermodynamics,

$$Q = \Delta U + W = (U_2 - U_1) + W$$
$$= mc_v(T_2 - T_1) + W$$
$$= 0.3 \times 0.75(433 - 313) - 30 = -3 \text{ kJ}$$

Hence, **heat 'rejected' during the process = 3 kJ. (Ans.)**

Note. Work, W has been taken –ve because it has been supplied from outside.

Example 2.4. *When a stationary mass of gas was compressed without friction at constant pressure its initial state of 0.4 m³ and 0.105 MPa was found to change to final state of 0.20 m³ and 0.105 MPa. There was a transfer of 42.5 kJ of heat from the gas during the process.*

How much did the internal energy of the gas change ?

Solution.

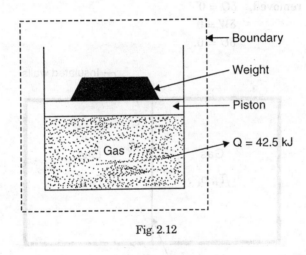

Fig. 2.12

Initial state
Pressure of gas, $p_1 = 0.105$ MPa
Volume of gas, $V_1 = 0.4$ m³

Final state
Pressure of gas, $p_2 = 0.105$ MPa
Volume of gas, $V_2 = 0.20$ m³
Process used : *Constant pressure*
Heat transferred, $Q = -42.5$ kJ
(–ve sign indicates that heat is rejected)

Change in internal energy, $\Delta U = U_2 - U_1$:
First law for a stationary system in a process gives

$$Q = \Delta U + W$$

or $Q_{1-2} = (U_2 - U_1) + W_{1-2}$ ...(i)

Here $W_{1-2} = \int_{V_1}^{V_2} pdV = p(V_2 - V_1)$

= 0.105(0.20 − 0.40) MJ = − 21 kJ   [∵ 1 MJ = 10^3 kJ]

Substituting this value of W_{1-2} in equation (i), we get

− 42.5 = $(U_2 − U_1)$ − 21

∴ $U_2 − U_1$ = − 42.5 + 21 = − 21.5 kJ

Hence **'decrease' in internal energy = 21.5 kJ.** **(Ans.)**

Example 2.5. *A container is divided into compartments by a partition. The container is completely insulated so that there is no heat transfer. One portion contains gas at temperature T_1 and pressure p_1 while the other portion also has the same gas but at temperature T_2 and pressure p_2.*

How will the First Law of Thermodynamics conclude the result if partition is removed ?

Solution. Refer to Fig. 2.13.

According to First Law of Thermodynamics,

$$\delta Q = \delta U + \delta W$$

When partition removed, $\delta Q = 0$

$\delta W = 0$

∴ $\delta U = 0.$

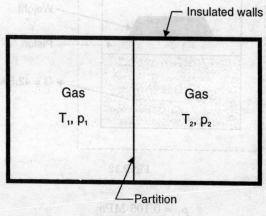

Fig. 2.13

Conclusion. *There is conservation of internal energy.*

Example 2.6. *Air enters a compressor at 10^5 Pa and 25°C having volume of 1.8 m^3/kg and is compressed to 5 × 10^5 Pa isothermally.*

Determine : (i) Work done ;

(ii) Change in internal energy ; and

(iii) Heat transferred.

Solution. Initial pressure of air, $p_1 = 10^5$ Pa

Initial temperature of air, T_1 = 25 + 273 = 298 K

Final pressure of air, $p_2 = 5 \times 10^5$ Pa

Final temperature of air, $T_2 = T_1$ = 298 K (isothermal process)

Since, it is a closed steady state process, we can write down the first law of thermodynamics as,

$$Q = (u_2 − u_1) + W \text{ ......per kg}$$

FIRST LAW OF THERMODYNAMICS

(i) *For isothermal process* :

$$W_{1-2} = \int_1^2 p \cdot dv = p_1 v_1 \log_e \left(\frac{p_1}{p_2}\right)$$

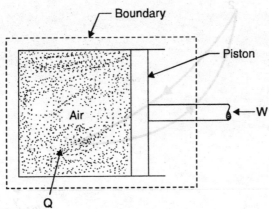

Fig. 2.14

as $\quad p_1 v_1 = p_2 v_2$ for isothermal process

$$\therefore \quad W_{1-2} = -10^5 \times 1.8 \log_e \left(\frac{1 \times 10^5}{5 \times 10^5}\right)$$
$$= -2.897 \times 10^5 = -289.7 \text{ kJ/kg}.$$

(–ve sign indicates that the work is supplied to the air)

∴ **Work done on the air = 289.7 kJ/kg. (Ans.)**

(ii) Since temperature is constant,

∴ $\quad u_2 - u_1 = 0$

∴ **Change in internal energy = zero. (Ans.)**

(iii) Again, $\quad Q_{1-2} = (u_2 - u_1) + W$
$$= 0 + (-289.7) = -289.7 \text{ kJ}$$

(–ve sign indicates that heat is lost from the system to the surroundings)

∴ **Heat rejected = 289.7 kJ/kg. (Ans.)**

Example 2.7. *A cylinder containing the air comprises the system. Cycle is completed as follows* :

(i) *82000 Nm of work is done by the piston on the air during compression stroke and 45 kJ of heat are rejected to the surroundings.*

(ii) *During expansion stroke 100000 Nm of work is done by the air on the piston.*

Calculate the quantity of heat added to the system.

Solution. Refer to Fig. 2.15.

Compression stroke. Process 1–2 :

Work done by the piston *on the air*, $W_{1-2} = -82000$ Nm $(= -82$ kJ$)$

Heat rejected to the system, $Q_{1-2} = -45$ kJ

Now, $\quad Q_{1-2} = (U_2 - U_1) + W$
$$-45 = (U_2 - U_1) + (-82)$$

∴ $\quad (U_2 - U_1) = 37$ kJ $\quad$ ...(i)

Expansion stroke. Process 2–1 :

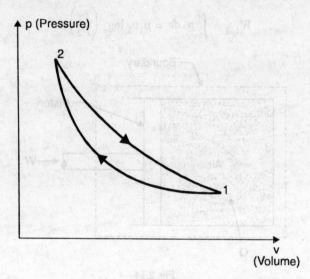

Fig. 2.15

Work done by air on the piston, W_{2-1} = 100000 N-m (= 100 kJ)

Now, $\qquad Q_{2-1} = (U_1 - U_2) + W$
$\qquad\qquad\qquad = -37 + 100$ kJ = 63 kJ

Hence, **quantity of heat added to the system = 63 kJ. (Ans.)**

Example 2.8. *The power developed by a turbine in a certain steam plant is 1200 kW. The heat supplied to the steam in the boiler is 3360 kJ/kg, the heat rejected by the system to cooling water in the condenser is 2520 kJ/kg and the feed pump work required to pump the condensate back into the boiler is 6 kW.*

Calculate the steam flow round the cycle in kg/s.

Solution. The power developed by the turbine = 1200 kW

The heat supplied to the steam in the boiler = 3360 kJ/kg

The heat rejected by the system to cooling water = 2520 kJ/kg

Feed pump work = 6 kW

Fig. 2.16 shows the cycle. A boundary is shown which encompasses the entire plant. Strictly, this boundary should be thought of as encompassing the working fluid only.

$$\oint dQ = 3360 - 2520 = 840 \text{ kJ/kg}$$

Let the system flow be in kg/s.

$\therefore \qquad\qquad \oint dQ = 840 \, \dot{m}$ kJ/s

$$\oint dW = 1200 - 6 = 1194 \text{ kJ/s}$$

FIRST LAW OF THERMODYNAMICS

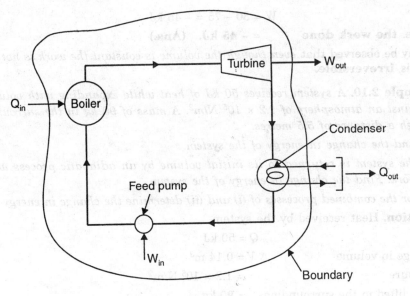

Fig. 2.16

But
$$\oint dQ = \oint dW$$

i.e., $\quad 840\, \dot{m} = 1194$

∴ $\quad \dot{m} = \dfrac{1194}{840} = 1.421$ kg/s

∴ **Steam flow round the cycle = 1.421 kg/s. (Ans.)**

Example 2.9. *A closed system of constant volume experiences a temperature rise of 25°C when a certain process occurs. The heat transferred in the process is 30 kJ. The specific heat at constant volume for the pure substance comprising the system is 1.2 kJ/kg°C, and the system contains 2.5 kg of this substance. Determine :*

(i) The change in internal energy ;

(ii) The work done.

Solution. Temperature rise, $(T_2 - T_1) = 25°C$

The heat transferred in the process, $Q = 30$ kJ

Specific heat at constant volume, $c_v = 1.2$ kJ/kg°C

Mass of the substance, $m = 2.5$ kg

Now, $\quad \Delta U = m \displaystyle\int_{T_1}^{T_2} c_v\, dT$

$\quad\quad\quad\quad = 2.5 \displaystyle\int_{T_1}^{T_2} 1.2\, dT = 3.0 \times (T_2 - T_1)$

$\quad\quad\quad\quad = 3.0 \times 25 = \mathbf{75\ kJ}$

Hence, **the change in internal energy is 75 kJ. (Ans.)**

According to the first law of thermodynamics,

$$Q = \Delta U + W$$
$$30 = 75 + W$$

$$W = 30 - 75 = -45 \text{ kJ}$$

Hence, **the work done** $= -45$ **kJ.** **(Ans.)**

It may be observed that *even though the volume is constant the work is not zero.* Clearly, the process is **irreversible**.

Example 2.10. *A system receives 50 kJ of heat while expanding with volume change of 0.14 m³ against an atmosphere of 1.2 × 10⁵ N/m². A mass of 90 kg in the surroundings is also lifted through a distance of 5.5 metres.*

(i) Find the change in energy of the system.

(ii) The system is returned to its initial volume by an adiabatic process which requires 110 kJ of work. Find the change in energy of the system.

(iii) For the combined processes of (i) and (ii) determine the change in energy of the system.

Solution. Heat received by the system,

$$Q = 50 \text{ kJ}$$

Change in volume $\quad \Delta V = 0.14 \text{ m}^3$

Pressure $\quad = 1.2 \times 10^5 \text{ N/m}^2$

Mass lifted in the surroundings $= 90$ kg

Distance through which lifted $= 5.5$ m

Work done during adiabatic process $= -110$ kJ.

(i) $\quad Q = \Delta E + W$ ...(i)

Now, $\quad W = p.\Delta V + W_{net}$

$$= \left(\frac{1.2 \times 10^5 \times 0.14}{1000}\right) + \left(\frac{90 \times 5.5 \times 9.8}{1000}\right) \text{ kJ}$$

$$= 16.8 + 4.85 = 21.65 \text{ kJ}$$

But [from (i)], $\quad \Delta E = Q - W$

$$= 50 - 21.65 = 28.35 \text{ kJ.} \quad \textbf{(Ans.)}$$

(ii) Since the process is adiabatic,

$$Q = 0$$

and $\quad \Delta E = -W$

$$= -(-110) = 110 \text{ kJ.} \quad \textbf{(Ans.)}$$

(iii) Change in internal energy,

$$\Delta E = Q - W$$

$$= 50 - [(-110) + 21.65] = \textbf{138.35 kJ.} \quad \textbf{(Ans.)}$$

☞ **Example 2.11.** *A fluid system undergoes a non-flow frictionless process following the pressure-volume relation as* $p = \frac{5}{V} + 1.5$, *where p is in bar and V is in m³. During the process the volume changes from 0.15 m³ to 0.05 m³ and the system rejects 45 kJ of heat. Determine :*

(i) Change in internal energy ;

(ii) Change in enthalpy.

Solution. Pressure-volume relation : $p = \frac{5}{V} + 1.5$

Initial volume, $\quad V_1 = 0.15 \text{ m}^3$

Final volume, $V_2 = 0.05 \text{ m}^3$
Heat *rejected* by the system, $Q = -45 \text{ kJ}$
Work done is given by,

$$W = \int_1^2 p \cdot dV = \int_{V_1}^{V_2} \left(\frac{5}{V} + 1.5\right) dV$$

$$= \int_{0.15}^{0.05} \left(\frac{5}{V} + 1.5\right) dV = \left[5 \log_e \frac{V_2}{V_1} + 1.5(V_2 - V_1)\right]_{0.15}^{0.05} \times 10^5 \text{ Nm}$$

$$= 10^5 \left[5 \log_e \frac{0.05}{0.15} + 1.5(0.05 - 0.15)\right] = 10^5(-5.49 - 0.15) \text{ Nm}$$

$$= -5.64 \times 10^5 \text{ Nm} = -5.64 \times 10^5 \text{ J} \quad [\because 1 \text{ Nm} = 1 \text{ J}]$$

$$= -564 \text{ kJ}.$$

(*i*) Applying the first law energy equation,
$$Q = \Delta U + W$$
$$-45 = \Delta U + (-564)$$
$\therefore \quad \Delta U = \textbf{519 kJ. (Ans.)}$

This shows that the internal energy is **increased**.

(*ii*) Change in enthalpy,
$$\Delta H = \Delta U + \Delta(pV)$$
$$= 519 \times 10^3 + (p_2 V_2 - p_1 V_1)$$

$$p_1 = \frac{5}{V_1} + 1.5 = \frac{5}{0.15} + 1.5 = 34.83 \text{ bar}$$

$$= 34.83 \times 10^5 \text{ N/m}^2$$

$$p_2 = \frac{5}{V_2} + 1.5 = \frac{5}{0.05} + 1.5$$

$$= 101.5 \text{ bar} = 101.5 \times 10^5 \text{ N/m}^2$$

$\therefore \quad \Delta H = 519 \times 10^3 + (101.5 \times 10^5 \times 0.05 - 34.83 \times 10^5 \times 0.15)$
$$= 519 \times 10^3 + 10^3(507.5 - 522.45)$$
$$= 10^3(519 + 507.5 - 522.45) = 504 \text{ kJ}$$

$\therefore$ **Change in enthalpy** $= \textbf{504 kJ. (Ans.)}$

☞ **Example 2.12.** *The following equation gives the internal energy of a certain substance*
$$u = 3.64 \, pv + 90$$
where u is kJ/kg, p is in kPa and v is in m^3/kg.

A system composed of 3.5 kg of this substance expands from an initial pressure of 500 kPa and a volume of 0.25 m^3 to a final pressure 100 kPa in a process in which pressure and volume are related by $pv^{1.25}$ = constant.

(i) If the expansion is quasi-static, find Q, ΔU and W for the process.

(ii) In another process, the same system expands according to the same pressure-volume relationship as in part (i), and from the same initial state to the same final state as in part (i), but the heat transfer in this case is 32 kJ. Find the work transfer for this process.

(iii) Explain the difference in work transfer in parts (i) and (ii).

Solution. Internal energy equation : $u = 3.64 \, pv + 90$
Initial volume, $V_1 = 0.25 \text{ m}^3$

Initial pressure, $p_1 = 500$ kPa
Final pressure, $p_2 = 100$ kPa
Process : $pv^{1.25}$ = constant.

(i) Now, $u = 3.64\, pv + 90$

$$\Delta u = u_2 - u_1$$
$$= 3.64\,(p_2 v_2 - p_1 v_1) \quad \text{...per kg}$$

$\therefore \quad \Delta U = 3.64\,(p_2 V_2 - p_1 V_1) \quad$ ...for 3.5 kg

Now, $p_1 V_1^{1.25} = p_2 V_2^{1.25}$

$$V_2 = V_1 \left(\frac{p_1}{p_2}\right)^{1/1.25} = 0.25 \left(\frac{500}{100}\right)^{1/1.25}$$

$$= 0.906 \text{ m}^3$$

$\therefore \quad \Delta U = 3.64\,(100 \times 10^3 \times 0.906 - 500 \times 10^3 \times 0.25)$ J $[\because\; 1$ Pa $= 1$ N/m$^2]$

$$= 3.64 \times 10^5 (0.906 - 5 \times 0.25) \text{ J}$$
$$= -3.64 \times 10^5 \times 0.344 \text{ J} = -125.2 \text{ kJ}$$

i.e., $\Delta U = -125.2$ kJ. **(Ans.)**

For a **quasi-static process**

$$W = \int p\,dV = \frac{p_1 V_1 - p_2 V_2}{n - 1}$$

$$= \frac{(500 \times 10^3 \times 0.25 - 100 \times 10^3 \times 0.906)}{(1.25 - 1)} = \frac{125 - 90.6}{0.25} \text{ kJ} = 137.6 \text{ kJ}$$

$\therefore \quad Q = \Delta U + W$
$\quad = -125.2 + 137.6 = 12.4$ kJ

i.e., **Q = 12.4 kJ. (Ans.)**

(ii) Here $Q = 32$ kJ

Since the end states are the same, ΔU would remain the same as in (i)

$\therefore \quad W = Q - \Delta U = 32 - (-125.2)$
$\quad = 157.2$ kJ. **(Ans.)**

(iii) The work in (ii) is **not equal** to $\int p\,dV$ since the process is **not quasi-static**. **(Ans.)**

☞**Example 2.13.** *The properties of a system, during a reversible constant pressure non-flow process at p = 1.6 bar, changed from $v_1 = 0.3$ m^3/kg, $T_1 = 20°C$ to $v_2 = 0.55$ m^3/kg, $T_2 = 260°C$. The specific heat of the fluid is given by*

$$c_p = \left(1.5 + \frac{75}{T + 45}\right) kJ/kg°C, \text{ where } T \text{ is in } °C.$$

Determine : (i) *Heat added/kg ;* (ii) *Work done/kg ;*
(iii) *Change in internal energy/kg ;* (iv) *Change in enthalpy/kg.*

Solution. Initial volume, $v_1 = 0.3$ m^3/kg
Initial temperature, $T_1 = 20°C$
Final volume, $v_2 = 0.55$ m^3/kg
Final temperature, $T_2 = 260°C$
Constant pressure, $p = 1.6$ bar

FIRST LAW OF THERMODYNAMICS

Specific heat at constant pressure, $c_p = \left(1.5 + \dfrac{75}{T+45}\right)$ kJ/kg°C

(*i*) The **heat added** per kg of fluid is given by

$$Q = \int_{T_1}^{T_2} c_p\, dT = \int_{20}^{260}\left(1.5 + \frac{75}{T+45}\right) dT$$

$$= \Big| 1.5\,T + 75 \log_e (T+45) \Big|_{20}^{260}$$

$$= 1.5\,(260 - 20) + 75 \times \log_e \left(\frac{260+45}{20+45}\right) = 475.94 \text{ kJ}$$

∴ **Heat added = 475.94 kJ/kg.** (Ans.)

(*ii*) The **work done** per kg of fluid is given by

$$W = \int_{v_1}^{v_2} p\,dv = p(v_2 - v_1) = 1.6 \times 10^5 (0.55 - 0.3) \text{ Nm}$$

$$= 40 \times 10^3 \text{ J} = 40 \text{ kJ}$$

∴ **Work done = 40 kJ/kg.** (Ans.)

(*iii*) **Change in internal energy,**

$$\Delta u = Q - W = 475.94 - 40 = \mathbf{435.94 \text{ kJ/kg}}. \quad \text{(Ans.)}$$

(*iv*) **Change in enthalpy,** (for **non-flow process**)

$$\Delta h = Q = 475.94 \text{ kJ/kg}. \quad \text{(Ans.)}$$

Example 2.14. *1 kg of gaseous CO_2 contained in a closed system undergoes a reversible process at constant pressure. During this process 42 kJ of internal energy is decreased. Determine the work done during the process.*

Take $c_p = 840$ J/kg°C and $c_v = 600$ J/kg°C.

Solution. Mass CO_2, $m = 1$ kg

Decrease in internal energy, $\Delta u = -42$ kJ $= -42 \times 10^3$ J

Specific heat at constant pressure, $c_p = 840$ J/kg°C

Specific heat at constant volume, $c_v = 600$ J/kg°C

Let, initial temperature of CO_2 $= T_1$

Final temperature of CO_2 $= T_2$

Now change in internal energy,

$$\Delta U = m \times c_v (T_2 - T_1)$$

$$-42 \times 10^3 = 1 \times 600 (T_2 - T_1)$$

∴ $T_2 - T_1 = -\dfrac{42 \times 10^3}{600} = -70°C$

The heat supplied or rejected,

$$Q = m c_p (T_2 - T_1)$$

$$= 1 \times 840 \times (-70) = -58800 \text{ J or} -58.8 \text{ kJ}$$

Applying first law to the process,
$$Q = \Delta U + W$$
$$-58.8 = -42 + W \text{ or } W = -16.8 \text{ kJ}$$

∴ **Work done during the process = – 16.8 kJ. (Ans.)**

☞**Example 2.15.** *A fluid is contained in a cylinder by a spring-loaded, frictionless piston so that the pressure in the fluid is a linear function of the volume ($p = a + bV$). The internal energy of the fluid is given by the following equation,*
$$U = 42 + 3.6 \, pV$$
where U is in kJ, p in kPa, and V in cubic metre. If the fluid changes from an initial state of 190 kPa, 0.035 m^3 to a final state of 420 kPa, 0.07 m^3, with no work other than that done on the piston, find the direction and magnitude of the work and heat transfer.

Solution. Relation between pressure and volume, $p = a + bV$.

Equation of internal energy : $U = 42 + 3.6pV$

Initial pressure, $p_1 = 190$ kPa
Initial volume, $V_1 = 0.035$ m^3
Final pressure, $p_2 = 420$ kPa
Final volume, $V_2 = 0.07$ m^3

The change in internal energy of the fluid during the process
$$U_2 - U_1 = (42 + 3.6 p_2 V_2) - (42 + 3.6 p_1 V_1)$$
$$= 3.6(p_2 V_2 - p_1 V_1)$$
$$= 3.6(4.2 \times 10^5 \times 0.07 - 1.9 \times 10^5 \times 0.035) \text{ J}$$
$$= 360(4.2 \times 0.07 - 1.9 \times 0.035) \text{ kJ}$$
$$= 81.9 \text{ kJ}$$

Now, $p = a + bV$
$$190 = a + b \times 0.035 \quad \ldots(i)$$
$$420 = a + b \times 0.07 \quad \ldots(ii)$$

Subtracting (i) from (ii), we get
$$230 = 0.035 \, b \text{ or } b = \frac{230}{0.035} = 6571 \text{ kN/}m^5$$

and $a = -40$ kN/m^2

Work transfer involved during the process
$$W_{1-2} = \int_{V_1}^{V_2} p \, dV = \int_{V_1}^{V_2} (a + bV) \, dV = a(V_2 - V_1) + b\left(\frac{V_2^2 - V_1^2}{2}\right)$$
$$= (V_2 - V_1)\left[a + \frac{b}{2}(V_1 + V_2)\right]$$
$$= (0.07 - 0.035)\left[-40 \text{ kN/}m^2 + \frac{6571}{2} \text{ kN/}m^5 (0.035 + 0.07)\right] = 10.67 \text{ kJ}$$

∴ **Work done by the system = 10.67 kJ. (Ans.)**

Heat transfer involved,
$$Q_{1-2} = (U_2 - U_1) + W_{1-2} = 81.9 + 10.67 = 92.57 \text{ kJ}.$$

92.57 kJ of heat flow into the system during the process. (Ans.)

FIRST LAW OF THERMODYNAMICS

Example 2.16. *90 kJ of heat are supplied to a system at a constant volume. The system rejects 95 kJ of heat at constant pressure and 18 kJ of work is done on it. The system is brought to original state by adiabatic process. Determine :*

   (i) *The adiabatic work ;*

   (ii) *The values of internal energy at all end states if initial value is 105 kJ.*

Solution. Refer to Fig. 2.17.

Heat supplied at constant volume   = 90 kJ

Heat rejected at constant pressure = – 95 kJ

Work done *on* the system   = – 18 kJ

Initial value of internal energy, U_l = 105 kJ

Process *l–m* (constant volume) :

$$W_{l-m} = 0$$
$$Q_{l-m} = 90 = U_m - U_l$$

∴ $U_m = U_l + 90 = 105 + 90$

   = 195 kJ

Process *m–n* (constant pressure) :

$$Q_{m-n} = (U_n - U_m) + W_{m-n}$$
$$-95 = (U_n - U_m) - 18$$

∴ $U_n - U_m = -77$ kJ

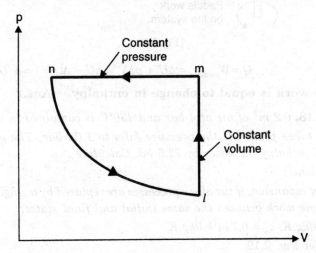

Fig. 2.17

∴ $U_n = 195 - 77 = 118$ kJ

Q_{n-l} = 0 being adiabatic process

∴ $\oint \delta Q = 90 - 95 = -5$ kJ

and $\oint \delta W = -18 + W_{n-l} = -5$

∴ $W_{n-l} = -5 + 18 = 13$ kJ

Hence, **$W_{n-l} = 13$ kJ ; $U_l = 105$ kJ ; $U_m = 195$ kJ ; $U_n = 118$ kJ.** **(Ans.)**

Example 2.17. *A movable frictionless piston closes a fully insulated cylinder on one side and offers a constant resistance during its motion. A paddle work is drawn into the cylinder and does work on the system.*

Prove that the paddle work is equal to change in enthalpy.

Solution. Refer Fig. 2.18.

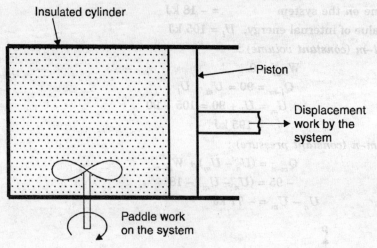

Fig. 2.18

$$Q = W_{paddle} = \Delta U + p\Delta V = \Delta U + \Delta(pV) = \Delta(U + pV) = \Delta H$$

Hence **paddle work is equal to change in enthalpy.** **(Ans.)**

☞**Example 2.18.** *0.2 m^3 of air at 4 bar and 130°C is contained in a system. A reversible adiabatic expansion takes place till the pressure falls to 1.02 bar. The gas is then heated at constant pressure till enthalpy increases by 72.5 kJ. Calculate :*

(i) The work done ;

(ii) The index of expansion, if the above processes are replaced by a single reversible polytropic process giving the same work between the same initial and final states.

Take $c_p = 1$ kJ/kg K, $c_v = 0.714$ kJ/kg K.

Solution. Refer Fig. 2.19.

Initial volume, $\quad\quad\quad\quad V_1 = 0.2$ m³

Initial pressure, $\quad\quad\quad\quad p_1 = 4$ bar $= 4 \times 10^5$ N/m²

Initial temperature, $\quad\quad\quad T_1 = 130 + 273 = 403$ K

Final pressure after adiabatic expansion,

$$p_2 = 1.02 \text{ bar} = 1.02 \times 10^5 \text{ N/m}^2$$

Increase in enthalpy during constant pressure process

$$= 72.5 \text{ kJ}.$$

FIRST LAW OF THERMODYNAMICS

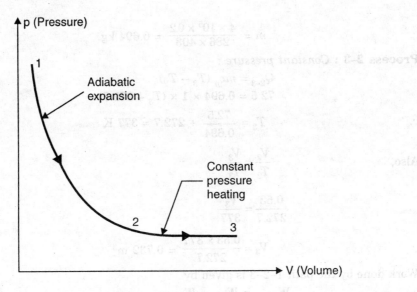

Fig. 2.19

(i) Work done :

Process 1–2 : *Reversibe adiabatic process :*

$$p_1 V_1^\gamma = p_2 V_2^\gamma$$

$$V_2 = V_1 \left(\frac{p_1}{p_2}\right)^{\frac{1}{\gamma}}$$

Also, $\quad\gamma = \dfrac{c_p}{c_v} = \dfrac{1}{0.714} = 1.4$

$\therefore\quad V_2 = 0.2 \times \left(\dfrac{4 \times 10^5}{1.02 \times 10^5}\right)^{\frac{1}{1.4}} = 0.53 \text{ m}^3$

Also, $\quad \dfrac{T_2}{T_1} = \left(\dfrac{p_2}{p_1}\right)^{\frac{\gamma-1}{\gamma}}$

$\therefore\quad T_2 = T_1 \left(\dfrac{p_2}{p_1}\right)^{\frac{\gamma-1}{\gamma}}$

$= 403 \left(\dfrac{1.02 \times 10^5}{4 \times 10^5}\right)^{\frac{1.4-1}{1.4}} = 272.7 \text{ K}$

Mass of the gas, $m = \dfrac{p_1 V_1}{RT_1}$ $\quad\quad\quad\quad [\because pV = mRT]$

where $\quad R = (c_p - c_v) = (1 - 0.714)$ kJ/kg K
$= 0.286$ kJ/kg K $= 286$ J/kg K or 286 Nm/kg K

$$\therefore \quad m = \frac{4 \times 10^5 \times 0.2}{286 \times 403} = 0.694 \text{ kg.}$$

Process 2-3 : *Constant pressure* :
$$Q_{2-3} = mc_p (T_3 - T_2)$$
$$72.5 = 0.694 \times 1 \times (T_3 - 272.7)$$

$$\therefore \quad T_3 = \frac{72.5}{0.694} + 272.7 = 377 \text{ K}$$

Also,
$$\frac{V_2}{T_2} = \frac{V_3}{T_3}$$

or
$$\frac{0.53}{272.7} = \frac{V_3}{377}$$

$$\therefore \quad V_3 = \frac{0.53 \times 377}{272.7} = 0.732 \text{ m}^3$$

Work done by the path 1–2–3 is given by
$$W_{1-2-3} = W_{1-2} + W_{2-3}$$
$$= \frac{p_1 V_1 - p_2 V_2}{\gamma - 1} + p_2 (V_3 - V_2)$$
$$= \frac{4 \times 10^5 \times 0.2 - 1.02 \times 10^5 \times 0.53}{1.4 - 1} + 1.02 \times 10^5 (0.732 - 0.53)$$
$$= \frac{10^5 (4 \times 0.2 - 1.02 \times 0.53)}{0.4} + 1.02 \times 10^5 (0.732 - 0.53)$$
$$= 64850 + 20604 = 85454 \text{ Nm or J}$$

Hence, **total work done** = **85454 Nm or J.** (Ans.)

(*ii*) **Index of expansion, n :**

If the work done by the polytropic process is the same,
$$W_{1-2-3} = W_{1-3} = \frac{p_1 V_1 - p_3 V_3}{n - 1}$$

$$85454 = \frac{4 \times 10^5 \times 0.2 - 1.02 \times 10^5 \times 0.732}{(n-1)} = \frac{5336}{n-1}$$

$$\therefore \quad n = \frac{5336}{85454} + 1$$

i.e., $\quad n = 1.062$

Hence, **value of index** = **1.062.** (Ans.)

Example 2.19. *The following is the equation which connects u, p and v for several gases*
$$u = a + bpv$$
where a and b are constants. Prove that for a reversible adiabatic process,
$$pv^\gamma = \text{constant, where } \gamma = \frac{b+1}{b}.$$

Solution. Considering a unit mass.

For a reversible adiabatic process, first law gives
$$0 = du + pdv$$

FIRST LAW OF THERMODYNAMICS

$\therefore \qquad \dfrac{du}{dv} = -p$ ...(i)

Also, $\qquad u = a + bpv$

$\therefore \qquad \dfrac{du}{dv} = \dfrac{d(a + bpv)}{dv} = bv\,\dfrac{dp}{dv} + bp$

$\qquad\qquad\quad = b\left(p + v \cdot \dfrac{dp}{dv}\right)$ ...(ii)

Equating (i) and (ii), we get

$$b\left(p + v \cdot \dfrac{dp}{dv}\right) = -p$$

$$bp + b \cdot v \cdot \dfrac{dp}{dv} = -p$$

$$bp + p + bv \cdot \dfrac{dp}{dv} = 0$$

$$p(b+1) + bv \cdot \dfrac{dp}{dv} = 0$$

Multiplying both sides by $\dfrac{dv}{bpv}$, we get

$$\left(\dfrac{b+1}{b}\right)\dfrac{dv}{v} + \dfrac{dp}{p} = 0$$

or $\qquad \dfrac{dp}{p} + \left(\dfrac{b+1}{b}\right)\dfrac{dv}{v} = 0$

$$d(\log_e p) + \left(\dfrac{b+1}{b}\right)d(\log_e v) = 0$$

Also, $\qquad \dfrac{b+1}{b} = \gamma$ ...(Given)

$\therefore \qquad d(\log_e p) + \gamma\, d(\log_e v) = 0$

Integrating, we get $\qquad pv^\gamma =$ constant.

☞ **Example 2.20.** *A cylinder contains $0.45\ m^3$ of a gas at $1 \times 10^5\ N/m^2$ and $80°C$. The gas is compressed to a volume of $0.13\ m^3$, the final pressure being $5 \times 10^5\ N/m^2$. Determine :*
  (i) *The mass of gas ;*
  (ii) *The value of index 'n' for compression ;*
  (iii) *The increase in internal energy of the gas ;*
  (iv) *The heat received or rejected by the gas during compression.*
  Take $\gamma = 1.4$, $R = 294.2\ J/kg°C$.

Solution. Initial volume of gas, $\quad V_1 = 0.45\ m^3$
Initial pressure of gas, $\qquad\quad p_1 = 1 \times 10^5\ N/m^2$
Initial temperature, $\qquad\qquad\quad T_1 = 80 + 273 = 353\ K$
Final volume after compression, $\ V_2 = 0.13\ m^3$
The final pressure, $\qquad\qquad\quad\ p_2 = 5 \times 10^5\ N/m^2$.

(i) To find mass 'm' using the relation

$$m = \frac{p_1 V_1}{RT_1} = \frac{1 \times 10^5 \times 0.45}{294.2 \times 353} = \mathbf{0.433 \text{ kg.}} \quad \textbf{(Ans.)}$$

(ii) To find index 'n' using the relation

$$p_1 V_1^n = p_2 V_2^n$$

or

$$\left(\frac{V_1}{V_2}\right)^n = \frac{p_2}{p_1}$$

$$\left(\frac{0.45}{0.13}\right)^n = \left(\frac{5 \times 10^5}{1 \times 10^5}\right) = 5$$

or
$$(3.46)^n = 5$$

Taking log on both sides, we get

$$n \log_e 3.46 = \log_e 5$$
$$n = \log_e 5 / \log_e 3.46 = \mathbf{1.296.} \quad \textbf{(Ans.)}$$

(iii) In a polytropic process,

$$\frac{T_2}{T_1} = \left(\frac{V_1}{V_2}\right)^{n-1} = \left(\frac{0.45}{0.13}\right)^{1.296-1} = 1.444$$

$$\therefore \quad T_2 = 353 \times 1.444 = 509.7 \text{ K}$$

Now, increase in internal energy,

$$\Delta U = m c_v (T_2 - T_1)$$

$$= 0.433 \times \frac{R}{(\gamma - 1)} (T_2 - T_1) \qquad \left[\because c_v = \frac{R}{(\gamma - 1)}\right]$$

$$= 0.433 \times \frac{294.2}{(1.4 - 1)1000} (509.7 - 353)$$

$$= \mathbf{49.9 \text{ kJ.}} \quad \textbf{(Ans.)}$$

(iv) $\quad Q = \Delta U + W$

Now, $\quad W = \dfrac{p_1 V_1 - p_2 V_2}{n - 1} = \dfrac{mR(T_1 - T_2)}{n - 1}$

$$= \frac{0.433 \times 294.2 (353 - 509.7)}{1.296 - 1}$$

$$= -67438 \text{ Nm or} -67438 \text{ J} = -67.44 \text{ kJ}$$

$\therefore \quad Q = 49.9 + (-67.44) = -17.54 \text{ kJ}$

$\therefore \quad$ **Heat rejected = 17.54 kJ. (Ans.)**

Example 2.21. *Air at 1.02 bar, 22°C, initially occupying a cylinder volume of 0.015 m^3, is compressed reversibly and adiabatically by a piston to a pressure of 6.8 bar. Calculate :*

(i) *The final temperature ;*

(ii) *The final volume ;*

(iii) *The work done.*

Solution. Initial pressure, $p_1 = 1.02$ bar

Initial temperature, $\quad T_1 = 22 + 273 = 295$ K

FIRST LAW OF THERMODYNAMICS

Initial volume, $V_1 = 0.015 \text{ m}^3$
Final pressure, $p_2 = 6.8$ bar
Law of compression : $pv^\gamma = C$

(i) Final temperature :
Using the relation,

$$\frac{T_2}{T_1} = \left(\frac{p_2}{p_1}\right)^{\frac{\gamma-1}{\gamma}}$$

$$\frac{T_2}{295} = \left(\frac{6.8}{1.02}\right)^{\frac{1.4-1}{1.4}} \qquad [\because \gamma \text{ for air} = 1.4]$$

$$\therefore \quad T_2 = 295 \left(\frac{6.8}{1.02}\right)^{\frac{1.4-1}{1.4}} = 507.24 \text{ K}$$

i.e., **Final temperature** = 507.24 − 273 = **234.24°C. (Ans.)**

(ii) Final volume :
Using the relation,

$$p_1 V_1^\gamma = p_2 V_2^\gamma$$

$$\frac{p_1}{p_2} = \left(\frac{V_2}{V_1}\right)^\gamma \quad \text{or} \quad \frac{V_2}{V_1} = \left(\frac{p_1}{p_2}\right)^{\frac{1}{\gamma}}$$

$$\therefore \quad V_2 = V_1 \times \left(\frac{p_1}{p_2}\right)^{\frac{1}{\gamma}} = 0.015 \times \left(\frac{1.02}{6.8}\right)^{\frac{1}{1.4}} = 0.00387 \text{ m}^3$$

i.e., **Final volume = 0.00387 m³. (Ans.)**

(iii) Work done :
Now, work done *on the air*,

$$W = \frac{mR(T_1 - T_2)}{(\gamma - 1)} \qquad \ldots(i)$$

where m is the mass of air and is found by the following relation,

$$pV = mRT$$

$$\therefore \quad m = \frac{p_1 V_1}{RT_1} = \frac{1.02 \times 10^5 \times 0.015}{0.287 \times 10^3 \times 295} \qquad [\because R \text{ for air} = 0.287 \times 10^3]$$
$$= 0.01807 \text{ kg}$$

$$\therefore \quad W = \frac{0.01807 \times 0.287 \times 10^3 (295 - 507.24)}{(1.4 - 1)} = -2751 \text{ J or} -2.751 \text{ kJ}$$

i.e., **Work done = 2.751 kJ. (Ans.)**

(−ve sign indicates that work is done on the air).

Example 2.22. *0.44 kg of air at 180°C expands adiabatically to three times its original volume and during the process, there is a fall in temperature to 15°C. The work done during the process is 52.5 kJ. Calculate c_p and c_v.*

Solution. Refer Fig. 2.20.

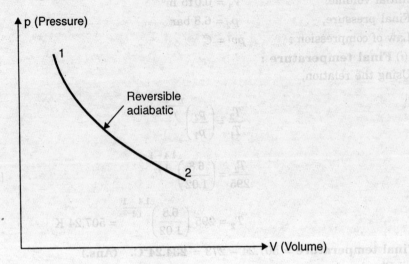

Fig. 2.20

Mass of air, $m = 0.44$ kg
Initial temperature, $T_1 = 180 + 273 = 453$ K
Ratio = $\dfrac{V_2}{V_1} = 3$
Final temperature, $T_2 = 15 + 273 = 288$ K
Work done during the process, $W_{1-2} = 52.5$ kJ
$c_p = ?$, $c_v = ?$

For adiabatic process, we have
$$\frac{T_2}{T_1} = \left(\frac{V_1}{V_2}\right)^{\gamma - 1}$$

$$\frac{288}{453} = \left(\frac{1}{3}\right)^{\gamma - 1} \quad \text{or} \quad 0.6357 = (0.333)^{\gamma - 1}$$

or Taking log on both sides, we get
$$\log_e (0.6357) = (\gamma - 1) \log_e (0.333)$$
$$-0.453 = (\gamma - 1) \times (-1.0996)$$
∴ $\gamma = \dfrac{0.453}{1.0996} + 1 = 1.41$

Also, $\dfrac{c_p}{c_v} = \gamma = 1.41$

Work done during adiabatic process,
$$W_{1-2} = \frac{mR(T_1 - T_2)}{\gamma - 1}$$

∴ $52.5 = \dfrac{0.44\, R(453 - 288)}{(1.41 - 1)}$

FIRST LAW OF THERMODYNAMICS

$$\therefore \quad R = \frac{52.5(1.41-1)}{0.44(453-288)} = 0.296$$

$$\therefore \quad c_p - c_v = 0.296 \qquad [\because R = c_p - c_v]$$

Also, $\quad \dfrac{c_p}{c_v} = 1.41 \quad$ or $\quad c_p = 1.41\, c_v$

$\therefore \quad 1.41\, c_v - c_p = 0.296$

or $\quad c_v = \mathbf{0.722\ kJ/kg\ K.}$ **(Ans.)**

and $\quad c_p = \mathbf{1.018\ kJ/kg\ K.}$ **(Ans.)**

☞ **Example 2.23.** *1 kg of ethane (perfect) gas is compressed from 1.1 bar, 27°C according to a law $pV^{1.3}$ = constant, until the pressure is 6.6 bar. Calculate the heat flow to or from the cylinder walls.*

Given : Molecular weight of ethane = 30, c_p = 1.75 kJ/kg K.

Solution. Mass of ethane gas, $\quad m = 1$ kg

Initial pressure, $\quad p_1 = 1.1$ bar

Initial temperature, $\quad T_1 = 27 + 273 = 300$ K

Final pressure, $\quad p_2 = 6.6$ bar

Law of compression, $\quad pV^{1.3} = C$

Quantity of heat transferred, Q :

Now, characteristic gas constant,

$$R = \frac{\text{Universal gas constant }(R_0)}{\text{Molecular weight }(M)}$$

$$= \frac{8314}{30} = 277.13\ \text{Nm/kg K} = 277.31\ \text{J/kg K}$$

$$= 0.277\ \text{kJ/kg K}$$

Also, $\quad c_p - c_v = R$

$\therefore \quad c_v = c_p - R = 1.75 - 0.277 = 1.473$ kJ/kg K

$$\gamma = \frac{c_p}{c_v} = \frac{1.75}{1.473} = 1.188$$

In case of a polytropic process,

$$\frac{T_2}{T_1} = \left(\frac{p_2}{p_1}\right)^{\frac{n-1}{n}} = \left(\frac{6.6}{1.1}\right)^{\frac{1.3-1}{1.3}} = 1.5119$$

$\therefore \quad T_2 = 300 \times 1.5119 = 453.6$ K

Now, work done, $\quad W = \dfrac{R(T_1 - T_2)}{n-1} = \dfrac{0.277(300-453.6)}{1.3-1} = -141.8$ kJ/kg

To find heat flow, using the relation,

$$Q = \left(\frac{\gamma - n}{\gamma - 1}\right) W = \left(\frac{1.188 - 1.3}{1.188 - 1}\right) \times -141.8 = +84.5\ \text{kJ/kg}$$

i.e., **Heat supplied** $= \mathbf{84.5\ kJ/kg.}$ **(Ans.)**

ADDITIONAL TYPICAL WORKED EXAMPLES

Example 2.24. *Derive the equation of state for a perfect gas.*

Solution. To derive the equation of state for a perfect gas let us consider a *unit mass* of a perfect gas to change its state in the following two successive processes (Fig. 2.21) :

(*i*) Process 1–2′ at constant pressure, and

(*ii*) Process 2′–2 at constant temperature.

For the *process 1–2′*, applying Charle's law, we get

$$\frac{v_1}{T_1} = \frac{v_2'}{T_2'}$$

and, since $T_2' = T_2$, we may write

$$\frac{v_1}{T_1} = \frac{v_2'}{T_2} \qquad \ldots(i)$$

For the *process 2′–2*, using Boyle's law, we have

$$p_2' v_2' = p_2 v_2$$

and, since $p_2' = p_1$

$$p_1 v_2' = p_2 v_2$$

i.e.,

$$v_2' = \frac{p_2 v_2}{p_1} \qquad \ldots(ii)$$

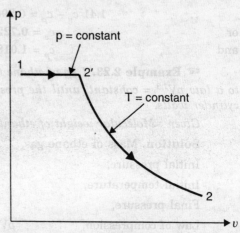

Fig. 2.21. Formulation of equation of state of a perfect gas.

Substituting the value of v_2' from Eqn. (*ii*) in Eqn. (*i*), we get

$$\frac{v_1}{T_1} = \frac{p_2 v_2}{p_1 T_2} \quad \text{or} \quad \frac{p_1 v_1}{T_1} = \frac{p_2 v_2}{T_2}$$

i.e.,

$$\frac{pv}{T} = \text{constant} \qquad \ldots(1)$$

The magnitude of this constant depends upon the particular gas and it is denoted by R, where R is called the **specific gas constant.** Then

$$\frac{pv}{T} = R$$

The equation of the state for a perfect gas is thus given by the equation

$$pv = RT \qquad \ldots(2)$$

or, for m kg, occupying V m³,

$$pV = mRT \qquad \ldots(3)$$

If the mass is chosen to be numerically equal to the molecular weight of the gas then 1 mole of the gas has been considered, *i.e.,* 1 kg mole of oxygen is 32 kg oxygen, or 1 kg mole of hydrogen is 2 kg hydrogen.

The equation may be written as

$$pV_0 = MRT \qquad \ldots(4)$$

where V_0 = Molar volume, and

M = Molecular weight of the gas.

FIRST LAW OF THERMODYNAMICS

Avogadro discovered that V_0 is the same for all gases at the same pressure and temperature and therefore it may be seen that $MR = $ a constant ; R_0 and thus

$$pV_0 = R_0 T \qquad \ldots(5)$$

R_0 is called the **molar or universal gas constant** and its value is

8.3143 kJ/kg mol K. (Ans.)

If there are n moles present then the ideal gas equation may be written as

$$pV = nR_0 T \qquad \ldots(6)$$

where V is the volume occupied by n moles at pressure p and temperature T.

Example 2.25. *0.1 m^3 of an ideal gas at 300 K and 1 bar is compressed adiabatically to 8 bar. It is then cooled at constant volume and further expanded isothermally so as to reach the condition from where it started. Calculate :*

(i) Pressure at the end of constant volume cooling.

(ii) Change in internal energy during constant volume process.

(iii) Net work done and heat transferred during the cycle.

Assume $c_p = 14.3$ kJ/kg K and $c_v = 10.2$ kJ/kg K.

Solution. Given : $V_1 = 0.1$ m^3 ; $T_1 = 300$ K ; $p_1 = 1$ bar ; $c_p = 14.3$ kJ/kg K ;
$c_v = 10.2$ kJ/kg K.

Refer to Fig. 2.22.

(i) **Pressure at the end of constant volume cooling, p_3 :**

$$\gamma = \frac{c_p}{c_v} = \frac{14.3}{10.2} = 1.402$$

Characteristic gas constant,

$$R = c_p - c_v = 14.3 - 10.2 = 4.1 \text{ kJ/kg K}$$

Considering *process 1-2*, we have

$$p_1 V_1^\gamma = p_2 V_2^\gamma$$

$$V_2 = V_1 \times \left(\frac{p_1}{p_2}\right)^{\frac{1}{\gamma}} = 0.1 \times \left(\frac{1}{8}\right)^{\frac{1}{1.402}} = 0.0227 \text{ m}^3$$

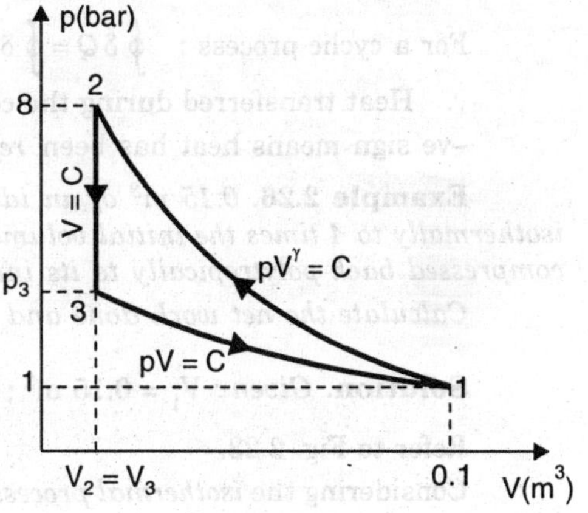

Also, $\quad \dfrac{T_2}{T_1} = \left(\dfrac{p_2}{p_1}\right)^{\frac{\gamma-1}{\gamma}} = \left(\dfrac{8}{1}\right)^{\frac{1.402-1}{1.402}} = 1.815$

or, $\quad T_2 = T_1 \times 1.815 = 300 \times 1.815 = 544.5$ K

Considering *process 3-1*, we have

$$p_3 V_3 = p_1 V_1$$

$\therefore \quad p_3 = \dfrac{p_1 V_1}{V_3} = \dfrac{1 \times 0.1}{0.0227} = $ **4.4 bar. (Ans.)** $\qquad (\because V_3 = V_2)$

Fig. 2.22

(ii) **Change in internal energy during constant volume process, $(U_3 - U_2)$:**

Mass of gas, $\quad m = \dfrac{p_1 V_1}{RT_1} = \dfrac{(1 \times 10^5) \times 0.1}{(4.1 \times 1000) \times 300} = 0.00813$ kg

$\therefore$ Change in internal energy during *constant volume process 2-3*,

$$U_3 - U_2 = mc_v(T_3 - T_2)$$
$$= 0.00813 \times 10.2 \,(300 - 544.5) \qquad (\because T_3 = T_1)$$
$$= \mathbf{-20.27 \text{ kJ. (Ans.)}} \qquad \text{(–ve sign means } decrease \text{ in internal energy)}$$

- During constant volume cooling process, temperature and hence internal energy is reduced. This *decrease in internal energy equals to heat flow to surroundings since work done is zero.*

(*iii*) **Net work done and heat transferred during the cycle :**

$$W_{1-2} = \frac{p_1V_1 - p_2V_2}{\gamma - 1} = \frac{mR(T_1 - T_2)}{\gamma - 1}$$

$$= \frac{0.00813 \times 4.1(300 - 544.5)}{1.402 - 1} = -20.27 \text{ kJ}$$

$W_{2-3} = 0$ ... since volume remains constant

$$W_{3-1} = p_3V_3 \log_e\left(\frac{V_1}{V_3}\right) = p_1V_1 \log_e\left(\frac{p_3}{p_1}\right)$$

$$= (1 \times 10^5) \times 0.1 \times \log_e\left(\frac{4.4}{1}\right) \qquad (\because p_3V_3 = p_1V_1)$$

$$= 14816 \text{ Nm (or J) or } 14.82 \text{ kJ}$$

∴ **Net work done** $= W_{1-2} + W_{2-3} + W_{3-1}$
$= (-20.27) + 0 + 14.82 = -$ **5.45 kJ**

–ve sign indicates that work has been done *on the system*. **(Ans.)**

For a cyclic process : $\oint \delta Q = \oint \delta W$

∴ Heat transferred during the complete cycle = **– 5.45 kJ**

–ve sign means heat has been *rejected i.e., lost from the system*. **(Ans.)**

Example 2.26. *0.15 m^3 of an ideal gas at a pressure of 15 bar and 550 K is expanded isothermally to 4 times the initial volume. It is then cooled to 290 K at constant volume and then compressed back polytropically to its initial state.*

Calculate the net work done and heat transferred during the cycle.

Solution. *Given :* $V_1 = 0.15 \text{ m}^3$; $p_1 = 15$ bar ; $T_1 = T_2 = 550$ K ; $\frac{V_2}{V_1} = 4$; $T_3 = 290$ K

Refer to Fig. 2.23.

Considering the *isothermal process 1–2*, we have

$$p_1V_1 = p_2V_2 \quad \text{or} \quad p_2 = \frac{p_1V_1}{V_2}$$

or $$p_2 = \frac{15 \times 0.15}{(4 \times 0.15)} = 3.75 \text{ bar}$$

Work done, $W_{1-2} = p_1V_1 \log_e\left(\frac{V_2}{V_1}\right)$

$$= (15 \times 10^5) \times 0.15 \times \log_e(4)$$
$$= 311916 \text{ J} = 311.9 \text{ kJ}$$

Considering *constant volume process 2–3*, we get

$$V_2 = V_3 = 4 \times 0.15 = 0.6 \text{ m}^3$$

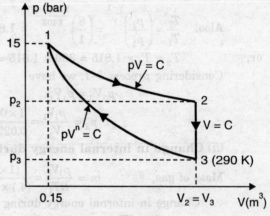

Fig. 2.23

FIRST LAW OF THERMODYNAMICS

$$\frac{p_2}{T_2} = \frac{p_3}{T_3}, \quad \text{or,} \quad p_3 = p_2 \times \frac{T_3}{T_2} = 3.75 \times \frac{290}{550} = 1.98 \text{ bar}$$

$W_{2-3} = 0$ ... since volume remains constant

Consider *polytropic process 3–1* :

$$p_3 V_3^n = p_1 V_1^n \quad \text{or} \quad \frac{p_1}{p_3} = \left(\frac{V_3}{V_1}\right)^n$$

Taking log on both sides, we get

$\log_e (p_1/p_3) = n \log_e (V_3/V_1)$

or
$$n = \frac{\log_e (p_1/p_3)}{\log_e (V_3/V_1)} = \frac{\log_e (15/1.98)}{\log_e (4)} = 1.46$$

$$W_{3-1} = \frac{p_3 V_3 - p_1 V_1}{n - 1} = \frac{1.98 \times 10^5 \times 0.6 - 15 \times 10^5 \times 0.15}{(1.46 - 1)}$$

$= -230869$ J or -230.87 kJ

∴ Net work done $= W_{1-2} + W_{2-3} + W_{3-1}$
$= 311.9 + 0 + (-230.87) = $ **81.03 kJ.** (Ans.)

For a cyclic process, $\oint \delta Q = \oint \delta W$

∴ Heat *transferred during the cycle* = **81.03 kJ.** (Ans.)

Example 2.27. *A system consisting of 1 kg of an ideal gas at 5 bar pressure and 0.02 m³ volume executes a cyclic process comprising the following three distinct operations : (i) Reversible expansion to 0.08 m³ volume 1.5 bar pressure, presuming pressure to be a linear function of volume (p = a + bV), (ii) Reversible cooling at constant pressure and (iii) Reversible hyperbolic compression according to law pV = constant. This brings the gas back to initial conditions.*

(i) Sketch the cycle on p-V diagram.

(ii) Calculate the work done in each process starting whether it is done on or by the system and evaluate the net cyclic work and heat transfer.

Solution. Given : $m = 1$ kg ; $p_1 = 5$ bar ; $V_1 = 0.02$ m³ ; $V_2 = 0.08$ m³ ; $p_2 = 1.5$ bar.

(*i*) **p-V diagram** : *p-V* diagram of the cycle is shown in Fig. 2.24.

(*ii*) **Work done and heat transfer :**

● *Process 1–2 (Linear law)* :

$p = a + bV$ ...(Given)

The values of constants a and b can be determined from the values of pressure and volume at the state points 1 and 2.

$5 = a + 0.02b$ ...(*i*)
$1.5 = a + 0.08b$ ...(*ii*)

From (*i*) and (*ii*) we get, $b = -58.33$ and $a = 6.167$

$$W_{1-2} = \int_1^2 p\, dV = \int_1^2 (a + bV)\, dV$$

$$= \int_1^2 (6.167 - 58.33 V)\, dV$$

$$= 10^5 \left| 6.167\, V - 58.33 \times \frac{V^2}{2} \right|_{0.02}^{0.08}$$

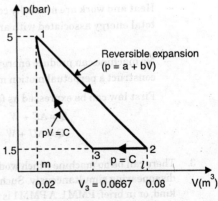

Fig. 2.24. *p-V* diagram.

$$= 10^5 \left| 6.167(0.08 - 0.02) - 58.33 \times \frac{(0.08^2 - 0.02^2)}{2} \right| \times 10^{-3} \text{ kJ} = \mathbf{19.5 \text{ kJ}}$$

This is work done **by the system. (Ans.)**

$$\left[\begin{array}{l} Alternatively: \ W_{1-2} = \text{area under the process line 1-2} \\ \qquad\qquad\qquad = \text{area of trapezium } 1\text{-}2\text{-}l\text{-}m \\ \qquad\qquad\qquad = \left[\dfrac{5 + 1.5}{2} \times 10^5 \right] \times (0.08 - 0.02) = 19.5 \text{ kJ} \end{array} \right]$$

- Process 2 – 3 (constant pressure):

$$p_3 = p_2 = 1.5 \text{ bar}$$

The volume V_3 can be worked out from the hyperbolic compression 3–1, as follows:

$$p_1 V_1 = p_3 V_3 \quad \text{or} \quad V_3 = \frac{p_1 V_1}{p_3} = \frac{5 \times 0.02}{1.5} = 0.0667 \text{ m}^3$$

$$\therefore \qquad W_{2-3} = p_2(V_3 - V_2) = 1.5 \times 10^5 (0.0667 - 0.08) \times 10^{-3} \text{ kJ} = -1.995 \text{ kJ}$$

- Process 3 – 1 (hyperbolic process):

$$W_{3-1} = p_3 V_3 \log_e \left(\frac{V_1}{V_3} \right)$$

$$= (10^5 \times 1.5) \times 0.0667 \log_e \left(\frac{0.02}{0.0667} \right) \times 10^{-3} \text{ kJ} = \mathbf{-12.05 \text{ kJ}}.$$

This is the work **done on the system. (Ans.)**

Net work done, $\quad W_{net} = W_{1-2} + W_{2-3} + W_{3-1}$

$$= 19.5 + (-1.995) + (-12.05) = \mathbf{5.445 \text{ kJ}}. \text{ (Ans.)}$$

Heat transferred during the complete cycle, $\oint \delta Q = \oint \delta W = \mathbf{5.455 \text{ kJ}}. \text{ (Ans.)}$

HIGHLIGHTS

1. *Internal energy* is the heat energy stored in a gas. The internal energy of a perfect gas is a function of *temperature* only.
2. First law of thermodynamics states:
   — Heat and work are mutually convertible but since energy can neither be created nor destroyed, the total energy associated with an energy conversion remains constant.

   Or

   — No machine can produce energy without corresponding expenditure of energy, *i.e.*, it is impossible to construct a perpetual motion machine of first kind.

   First law can be expressed as follows:

   $$Q = \Delta E + W$$
   $$Q = \Delta U + W \quad \text{... if electric, magnetic, chemical energies are absent and changes in potential and kinetic energies are neglected.}$$

3. There can be no machine which would continuously supply mechanical work without some form of energy disappearing simultaneously. Such a fictitious machine is called a perpetual motion machine of the first kind, or in brief, PMM1. A PMM1 is thus impossible.
4. The energy of an isolated system is always constant.

FIRST LAW OF THERMODYNAMICS

5. In case of
   (i) **Reversible constant volume process** (v = constant)
   $$\Delta u = c_v(T_2 - T_1)\,;\ W = 0\,;\ Q = c_v(T_2 - T_1)$$
   (ii) **Reversible constant pressure process** (p = constant)
   $$\Delta u = c_v(T_2 - T_1)\,;\ W = p(v_2 - v_1)\,;\ Q = c_p(T_2 - T_1)$$
   (iii) **Reversible temperature or isothermal process** (pv = constant)
   $$\Delta u = 0,\ W = p_1 V_1 \log_e r,\ Q = W$$
   where r = expansion or compression ratio.
   (iv) **Reversible adiabatic process** (pv^γ = constant)
   $$\pm \Delta u = \mp W = \frac{R(T_1 - T_2)}{\gamma - 1}\,;\ Q = 0\,;\ \frac{T_2}{T_1} = \left(\frac{v_1}{v_2}\right)^{\gamma - 1} = \left(\frac{p_2}{p_1}\right)^{\frac{\gamma - 1}{\gamma}}$$
   (v) **Polytropic reversible process** (pv^n = constant)
   $$\Delta u = c_v(T_2 - T_1)\,;\ W = \frac{R(T_1 - T_2)}{n - 1}\,;\ Q = \Delta u + W\,;$$
   and
   $$\frac{T_2}{T_1} = \left(\frac{v_1}{v_2}\right)^{n - 1} = \left(\frac{p_2}{p_1}\right)^{\frac{n - 1}{n}}\ \ \text{and}\ \ Q = \left(\frac{\gamma - n}{n - 1}\right) \times W.$$

OBJECTIVE TYPE QUESTIONS

Choose the Correct Answer:

1. If all the variables of a stream are independent of time it is said to be in
   (a) steady flow
   (b) unsteady flow
   (c) uniform flow
   (d) closed flow
   (e) constant flow.
2. A control volume refers to
   (a) a fixed region in space
   (b) a specified mass
   (c) an isolated system
   (d) a reversible process only
   (e) a closed system.
3. Internal energy of a perfect gas depends on
   (a) temperature, specific heats and pressure
   (b) temperature, specific heats and enthalpy
   (c) temperature, specific heats and entropy
   (d) temperature only.
4. In reversible polytropic process
   (a) true heat transfer occurs
   (b) the entropy remains constant
   (c) the enthalpy remains constant
   (d) the internal energy remains constant
   (e) the temperature remains constant.
5. An isentropic process is always
   (a) irreversible and adiabatic
   (b) reversible and isothermal
   (c) frictionless and irreversible
   (d) reversible and adiabatic
   (e) none of the above.
6. The net work done per kg of gas in a polytropic process is equal to
   (a) $p_1 v_1 \log_e \dfrac{v_2}{v_1}$
   (b) $p_1(v_1 - v_2)$
   (c) $p_2 \left(v_2 - \dfrac{v_1}{v_2}\right)$
   (d) $\dfrac{p_1 v_1 - p_2 v_2}{n - 1}$
   (e) $\dfrac{p_2 v_1 - p_2 v_2}{n - 1}$.

7. Steady flow occurs when
   (a) conditions do not change with time at any point
   (b) conditions are the same at adjacent points at any instant
   (c) conditions change steadily with the time
   (d) $\left(\dfrac{\partial v}{\partial t}\right)$ is constant.
8. A reversible process requires that
   (a) there be no heat transfer
   (b) newton's law of viscosity be satisfied
   (c) temperature of system and surroundings be equal
   (d) there be no viscous or coloumb friction in the system
   (e) heat transfer occurs from surroundings to system only.
9. The first law of thermodynamics for steady flow
   (a) accounts for all energy entering and leaving a control volume
   (b) is an energy balance for a specified mass of fluid
   (c) is an expression of the conservation of linear momentum
   (d) is primarily concerned with heat transfer
   (e) is restricted in its application to perfect gases.
10. The characteristic equation of gases $pV = mRT$ holds good for
    (a) monoatomic gases
    (b) diatomic gas
    (c) real gases
    (d) ideal gases
    (e) mixture of gases.
11. A gas which obeys kinetic theory perfectly is known as
    (a) monoatomic gas
    (b) diatomic gas
    (c) real gas
    (d) pure gas
    (e) perfect gas.
12. Work done in a free expansion process is
    (a) zero
    (b) minimum
    (c) maximum
    (d) positive
    (e) negative.
13. Which of the following is not a property of the system?
    (a) Temperature
    (b) Pressure
    (c) Specific volume
    (d) Heat
    (e) None of the above.
14. In the polytropic process equation $pv^n = $ constant, if $n = 0$, the process is termed as
    (a) constant volume
    (b) constant pressure
    (c) constant temperature
    (d) adiabatic
    (e) isothermal.
15. In the polytropic process equation $pv^n = $ constant, if n is infinitely large, the process is termed as
    (a) constant volume
    (b) constant pressure
    (c) constant temperature
    (d) adiabatic
    (e) isothermal.
16. The processes or systems that do not involve heat are called
    (a) isothermal processes
    (b) equilibrium processes
    (c) thermal processes
    (d) steady processes
    (e) adiabatic processes.
17. In a reversible adiabatic process the ratio (T_1/T_2) is equal to
    (a) $\left(\dfrac{p_1}{p_2}\right)^{\frac{\gamma-1}{\gamma}}$
    (b) $\left(\dfrac{v_1}{v_2}\right)^{\frac{\gamma-1}{\gamma}}$
    (c) $(v_1 v_2)^{\frac{\gamma-1}{2\gamma}}$
    (d) $\left(\dfrac{v_2}{v_1}\right)^{\gamma}$.

FIRST LAW OF THERMODYNAMICS

18. In isothermal process
    (a) temperature increases gradually
    (b) volume remains constant
    (c) pressure remains constant
    (d) enthalpy change is maximum
    (e) change in internal energy is zero.
19. During throttling process
    (a) internal energy does not change
    (b) pressure does not change
    (c) entropy does not change
    (d) enthalpy does not change
    (e) volume change is negligible.
20. When a gas is to be stored, the type of compression that would be ideal is
    (a) isothermal
    (b) adiabatic
    (c) polytropic
    (d) constant volume
    (e) none of the above.
21. If a process can be stopped at any stage and reversed so that the system and surroundings are exactly restored to their initial states, it is known as
    (a) adiabatic process
    (b) isothermal process
    (c) ideal process
    (d) frictionless process
    (e) energyless process.
22. The state of a substance whose evaporation from its liquid state is complete, is known as
    (a) vapour
    (b) perfect gas
    (c) air
    (d) steam.
23. In SI units, the value of the universal gas constant is
    (a) 0.8314 J/mole/K
    (b) 8.314 J/mole/K
    (c) 83.14 J/mole/K
    (d) 831.4 J/mole/K
    (e) 8314 J/mole/K.
24. When the gas is heated at constant pressure, the heat supplied
    (a) increases the internal energy of the gas
    (b) increases the temperature of the gas
    (c) does some external work during expansion
    (d) both (b) and (c)
    (e) none of the above.
25. The gas constant (R) is equal to the
    (a) sum of two specific heats
    (b) difference of two specific heats
    (c) product of two specific heats
    (d) ratio of two specific heats.
26. The heat absorbed or rejected during a polytropic process is
    (a) $\left(\dfrac{\gamma - n}{\gamma - 1}\right) \times$ work done
    (b) $\left(\dfrac{\gamma - n}{\gamma - 1}\right)^2 \times$ work done
    (c) $\left(\dfrac{\gamma - n}{\gamma - 1}\right)^{1/2} \times$ work done
    (d) $\left(\dfrac{\gamma - n}{\gamma - 1}\right)^3 \times$ work done.

Answers

1. (a)	2. (a)	3. (d)	4. (a)	5. (d)	6. (d)	7. (a)
8. (d)	9. (a)	10. (c)	11. (e)	12. (a)	13. (d)	14. (b)
15. (a)	16. (e)	17. (a)	18. (e)	19. (d)	20. (a)	21. (c)
22. (b)	23. (e)	24. (d)	25. (b)	26. (a).		

THEORETICAL QUESTIONS

1. Define 'internal energy' and prove that it is a property of a system.
2. Explain the First Law of Thermodynamics as referred to closed systems undergoing a cyclic change.
3. State the First Law of Thermodynamics and prove that for a non-flow process, it leads to the energy equation $Q = \Delta U + W$.

4. What is the mechanical equivalent of heat ? Write down its value when heat is expressed in kJ and work is expressed in Nm.
5. What do you mean by "Perpetual motion machine of first kind-PMM 1" ?
6. Why only in constant pressure non-flow process, the enthalpy change is equal to heat transfer ?
7. Prove that the rate of change of heat interchange per unit change of volume when gas is compressed or expanded is given by $\dfrac{\gamma - n}{\gamma - 1} \times \dfrac{pdv}{J}$.

UNSOLVED EXAMPLES

Closed Systems

1. In a cyclic process, heat transfers are + 14.7 kJ, – 25.2 kJ, – 3.56 kJ and + 31.5 kJ. What is the net work for this cyclic process ? [Ans. 17.34 kJ]
2. A domestic refrigerator is loaded with food and the door closed. During a certain period the machine consumes 1 kWh of energy and the internal energy of the system drops by 5000 kJ. Find the net heat transferred in the system. [Ans. – 8.6 MJ]
3. 1.5 kg of liquid having a constant specific heat of 2.5 kJ/kg°C is stirred in a *well-insulated* chamber causing the temperature to rise by 15°C. Find :
   (i) Change in internal energy, and
   (ii) Work done for the process. [Ans. (i) 56.25 kJ ; (ii) W = – 56.25 kJ]
4. A system is composed of a stone having a mass of 10 kg and a bucket containing 100 kg of water. Initially the stone and water are at the same temperature, the stone then falls into the water. Determine ΔU, ΔKE, ΔPE, ΔQ and ΔW for the following cases :
   (i) At the instant the stone is about to enter the water.
   (ii) Just after the stone comes to rest in the bucket.
   $\begin{bmatrix} \text{Ans. } (i) \ \Delta Q = \Delta W = \Delta E = 0, \Delta KE = 4.184 \text{ kJ}, \Delta PE = -4.184 \text{ kJ} ; \\ (ii) \ \Delta Q = 0, \Delta W = 0, \Delta KE = 0, \Delta U = +4.184 \text{ kJ}, \Delta PE = -4.184 \text{ kJ} \end{bmatrix}$
5. A closed system of constant volume experiences a temperature rise of 20°C when a certain process occurs. The heat transferred in the process is 18 kJ. The specific heat at constant volume for the pure substance comprising the system is 1.2 kJ/kg°C, and the system contains 2 kg of this substance. Determine the change in internal energy and the work done. [Ans. ΔU = 48 kJ ; W = – 30 kJ]
6. A stationary mass of gas is compressed without friction from an initial state of 2 m³ and 2×10^5 N/m² to a final state of 1 m³ and 2×10^5 N/m², the pressure remaining the same. There is a transfer of 360 kJ of heat from the gas during the process. How much does the internal energy of the gas change ? [Ans. ΔU = – 160 kJ]
7. The internal energy of a certain substance is given by the following equation :
   $$u = pv + 84$$
   where u is given in kJ/kg, p is in kPa and v is in m³/kg.
   A system composed of 3 kg of this substance expands from an initial pressure of 500 kPa and a volume of 0.22 m³ to a final pressure 100 kPa in a process in which pressure and volume are related by $pv^{1.2}$ = constant.
   (i) If the expansion is quasi-static, find Q, ΔU and W for the process.
   (ii) In another process the same system expands according to the same pressure-volume relationship as in part (i) and from the same initial state to the same final state as in part (i) but the heat transfer in this case is 30 kJ. Find the work transfer for this process.
   (iii) Explain the difference in work transfer in parts (i) and (ii).
   $\begin{bmatrix} \text{Ans. } (i) \ \Delta U = -91 \text{ kJ}, W = 127.5 \text{ kJ}, Q = 36.5 \text{ kJ} ; \\ (ii) \ W = 121 \text{ kJ} ; (iii) \text{ The work in } (ii) \text{ in not equal} \\ \text{to } \int pdV \text{ since the process is not quasi-static.} \end{bmatrix}$

FIRST LAW OF THERMODYNAMICS

8. A fluid is contained in a cylinder by a spring-loaded, frictionless piston so that the pressure in the fluid is linear function of the volume ($p = a + bV$). The internal energy of the fluid is given by the following equation
$$U = 34 + 3.15 pV$$
where U is in kJ, p in kPa and V in cubic metre. If the fluid changes from an initial state of 170 kPa, 0.03 m³ to a final state of 400 kPa, 0.06 m³, with no work other than that done on the piston, find the direction and magnitude of the work and heat transfer.

[**Ans.** $W_{1-2} = 10.35$ kJ ;
$Q_{1-2} = 69.85$ kJ (heat flows into the system during the process)]

9. A piston cylinder arrangement has a gas in the cylinder space. During a constant pressure expansion to a larger volume the work effect for the gas are 1.6 kJ, the heat added to the gas and cylinder arrangement is 3.2 kJ and the friction between the piston and cylinder wall amounts to 0.24 kJ. Determine the change in internal energy of the entire apparatus (gas, cylinder, piston). [**Ans.** 1.84 kJ]

10. A system receives 42 kJ of heat while expanding with volume change of 0.123 m³ against an atmosphere of 12 N/cm². A mass of 80 kg in the surroundings is also lifted through a distance of 6 metres.
    (i) Find the change in energy of the system.
    (ii) The system is returned to its initial volume by an adiabatic process which requires 100 kJ of work. Find the change in energy of system.
    (iii) Determine the total change in energy of the system. [**Ans.** (i) 22.54 kJ ; (ii) 100 kJ ; (iii) 122.54 kJ]

11. A thermally insulated battery is being discharged at atmospheric pressure and constant volume. During a 1 hour test it is found that a current of 50 A and 2 V flows while the temperature increases from 20°C to 32.5°C. Find the change in internal energy of the cell during the period of operation. [**Ans.** -36×10^4 J]

12. In a certain steam plant the turbine develops 1000 kW. The heat supplied to the steam in the boiler is 2800 kJ/kg, the heat received by the system from cooling water in the condenser is 2100 kJ/kg and the feed pump work required to pump the condensate back into the boiler is 5 kW. Calculate the steam flow round the cycle in kg/s. [**Ans.** 1.421 kg/s]

13. In the compression stroke of an internal-combustion engine the heat rejected to the cooling water is 45 kJ/kg and the work input is 90 kJ/kg. Calculate the change in internal energy of the working fluid stating whether it is a gain or a loss. [**Ans.** 45 kJ/kg (gain)]

14. 85 kJ of heat are supplied to a system at constant volume. The system rejects 90 kJ of heat at constant pressure and 20 kJ of work is done on it. The system is brought to its original state by adiabatic process. Determine the adiabatic work. Determine also the value of internal energy at all end states if initial value is 100 kJ. [**Ans.** $W = 15$ kJ ; $U_1 = 100$ kJ ; $U_2 = 185$ kJ ; $U_3 = 115$ kJ]

15. A closed system undergoes a reversible process at a constant pressure process of 3.5 bar and its volume changes from 0.15 m³ to 0.06 m³. 25 kJ of heat is rejected by the system during the process. Determine the change in internal energy of the system. [**Ans.** 6.5 kJ (increase)]

16. An air compressor takes in air at 10^5 Pa and 27°C having volume of 1.5 m³/kg and compresses it to 4.5×10^5 Pa. Find the work done, heat transfer and change in internal energy if the compression is isothermal.
[**Ans.** -225 kJ ; -225 kJ ; $\Delta U = 0$]

17. A cylinder fitted with piston contains 0.2 kg of N_2 at 100 kPa and 30°C. The piston is moved compressing N_2 until the pressure becomes 1 MPa and temperature becomes 150°C. The work done during the process is 20 kJ. Determine the heat transferred from N_2 to the surroundings. Take $c_v = 0.75$ kJ/kg K for N_2.
[**Ans.** -2 kJ]

18. A closed system consisting of 1 kg of gaseous CO_2 undergoes a reversible process at constant pressure causing a decrease of 30 kJ in internal energy. Determine the work done during the process. Take $c_p = 840$ J/kg°C and $c_v = 600$ J/kg°C. [**Ans.** -12 kJ]

19. The specific heat at constant pressure of one kg fluid undergoing a non-flow constant pressure process is given by
$$c_p = \left[2.5 + \frac{40}{T + 20} \right] \text{kg/kg°C}$$
where T is in °C.

The pressure during the process is maintained at 2 bar and volume changes from 1 m³ to 1.8 m³ and temperature changes from 50°C to 450°C. Determine :
(i) Heat added (ii) Work done
(iii) Change in internal energy (iv) Change in enthalpy.
[**Ans.** (i) 1076 kJ ; (ii) 160 kJ ; (iii) 916 kJ ; (iv) 1076 kJ]

20. 1 kg of nitrogen (molecular weight 28) is compressed reversibly and isothermally from 1.01 bar, 20°C to 4.2 bar. Calculate the work done and the heat flow during the process. Assume nitrogen to be a perfect gas.
[**Ans.** $W = 124$ kJ/kg ; $Q = -124$ kJ/kg]

21. Air at 1.02 bar, 22°C, initially occupying a cylinder volume of 0.015 m³, is compressed reversibly and adiabatically by a piston to a pressure of 6.8 bar. Calculate :
(i) The final temperature (ii) The final volume
(iii) The work done on the mass of air in the cylinder. [**Ans.** (i) 234.5°C ; (ii) 0.00388 m³ ; (iii) 2.76 kJ]

22. 1 kg of a perfect gas is compressed from 1.1 bar, 27°C according to a law $pv^{1.3}$ = constant, until the pressure is 6.6 bar. Calculate the heat flow to or from the cylinder walls,
(i) When the gas is ethane (molecular weight 30), which has
$c_p = 1.75$ kJ/kg K.
(ii) When the gas is argon (molecular weight 40), which has
$c_p = 0.515$ kJ/kg K. [**Ans.** (i) 84.5 kJ/kg ; (ii) – 59.4 kJ/kg]

23. 1 kg of air at 1 bar, 15°C is compressed reversibly and adiabatically to a pressure of 4 bar. Calculate the final temperature and the work done on the air. [**Ans.** 155°C ; 100.5 kJ/kg]

24. A certain perfect gas is compressed reversibly from 1 bar, 17°C to a pressure of 5 bar in a perfectly thermally insulated cylinder, the final temperature being 77°C. The work done on the gas during the compression is 45 kJ/kg. Calculate γ, c_v, R and the molecular weight of the gas.
[**Ans.** 1.132 ; 0.75 kJ/kg K ; 0.099 kJ/kg K ; 84]

25. 1 kg of air at 1.02 bar, 20°C is compressed reversibly according to a law $pv^{1.3}$ = constant, to a pressure of 5.5 bar. Calculate the work done on the air and heat flow to or from the cylinder walls during the compression. [**Ans.** 133.5 kJ/kg ; – 33.38 kJ/kg]

26. 0.05 kg of carbon dioxide (molecular weight 44), occupying a volume of 0.03 m³ at 1.025 bar, is compressed reversibly until the pressure is 6.15 bar. Calculate final temperature, the work done on the CO_2, the heat flow to or from the cylinder walls,
(i) When the process is according to law $pv^{1.4}$ = constant,
(ii) When the process is isothermal,
(iii) When the process takes place in a perfectly thermally insulated cylinder.
Assume CO_2 to be a perfect gas, and take $\gamma = 1.3$.
$$\left[\textbf{Ans.}\ 270°C\ ;\ 5.138\text{ kJ}\ ;\ 1.713\text{ kJ}\ ;\ 52.6°C\ ;\ 5.51\text{ kJ}\ ;\\ -5.51\text{ kJ}\ ;\ 219°C\ ;\ 5.25\text{ kJ}\ ;\ 0\text{ kJ} \right]$$

27. Oxygen (molecular weight 32) is compressed reversibly and polytropically in a cylinder from 1.05 bar, 15°C to 4.2 bar in such a way that one-third of the work input is rejected as heat to the cylinder walls. Calculate the final temperature of the oxygen.
Assume oxygen to be a perfect gas and take $c_v = 0.649$ kJ/kg K. [**Ans.** 113°C]

28. A cylinder contains 0.5 m³ of a gas at 1×10^5 N/m² and 90°C. The gas is compressed to a volume of 0.125 m³, the final pressure being 6×10^5 N/m². Determine :
(i) The mass of gas.
(ii) The value of index 'n' for compression.
(iii) The increase in internal energy of gas.
(iv) The heat received or rejected by the gas during compression.
($\gamma = 1.4, R = 294.2$ Nm/kg°C). [**Ans.** 0.468 kg ; 1.292 ; 62.7 kJ ; – 22.67 kJ]

3

Gas Power Cycles

3.1. Definition of a cycle. 3.2. Air standard efficiency. 3.3. The Carnot cycle. 3.4. Constant volume or Otto cycle. 3.5. Constant pressure or Diesel cycle—Highlights—Objective Type Questions—Theoretical Questions—Unsolved Examples.

3.1. DEFINITION OF A CYCLE

A **cycle** is defined as a *repeated series of operations occurring in a certain order*. It may be repeated by repeating the processes in the same order. The cycle may be of *imaginary perfect engine or actual engine*. The former is called **ideal cycle** and the latter **actual cycle**. In ideal cycle all accidental heat losses are prevented and the working substance is assumed to behave like a perfect working substance.

3.2. AIR STANDARD EFFICIENCY

To compare the effects of different cycles, it is of paramount importance that the effect of the calorific value of the fuel is altogether eliminated and this can be achieved by considering air (which is assumed to behave as a perfect gas) as the working substance in the engine cylinder. *The efficiency of engine using air as the working medium is known as an* **"Air standard efficiency"**. This efficiency is oftenly called **ideal efficiency.**

The actual efficiency of a cycle is always *less* than the air-standard efficiency of that cycle under ideal conditions. This is taken into account by introducing a new term **"Relative efficiency"** which is defined as ;

$$\eta_{relative} = \frac{\text{Actual thermal efficiency}}{\text{Air standard efficiency}} \qquad \ldots(3.1)$$

The analysis of all air standard cycles is based upon the following *assumptions* :

Assumptions :

1. The gas in the engine cylinder is a *perfect gas i.e.*, it obeys the gas laws and has constant specific heats.
2. The physical constants of the gas in the cylinder are the same as those of air at moderate temperatures *i.e.*, the molecular weight of cylinder gas is 29.
   c_p = 1.005 kJ/kg K, c_p = 0.718 kJ/kg K.
3. The compression and expansion processes are adiabatic and they take place without internal friction, *i.e.*, these processes are *isentropic*.
4. No chemical reaction takes place in the cylinder. Heat is supplied or rejected by bringing a hot body or a cold body in contact with cylinder at appropriate points during the process.
5. The cycle is considered closed with the same 'air' always remaining in the cylinder to repeat the cycle.

3.3. THE CARNOT CYCLE

This cycle has the *highest possible efficiency* and consists of four simple operations namely,
(a) Isothermal expansion
(b) Adiabatic expansion
(c) Isothermal compression
(d) Adiabatic compression.

The condition of the Carnot cycle may be imagined to occur in the following way :

One kg of a air is enclosed in the cylinder which (except at the end) is made of perfect non-conducting material. A source of heat 'H' is supposed to provide unlimited quantity of heat, non-conducting cover 'C' and a sump 'S' which is of infinite capacity so that its temperature remains unchanged irrespective of the fact how much heat is supplied to it. The temperature of source H is T_1 and the same is of the working substance. The working substance while rejecting heat to sump 'S' has the temperature. T_2 i.e., the same as that of sump S.

Following are the *four stages* of the Carnot cycle. Refer Fig. 3.1 (a).

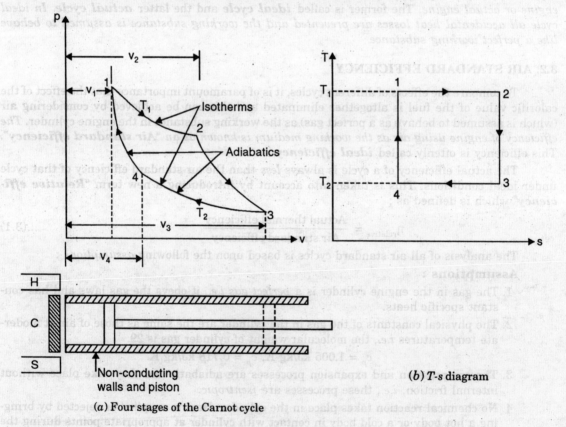

(a) Four stages of the Carnot cycle   (b) T-s diagram

Fig. 3.1. Carnot cycle.

Stage (1). Line 1–2 [Fig. 3.1 (a)] represents the isothermal expansion which takes place at temperature T_1 when source of heat H is applied to the end of cylinder. Heat supplied in this case is given by $RT_1 \log_e r$ and where r is the ratio of expansion.

GAS POWER CYCLES

Stage (2). Line 2–3 represents the application of non-conducting cover to the end of the cylinder. This is followed by the adiabatic expansion and the temperature falls from T_1 to T_2.

Stage (3). Line 3–4 represents the isothermal compression which takes place when sump 'S' is applied to the end of cylinder. Heat is rejected during this operation whose value is given by $RT_2 \log_e r$, where r is the ratio of compression.

Stage (4). Line 4–1 represents repeated application of non-conducting cover and adiabatic compression due to which temperature increases from T_2 to T_1.

It may be noted that ratio of expansion during isotherm 1-2 and ratio of compression during isotherm 3-4 must be equal to get a closed cycle.

Fig. 3.1 (b) represents the Carnot cycle on T-s coordinates.

Now according to law of conservation of energy,

Heat supplied $\quad$ = Work done + Heat rejected

Work done $\quad$ = Heat supplied – Heat rejected

$\quad\quad\quad\quad\quad$ = $RT_1 . \log_e r - RT_2 \log_e r$

Efficiency of cycle $\quad = \dfrac{\text{Work done}}{\text{Heat supplied}} = \dfrac{R \log_e r (T_1 - T_2)}{RT_1 . \log_e r}$

$\quad\quad\quad\quad\quad\quad\quad = \dfrac{T_1 - T_2}{T_1}$ ...(3.2)

From this equation, it is quite obvious that if temperature T_2 decreases efficiency increases and it becomes 100% if T_2 becomes absolute zero which, of course is impossible to attain. Further more *it is not possible to produce an engine that should work on Carnot's cycle as it would necessitate the piston to travel very slowly during first portion of the forward stroke (isothermal expansion) and to travel more quickly during the remainder of the stroke (adiabatic expansion) which however is not practicable.*

Example 3.1. *A Carnot engine working between 400°C and 40°C produces 130 kJ of work. Determine :*

(i) *The engine thermal efficiency.*
(ii) *The heat added.*
(iii) *The entropy changes during heat rejection process.*

Solution. Temperature, $\quad T_1 = T_2 = 400 + 273 = 673$ K
Temperature, $\quad\quad\quad\quad\quad T_3 = T_4 = 40 + 273 = 313$ K
Work produced, $\quad\quad\quad\quad W = 130$ kJ.

(i) **Engine thermal efficiency, η_{th} :**

$$\eta_{th} = \dfrac{673 - 313}{673} = 0.535 \text{ or } 53.5\%. \text{ (Ans.)}$$

(ii) **Heat added :**

$$\eta_{th} = \dfrac{\text{Work done}}{\text{Heat added}}$$

i.e., $\quad\quad\quad\quad 0.535 = \dfrac{130}{\text{Heat added}}$

$\therefore\quad$ Heat added $\quad = \dfrac{130}{0.535} = 243$ kJ. **(Ans.)**

(iii) **Entropy change during the heat rejection process, $(S_3 - S_4)$:**

Heat rejected $\quad\quad\quad$ = Heat added – Work done
$\quad\quad\quad\quad\quad\quad\quad$ = 243 – 130 = 113 kJ

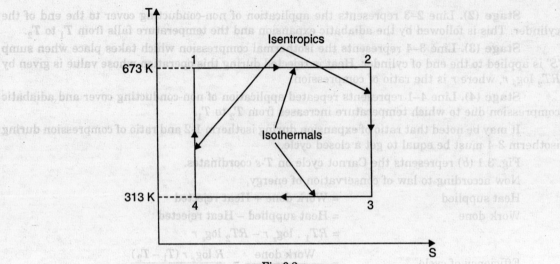
Fig. 3.2

Heat rejected $= T_3 (S_3 - S_4) = 113$

$\therefore (S_3 - S_4) = \dfrac{113}{T_3} = \dfrac{113}{313} = 0.361$ kJ/K. (Ans.)

☞**Example 3.2.** *In a Carnot cycle, the maximum pressure and temperature are limited to 18 bar and 410°C. The ratio of isentropic compression is 6 and isothermal expansion is 1.5. Assuming the volume of the air at the beginning of isothermal expansion as 0.18 m³, determine :*

(i) *The temperature and pressures at main points in the cycle.*
(ii) *Change in entropy during isothermal expansion.*
(iii) *Mean thermal efficiency of the cycle.*
(iv) *Mean effective pressure of the cycle.*
(v) *The theoretical power if there are 210 working cycles per minute.*

Solution. Refer to Fig. 3.3.
Maximum pressure, $p_1 = 18$ bar
Maximum temperature, $T_1 = (T_2) = 410 + 273 = 683$ K

Ratio of isentropic (or adiabatic) compression, $\dfrac{V_4}{V_1} = 6$

Ratio of isothermal expansion, $\dfrac{V_2}{V_1} = 1.5$.

Volume of the air at the beginning of isothermal expansion, $V_1 = 0.18$ m³.

(i) **Temperatures and pressures at the main points in the cycle :**

For the *isentropic process* 4–1 :

$$\dfrac{T_1}{T_4} = \left(\dfrac{V_4}{V_1}\right)^{\gamma-1} = (6)^{1.4-1} = (6)^{0.4} = 2.05$$

$\therefore T_4 = \dfrac{T_1}{2.05} = \dfrac{683}{2.05} = 333.2$ K $= T_3$

GAS POWER CYCLES

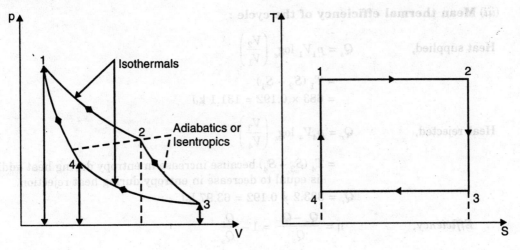

Fig. 3.3

Also, $\quad \dfrac{p_1}{p_4} = \left(\dfrac{V_4}{V_1}\right)^\gamma = (6)^{1.4} = 12.29$

$\therefore \quad p_4 = \dfrac{p_1}{12.29} = \dfrac{18}{12.29} = 1.46 \text{ bar}$

For the *isothermal process* 1–2 :

$p_1 V_1 = p_2 V_2$

$p_2 = \dfrac{p_1 V_1}{V_2} = \dfrac{18}{1.5} = 12 \text{ bar}$

For *isentropic process* 2–3, we have :

$p_2 V_2^\gamma = p_3 V_3^\gamma$

$p_3 = p_2 \times \left(\dfrac{V_2}{V_3}\right)^\gamma = 12 \times \left(\dfrac{V_1}{V_4}\right)^\gamma \quad \left[\because \dfrac{V_4}{V_1} = \dfrac{V_3}{V_2}\right]$

$= 12 \times \left(\dfrac{1}{6}\right)^{1.4} = \textbf{0.97 bar. (Ans.)}$

Hence $\quad \begin{array}{ll} p_1 = \textbf{18 bar} & T_1 = T_2 = \textbf{683 K} \\ p_2 = \textbf{12 bar} & \\ p_3 = \textbf{0.97 bar} & T_3 = T_4 = \textbf{333.2 K} \\ p_4 = \textbf{1.46 bar} & \end{array} \quad \textbf{(Ans.)}$

(*ii*) **Change in entropy :**

Change in entropy during isothermal expansion,

$S_2 - S_1 = mR \log_e \left(\dfrac{V_2}{V_1}\right) = \dfrac{p_1 V_1}{T_1} \log_e \left(\dfrac{V_2}{V_1}\right) \quad \begin{bmatrix} \because \ pV = mRT \\ \text{or} \ \ mR = \dfrac{pV}{T} \end{bmatrix}$

$= \dfrac{18 \times 10^5 \times 0.18}{10^3 \times 683} \cdot \log_e (1.5) = \textbf{0.192 kJ/K. (Ans.)}$

(*iii*) **Mean thermal efficiency of the cycle :**

Heat supplied, $Q_s = p_1 V_1 \log_e \left(\dfrac{V_2}{V_1}\right)$

$= T_1 (S_2 - S_1)$
$= 683 \times 0.192 = 131.1 \text{ kJ}$

Heat rejected, $Q_r = p_4 V_4 \log_e \left(\dfrac{V_3}{V_4}\right)$

$= T_4 (S_3 - S_4)$ because increase in entropy during heat addition is equal to decrease in entropy during heat rejection.

∴ $Q_r = 333.2 \times 0.192 = 63.97 \text{ kJ}$

∴ Efficiency, $\eta = \dfrac{Q_s - Q_r}{Q_s} = 1 - \dfrac{Q_r}{Q_s}$

$= 1 - \dfrac{63.97}{131.1} = 0.512$ or **51.2%. (Ans.)**

(*iv*) **Mean effective pressure of the cycle, p_m :**

The mean effective pressure of the cycle is given by

$$p_m = \dfrac{\text{Work done per cycle}}{\text{Stroke volume}}$$

$\dfrac{V_3}{V_1} = 6 \times 1.5 = 9$

Stroke volume, $V_s = V_3 - V_1 = 9V_1 - V_1 = 8V_1 = 8 \times 0.18 = 1.44 \text{ m}^3$

∴ $p_m = \dfrac{(Q_s - Q_r) \times J}{V_s} = \dfrac{(Q_s - Q_r) \times 1}{V_s}$ (∵ $J = 1$)

$= \dfrac{(131.1 - 63.97) \times 10^3}{1.44 \times 10^5} = \textbf{0.466 bar. (Ans.)}$

(*v*) **Power of the engine, P :**

Power of the engine working on this cycle is given by

$P = (131.1 - 63.97) \times (210/60) = \textbf{234.9 kW. (Ans.)}$

Example 3.3. *A reversible engine converts one-sixth of the heat input into work. When the temperature of the sink is reduced by 70°C, its efficiency is doubled. Find the temperature of the source and the sink.*

Solution. Let, T_1 = temperature of the source (K), and
T_2 = temperature of the sink (K)

First case :

$$\dfrac{T_1 - T_2}{T_1} = \dfrac{1}{6}$$

i.e., $6T_1 - 6T_2 = T_1$

or $5T_1 = 6T_2$ or $T_1 = 1.2 T_2$ ...(*i*)

Second case :

$$\dfrac{T_1 - [T_2 - (70 + 273)]}{T_1} = \dfrac{1}{3}$$

GAS POWER CYCLES

$$\frac{T_1 - T_2 + 343}{T_1} = \frac{1}{3}$$

$$3T_1 - 3T_2 + 1029 = T_1$$
$$2T_1 = 3T_2 - 1029$$
$$2 \times (1.2 T_2) = 3T_2 - 1029 \qquad (\because T_1 = 1.2 T_2)$$
$$2.4 T_2 = 3T_2 - 1029$$

or
$$0.6 T_2 = 1029$$

∴
$$T_2 = \frac{1029}{0.6} = \textbf{1715 K} \text{ or } \textbf{1442°C. (Ans.)}$$

and
$$T_1 = 1.2 \times 1715 = \textbf{2058 K} \text{ or } \textbf{1785°C. (Ans.)}$$

Example 3.4. *An inventor claims that a new heat cycle will develop 0.4 kW for a heat addition of 32.5 kJ/min. The temperature of heat source is 1990 K and that of sink is 850 K. Is his claim possible ?*

Solution. Temperature of heat source, $T_1 = 1990$ K
Temperature of sink, $T_2 = 850$ K
Heat supplied, $= 32.5$ kJ/min
Power developed by the engine, $P = 0.4$ kW

The *most efficient engine is one that works on Carnot cycle*

$$\eta_{carnot} = \frac{T_1 - T_2}{T_1} = \frac{1990 - 850}{1990} = 0.573 \text{ or } 57.3\%$$

Also, thermal efficiency of the engine,

$$\eta_{th} = \frac{\text{Work done}}{\text{Heat supplied}} = \frac{0.4}{(32.5/60)} = \frac{0.4 \times 60}{32.5}$$
$$= 0.738 \text{ or } 73.8\%$$

which is not feasible as no engine can be more efficient than that working on Carnot cycle.

Hence claims of the inventor is **not true. (Ans.)**

3.4. CONSTANT VOLUME OR OTTO CYCLE

This cycle is so named as it was conceived by 'Otto'. On this cycle, petrol, gas and many types of oil engines work. It is the standard of comparison for internal combustion engines.

Fig. 3.4 (a) and (b) shows the theoretical p-V diagram and T-s diagrams of this cycle respectively.

- The point 1 represents that cylinder is full of air with volume V_1, pressure p_1 and absolute temperature T_1.
- Line 1–2 represents the *adiabatic compression* of air due to which p_1, V_1 and T_1 change to p_2, V_2 and T_2 respectively.
- Line 2–3 shows the *supply of heat* to the air *at constant volume* so that p_2 and T_2 change to p_3 and T_3 (V_3 being the same as V_2).
- Line 3–4 represents the *adiabatic expansion* of the air. During expansion p_3, V_3 and T_3 change to a final value of p_4, V_4 or V_1 and T_4 respectively.
- Line 4–1 shows the *rejection of heat* by air at *constant volume* till original state (point 1) reaches.

Consider **1 kg of air** (working substance) :

Heat supplied at constant volume = $c_v(T_3 - T_2)$.

Heat rejected at constant volume = $c_v(T_4 - T_1)$.

But, work done = Heat supplied – Heat rejected

$= c_v(T_3 - T_2) - c_v(T_4 - T_1)$

$\therefore$ Efficiency $= \dfrac{\text{Work done}}{\text{Heat supplied}} = \dfrac{c_v(T_3 - T_2) - c_v(T_4 - T_1)}{c_v(T_3 - T_2)}$

$= 1 - \dfrac{T_4 - T_1}{T_3 - T_2}$ ...(i)

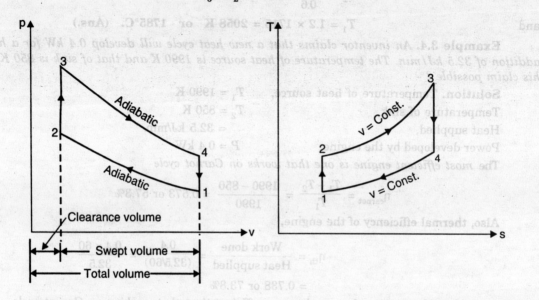

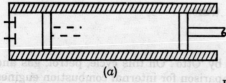

Fig. 3.4

Let compression ratio, $r_c \,(= r) = \dfrac{v_1}{v_2}$

and expansion ratio, $r_e \,(= r) = \dfrac{v_4}{v_3}$

(These *two ratios are same* in this cycle)

As $\dfrac{T_2}{T_1} = \left(\dfrac{v_1}{v_2}\right)^{\gamma - 1}$

Then $T_2 = T_1 \cdot (r)^{\gamma - 1}$

GAS POWER CYCLES

Similarly,
$$\frac{T_3}{T_4} = \left(\frac{v_4}{v_3}\right)^{\gamma-1}$$

or
$$T_3 = T_4 \cdot (r)^{\gamma-1}$$

Inserting the values of T_2 and T_3 in equation (i), we get

$$\eta_{otto} = 1 - \frac{T_4 - T_1}{T_4 \cdot (r)^{\gamma-1} - T_1 \cdot (r)^{\gamma-1}} = 1 - \frac{T_4 - T_1}{r^{\gamma-1}(T_4 - T_1)}$$

$$= 1 - \frac{1}{(r)^{\gamma-1}} \qquad \ldots(3.3)$$

*This expression is known as the **air standard efficiency of the Otto cycle.***

It is clear from the above expression that efficiency increases with the increase in the value of r, which means we can have maximum efficiency by increasing r to a considerable extent, but due to practical difficulties its value is limited to about 8.

The *net work done per kg* in the Otto cycle can also be expressed in terms of p, v. If p is expressed in bar *i.e.*, 10^5 N/m^2, then work done

$$W = \left(\frac{p_3 v_3 - p_4 v_4}{\gamma - 1} - \frac{p_2 v_2 - p_1 v_1}{\gamma - 1}\right) \times 10^2 \text{ kJ} \qquad \ldots(3.4)$$

Also
$$\frac{p_3}{p_4} = r^\gamma = \frac{p_2}{p_1}$$

$$\therefore \quad \frac{p_3}{p_2} = \frac{p_4}{p_1} = r_p$$

where r_p stands for *pressure ratio*.

and
$$v_1 = rv_2 = v_4 = rv_3 \qquad \left[\because \frac{v_1}{v_2} = \frac{v_4}{v_3} = r\right]$$

$\therefore$
$$W = \frac{1}{\gamma - 1}\left[p_4 v_4 \left(\frac{p_3 v_3}{p_4 v_4} - 1\right) - p_1 v_1 \left(\frac{p_2 v_2}{p_1 v_1} - 1\right)\right]$$

$$= \frac{1}{\gamma - 1}\left[p_4 v_4 \left(\frac{p_3}{p_4 r} - 1\right) - p_1 v_1 \left(\frac{p_2}{p_1 r} - 1\right)\right]$$

$$= \frac{v_1}{\gamma - 1}\left[p_4 (r^{\gamma-1} - 1) - p_1 (r^{\gamma-1} - 1)\right]$$

$$= \frac{v_1}{\gamma - 1}\left[(r^{\gamma-1} - 1)(p_4 - p_1)\right]$$

$$= \frac{p_1 v_1}{\gamma - 1}\left[(r^{\gamma-1} - 1)(r_p - 1)\right] \qquad \ldots[3.4\,(a)]$$

Mean effective pressure (p_m) is given by:

$$p_m = \left[\left(\frac{p_3 v_3 - p_4 v_4}{\gamma - 1} - \frac{p_2 v_2 - p_1 v_1}{\gamma - 1}\right) \div (v_1 - v_2)\right] \text{bar} \qquad \ldots(3.5)$$

Also
$$p_m = \frac{\left[\frac{p_1 v_1}{\gamma - 1}(r^{\gamma-1} - 1)(r_p - 1)\right]}{(v_1 - v_2)}$$

$$= \frac{\frac{p_1 v_1}{\gamma - 1}[(r^{\gamma-1} - 1)(r_p - 1)]}{v_1 - \frac{v_1}{r}}$$

$$= \frac{\frac{p_1 v_1}{\gamma - 1}[(r^{\gamma-1} - 1)(r_p - 1)]}{v_1\left(\frac{r-1}{r}\right)}$$

i.e.,
$$p_m = \frac{p_1 r[(r^{\gamma-1} - 1)(r_p - 1)]}{(\gamma - 1)(r - 1)} \qquad \ldots(3.6)$$

Example 3.5. *The efficiency of an Otto cycle is 60% and γ = 1.5. What is the compression ratio ?*

Solution. Efficiency of Otto cycle, η = 60%
Ratio of specific heats, γ = 1.5
Compression ratio, r = ?
Efficiency of Otto cycle is given by,

$$\eta_{otto} = 1 - \frac{1}{(r)^{\gamma-1}}$$

$$0.6 = 1 - \frac{1}{(r)^{1.5-1}}$$

or $\quad \frac{1}{(r)^{0.5}} = 0.4 \quad \text{or} \quad (r)^{0.5} = \frac{1}{0.4} = 2.5 \quad \text{or} \quad r = 6.25$

Hence, *compression ratio* = **6.25.** (Ans.)

Example 3.6. *An engine of 250 mm bore and 375 mm stroke works on Otto cycle. The clearance volume is 0.00263 m³. The initial pressure and temperature are 1 bar and 50°C. If the maximum pressure is limited to 25 bar, find the following :*

(i) The air standard efficiency of the cycle.

(ii) The mean effective pressure for the cycle.

Assume the ideal conditions.

Solution. Bore of the engine, $D = 250$ mm $= 0.25$ m
Stroke of the engine, $L = 375$ mm $= 0.375$ m
Clearance volume, $V_c = 0.00263$ m³
Initial pressure, $p_1 = 1$ bar
Initial temperature, $T_1 = 50 + 273 = 323$ K
Maximum pressure, $p_3 = 25$ bar
Swept volume, $V_s = \pi/4\, D^2 L = \pi/4 \times 0.25^2 \times 0.375 = 0.0184$ m³

Compression ratio, $r = \dfrac{V_s + V_c}{V_c} = \dfrac{0.0184 + 0.00263}{0.00263} = 8.$

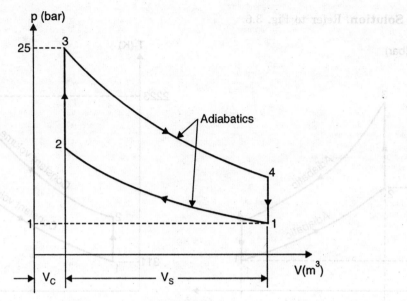

Fig. 3.5

(i) Air standard efficiency :

The air standard efficiency of Otto cycle is given by

$$\eta_{otto} = 1 - \frac{1}{(r)^{\gamma-1}} = 1 - \frac{1}{(8)^{1.4-1}} = 1 - \frac{1}{(8)^{0.4}}$$

$$= 1 - 0.435 = \mathbf{0.565} \text{ or } \mathbf{56.5\%.} \quad \textbf{(Ans.)}$$

(ii) Mean effective pressure, p_m :

For adiabatic (or isentropic) process 1–2

$$p_1 V_1^\gamma = p_2 V_2^\gamma$$

or

$$p_2 = p_1 \left(\frac{V_1}{V_2}\right)^\gamma = 1 \times (r)^{1.4} = 1 \times (8)^{1.4} = 18.38 \text{ bar}$$

∴ Pressure ratio, $r_p = \dfrac{p_3}{p_2} = \dfrac{25}{18.38} = 1.36$

The mean effective pressure is given by

$$p_m = \frac{p_1 r[(r^{\gamma-1} - 1)(r_p - 1)]}{(\gamma - 1)(r - 1)} = \frac{1 \times 8 \, [\{(8)^{1.4-1} - 1\}(1.36 - 1)]}{(1.4 - 1)(8 - 1)}$$

...[Eqn. (3.6)]

$$= \frac{8 \, (2.297 - 1)(0.36)}{0.4 \times 7} = 1.334 \text{ bar}$$

Hence *mean effective pressure* = **1.334 bar.** **(Ans.)**

Example 3.7. *In a constant volume 'Otto cycle', the pressure at the end of compression is 15 times that at the start, the temperature of air at the beginning of compression is 38°C and maximum temperature attained in the cycle is 1950°C. Determine :*

(i) *Compression ratio.*   (ii) *Thermal efficiency of the cycle.*

(iii) *Work done.*

Take γ for air = 1.4.

Solution. Refer to Fig. 3.6.

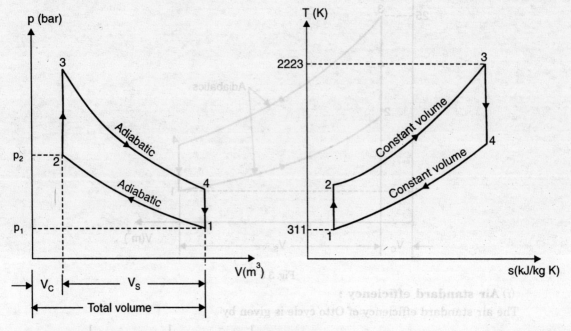

Fig. 3.6

Initial temperature, $T_1 = 38 + 273 = 311$ K
Maximum temperature, $T_3 = 1950 + 273 = 2223$ K.

(i) Compression ratio, r :

For *adiabatic compression 1–2*,
$$p_1 V_1^\gamma = p_2 V_2^\gamma$$

or
$$\left(\frac{V_1}{V_2}\right)^\gamma = \frac{p_2}{p_1}$$

But $\dfrac{p_2}{p_1} = 15$ ...(given)

∴ $(r)^\gamma = 15$ $\left[\because r = \dfrac{V_1}{V_2}\right]$

or $(r)^{1.4} = 15$

or $r = (15)^{\frac{1}{1.4}} = (15)^{0.714} = 6.9$

Hence *compression ratio* = **6.9. (Ans.)**

(ii) Thermal efficiency :

Thermal efficiency, $\eta_{th} = 1 - \dfrac{1}{(r)^{\gamma-1}} = 1 - \dfrac{1}{(6.9)^{1.4-1}}$

= **0.538 or 53.8%. (Ans.)**

GAS POWER CYCLES

(*iii*) **Work done :**

Again, for *adiabatic compression 1–2*,

$$\frac{T_2}{T_1} = \left(\frac{V_1}{V_2}\right)^{\gamma-1} = (r)^{\gamma-1} = (6.9)^{1.4-1} = (6.9)^{0.4} = 2.16$$

or $\quad T_2 = T_1 \times 2.16 = 311 \times 2.16 = 671.7$ K or 398.7°C

For *adiabatic expansion process 3–4*

$$\frac{T_3}{T_4} = \left(\frac{V_4}{V_3}\right)^{\gamma-1} = (r)^{\gamma-1} = (6.9)^{0.4} = 2.16$$

or $\quad T_4 = \dfrac{T_3}{2.16} = \dfrac{2223}{2.16} = 1029$ K or 756°C

Heat supplied *per kg of air*

$= c_v(T_3 - T_2) = 0.717(2223 - 671.7)$
$= 1112.3$ kJ/kg or air

$$\left[c_v = \frac{R}{\gamma-1} = \frac{0.287}{1.4-1}\right.$$
$$\left. = 0.717 \text{ kJ/kg K}\right]$$

Heat rejected per kg of air $\quad = c_v(T_4 - T_1) = 0.717(1029 - 311)$
$\quad\quad = 514.8$ kJ/kg of air

∴ Work done per kg of air $\quad=$ Heat supplied – heat rejected
$\quad\quad = 1112.3 - 514.8$
$\quad\quad = $ **597.5 kJ or 597500 Nm.** (**Ans.**)

☞**Example 3.8.** *An engine working on Otto cycle has a volume of 0.45 m³, pressure 1 bar and temperature 30°C at the beginning of compression stroke. At the end of compression stroke, the pressure is 11 bar. 210 kJ of heat is added at constant volume. Determine :*

(*i*) *Pressures, temperatures and volumes at salient points in the cycle.*
(*ii*) *Percentage clearance.*
(*iii*) *Efficiency.*
(*iv*) *Mean effective pressure.*
(*v*) *Ideal power developed by the engine if the number of working cycles per minute is 210. Assume the cycle is reversible.*

Solution. Refer Fig. 3.7
Volume, $\quad V_1 = 0.45$ m³
Initial pressure, $\quad p_1 = 1$ bar
Initial temperature, $\quad T_1 = 30 + 273 = 303$ K
Pressure at the end of compression stroke, $\quad p_2 = 11$ bar
Heat added at constant volume $\quad = 210$ kJ
Number of working cycles/min. $\quad = 210.$

(*i*) **Pressures, temperatures and volumes at salient points :**
For *adiabatic compression 1–2*
$$p_1 V_1^\gamma = p_2 V_2^\gamma$$

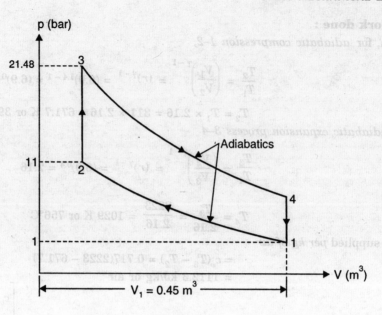

Fig. 3.7

or
$$\frac{p_2}{p_1} = \left(\frac{V_1}{V_2}\right)^\gamma = (r)^\gamma \quad \text{or} \quad r = \left(\frac{p_2}{p_1}\right)^{\frac{1}{\gamma}} = \left(\frac{11}{1}\right)^{\frac{1}{1.4}} = (11)^{0.714} = 5.5$$

Also
$$\frac{T_2}{T_1} = \left(\frac{V_1}{V_2}\right)^{\gamma-1} = (r)^{\gamma-1} = (5.5)^{1.4-1} = 1.977 \simeq 1.98$$

∴ $T_2 = T_1 \times 1.98 = 303 \times 1.98 = $ **600 K.** (Ans.)

Applying gas laws to points 1 and 2

$$\frac{p_1 V_1}{T_1} = \frac{p_2 V_2}{T_2}$$

∴ $V_2 = \frac{T_2}{T_1} \times \frac{p_1}{p_2} \times V_1 = \frac{600 \times 1 \times 0.45}{303 \times 11} = $ **0.081 m³.** (Ans.)

The heat supplied during the *process 2–3* is given by :

$$Q_s = m\, c_v\, (T_3 - T_2)$$

where
$$m = \frac{p_1 V_1}{R T_1} = \frac{1 \times 10^5 \times 0.45}{287 \times 303} = 0.517 \text{ kg}$$

∴ $210 = 0.517 \times 0.71 (T_3 - 600)$

or
$$T_3 = \frac{210}{0.517 \times 0.71} + 600 = \textbf{1172 K.} \quad \text{(Ans.)}$$

For the *constant volume process 2–3*

$$\frac{p_3}{T_3} = \frac{p_2}{T_2}$$

∴ $p_3 = \frac{T_3}{T_2} \times p_2 = \frac{1172}{600} \times 11 = $ **21.48 bar.** (Ans.)

$V_3 = V_2 = $ **0.081 m³.** (Ans.)

For the *adiabatic (or isentropic) process 3–4*

$$p_3 V_3^\gamma = p_4 V_4^\gamma$$

$$p_4 = p_3 \times \left(\frac{V_3}{V_4}\right)^\gamma = p_3 \times \left(\frac{1}{r}\right)^\gamma$$

$$= 21.48 \times \left(\frac{1}{5.5}\right)^{1.4} = \textbf{1.97 bar. (Ans.)}$$

Also
$$\frac{T_4}{T_3} = \left(\frac{V_3}{V_4}\right)^{\gamma-1} = \left(\frac{1}{r}\right)^{\gamma-1} = \left(\frac{1}{5.5}\right)^{1.4-1} = 0.505$$

∴ $T_4 = 0.505\, T_3 = 0.505 \times 1172 = \textbf{591.8 K. (Ans.)}$
$V_4 = V_1 = \textbf{0.45 m}^3\textbf{. (Ans.)}$

(*ii*) **Percentage clearance :**

Percentage clearance

$$= \frac{V_c}{V_s} = \frac{V_2}{V_1 - V_2} \times 100 = \frac{0.081}{0.45 - 0.081} \times 100$$

$$= \textbf{21.95\%. (Ans.)}$$

(*iii*) **Efficiency :**

The heat rejected per cycle is given by

$$Q_r = m c_v (T_4 - T_1)$$
$$= 0.517 \times 0.71 (591.8 - 303) = 106 \text{ kJ}$$

The air-standard efficiency of the cycle is given by

$$\eta_{otto} = \frac{Q_s - Q_r}{Q_s} = \frac{210 - 106}{210} = \textbf{0.495 or 49.5\%. (Ans.)}$$

$$\left[\begin{array}{l}\text{Alternatively :} \\ \qquad \eta_{otto} = 1 - \frac{1}{(r)^{\gamma-1}} = 1 - \frac{1}{(5.5)^{1.4-1}} = \textbf{0.495 or 49.5\%. (Ans.)}\end{array}\right]$$

(*iv*) **Mean effective pressure, p_m :**

The mean effective pressure is given by

$$p_m = \frac{W \text{ (work done)}}{V_s \text{ (swept volume)}} = \frac{Q_s - Q_r}{(V_1 - V_2)}$$

$$= \frac{(210 - 106) \times 10^3}{(0.45 - 0.081) \times 10^5} = \textbf{2.818 bar. (Ans.)}$$

(*v*) **Power developed, P :**

Power developed, P = Work done per second
$\qquad$ = Work done per cycle × number of cycles per second
$\qquad$ = $(210 - 106) \times (210/60)$ = **364 kW. (Ans.)**

3.5. CONSTANT PRESSURE OR DIESEL CYCLE

This cycle was introduced by Dr. R. Diesel in 1897. It differs from Otto cycle in that *heat is supplied at constant pressure instead of at constant volume*. Fig. 3.8 (*a* and *b*) shows the *p-v* and *T-s* diagrams of this cycle respectively.

This cycle comprises of the following *operations* :
   (*i*) 1–2......*Adiabatic compression.*
   (*ii*) 2–3......*Addition of heat at constant pressure.*
   (*iii*) 3–4......*Adiabatic expansion.*
   (*iv*) 4–1......*Rejection of heat at constant volume.*

Point 1 represents that the cylinder is full of air. Let p_1, V_1 and T_1 be the corresponding pressure, volume and absolute temperature. The piston then compresses the air adiabatically (*i.e.*, pV^γ = constant) till the values become p_2, V_2 and T_2 respectively (at the end of the stroke) at point 2. Heat is then added from a hot body at a constant pressure. During this addition of heat let volume increases from V_2 to V_3 and temperature T_2 to T_3, corresponding to point 3. This point (3) is called the *point of cut off*. The air then expands adiabatically to the conditions p_4, V_4 and T_4 respectively corresponding to point 4. Finally, the air rejects the heat to the cold body at constant volume till the point 1 where it returns to its original state.

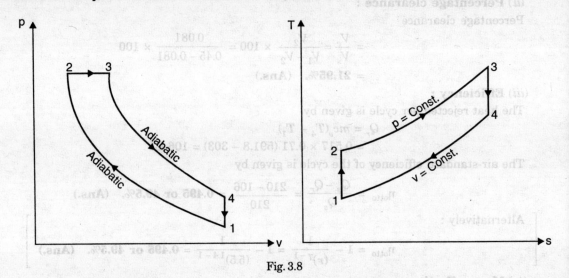

Fig. 3.8

Consider 1 kg of air.
Heat supplied at constant pressure = $c_p(T_3 - T_2)$
Heat rejected at constant volume = $c_v(T_4 - T_1)$
Work done = Heat supplied – heat rejected
          = $c_p(T_3 - T_2) - c_v(T_4 - T_1)$

∴ $\eta_{\text{diesel}} = \dfrac{\text{Work done}}{\text{Heat supplied}}$

$= \dfrac{c_p(T_3 - T_2) - c_v(T_4 - T_1)}{c_p(T_3 - T_2)}$

$= 1 - \dfrac{(T_4 - T_1)}{\gamma(T_3 - T_2)}$ ...(*i*) $\left[\because \dfrac{c_p}{c_v} = \gamma\right]$

Let compression ratio, $r = \dfrac{v_1}{v_2}$, and cut-off ratio, $\rho = \dfrac{v_3}{v_2}$ *i.e.*, $\dfrac{\text{Volume at cut-off}}{\text{Clearance volume}}$

GAS POWER CYCLES

Now, during *adiabatic compression 1–2*,

$$\frac{T_2}{T_1} = \left(\frac{v_1}{v_2}\right)^{\gamma-1} = (r)^{\gamma-1} \quad \text{or} \quad T_2 = T_1 \cdot (r)^{\gamma-1}$$

During *constant pressure process 2–3*,

$$\frac{T_3}{T_2} = \frac{v_3}{v_2} = \rho \quad \text{or} \quad T_3 = \rho \cdot T_2 = \rho \cdot T_1 \cdot (r)^{\gamma-1}$$

During *adiabatic expansion 3–4*,

$$\frac{T_3}{T_4} = \left(\frac{v_4}{v_3}\right)^{\gamma-1}$$

$$= \left(\frac{r}{\rho}\right)^{\gamma-1} \quad \left(\because \frac{v_4}{v_3} = \frac{v_1}{v_3} = \frac{v_1}{v_2} \times \frac{v_2}{v_3} = \frac{r}{\rho}\right)$$

$$\therefore \quad T_4 = \frac{T_3}{\left(\dfrac{r}{\rho}\right)^{\gamma-1}} = \frac{\rho \cdot T_1 (r)^{\gamma-1}}{\left(\dfrac{r}{\rho}\right)^{\gamma-1}} = T_1 \cdot \rho^{\gamma}$$

By inserting values of T_2, T_3 and T_4 in eqn. (*i*), we get

$$\eta_{\text{diesel}} = 1 - \frac{(T_1 \cdot \rho^{\gamma} - T_1)}{\gamma (\rho \cdot T_1 \cdot (r)^{\gamma-1} - T_1 \cdot (r)^{\gamma-1})} = 1 - \frac{(\rho^{\gamma} - 1)}{\gamma (r)^{\gamma-1} (\rho - 1)}$$

or

$$\eta_{\text{diesel}} = 1 - \frac{1}{\gamma (r)^{\gamma-1}} \left[\frac{\rho^{\gamma} - 1}{\rho - 1}\right] \qquad \ldots(3.7)$$

It may be observed that eqn. (3.7) for efficiency of diesel cycle is different from that of the Otto cycle only in bracketed factor. This factor is always greater than unity, because $\rho > 1$. Hence *for a given compression ratio, the Otto cycle is more efficient.*

The *net work* for diesel cycle can be expressed in terms of pv as follows :

$$W = p_2(v_3 - v_2) + \frac{p_3 v_3 - p_4 v_4}{\gamma - 1} - \frac{p_2 v_2 - p_1 v_1}{\gamma - 1}$$

$$= p_2 (\rho v_2 - v_2) + \frac{p_3 \rho v_2 - p_4 r v_2}{\gamma - 1} - \frac{p_2 v_2 - p_1 r v_1}{\gamma - 1}$$

$$\left[\begin{array}{l} \because \dfrac{v_3}{v_2} = \rho \quad \therefore v_3 = \rho v_2 \text{ and } \dfrac{v_1}{v_2} = r \quad \therefore v_1 = r v_2 \\ \text{But } v_4 = v_1 \quad \therefore \quad v_4 = r v_2 \end{array}\right]$$

$$= p_2 v_2 (\rho - 1) + \frac{p_3 \rho v_2 - p_4 r v_2}{\gamma - 1} - \frac{p_2 v_2 - p_1 r v_2}{\gamma - 1}$$

$$= \frac{v_2 [p_2 (\rho - 1)(\gamma - 1) + p_3 \rho - p_4 r - (p_2 - p_1 r)]}{\gamma - 1}$$

$$= \frac{v_2 \left[p_2 (\rho - 1)(\gamma - 1) + p_3 \left(\rho - \dfrac{p_4 r}{p_3} \right) - p_2 \left(1 - \dfrac{p_1 r}{p_2} \right) \right]}{\gamma - 1}$$

$$= \frac{p_2 v_2 [(\rho-1)(\gamma-1) + \rho - \rho^\gamma \cdot r^{1-\gamma} - (1 - r^{1-\gamma})]}{\gamma - 1}$$

$$\left[\because \frac{p_4}{p_3} = \left(\frac{v_3}{v_4}\right)^\gamma = \left(\frac{\rho}{r}\right)^\gamma = \rho^\gamma r^{-\gamma} \right]$$

$$= \frac{p_1 v_1 r^{\gamma-1} [(\rho-1)(\gamma-1) + \rho - \rho^\gamma r^{1-\gamma} - (1 - r^{1-\gamma})]}{\gamma - 1}$$

$$\left[\because \frac{p_2}{p_1} = \left(\frac{v_1}{v_2}\right)^\gamma \text{ or } p_2 = p_1 \cdot r^\gamma \text{ and } \frac{v_1}{v_2} = r \text{ or } v_2 = v_1 r^{-1} \right]$$

$$= \frac{p_1 v_1 r^{\gamma-1} [\gamma(\rho-1) - r^{1-\gamma}(\rho^\gamma - 1)]}{(\gamma-1)} \qquad \ldots(3.8)$$

Mean effective pressure p_m is given by :

$$p_m = \frac{p_1 v_1 r^{\gamma-1} [\gamma(\rho-1) - r^{1-\gamma}(\rho^\gamma - 1)]}{(\gamma-1) v_1 \left(\dfrac{r-1}{r}\right)}$$

or

$$\mathbf{p_m} = \frac{p_1 r^\gamma [\gamma(\rho-1) - r^{1-\gamma}(\rho^\gamma - 1)]}{(\gamma-1)(r-1)} \qquad \ldots(3.9)$$

Example 3.9. *A diesel engine has a compression ratio of 15 and heat addition at constant pressure takes place at 6% of stroke. Find the air standard efficiency of the engine.*

Take γ for air as 1.4.

Solution. Refer Fig. 3.9.

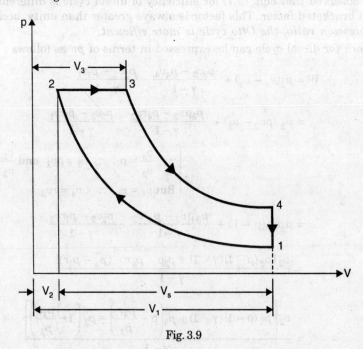

Fig. 3.9

Compression ratio, $r \left(= \dfrac{V_1}{V_2}\right) = 15$

γ for air = 1.4

Air standard efficiency of diesel cycle is given by

$$\eta_{\text{diesel}} = 1 - \dfrac{1}{\gamma(r)^{\gamma-1}} \left[\dfrac{\rho^\gamma - 1}{\rho - 1} \right] \qquad \ldots(i)$$

where ρ = cut-off ratio = $\dfrac{V_3}{V_2}$

But $\quad V_3 - V_2 = \dfrac{6}{100} V_s \quad (V_s = \text{stroke volume})$

$\qquad\qquad\qquad = 0.06 (V_1 - V_2) = 0.06 (15 V_2 - V_2)$

$\qquad\qquad\qquad = 0.84 V_2 \text{ or } V_3 = 1.84 V_2$

$\therefore \qquad \rho = \dfrac{V_3}{V_2} = \dfrac{1.84 V_2}{V_2} = 1.84$

Putting the value in eqn. (i), we get

$$\eta_{\text{diesel}} = 1 - \dfrac{1}{1.4 (15)^{1.4-1}} \left[\dfrac{(1.84)^{1.4} - 1}{1.84 - 1} \right]$$

$\qquad\qquad\qquad = 1 - 0.2417 \times 1.605 = \mathbf{0.612 \text{ or } 61.2\%}.$ **(Ans.)**

Example 3.10. *The stroke and cylinder diameter of a compression ignition engine are 250 mm and 150 mm respectively. If the clearance volume is 0.0004 m³ and fuel injection takes place at constant pressure for 5 per cent of the stroke determine the efficiency of the engine. Assume the engine working on the diesel cycle.*

Solution. Refer Fig. 3.9.

Length of stroke, $\qquad L$ = 250 mm = 0.25 m

Diameter of cylinder, $\qquad D$ = 150 mm = 0.15 m

Clearance volume, $\qquad V_2$ = 0.0004 m³

Swept volume, $\qquad V_s = \pi/4\ D^2 L = \pi/4 \times 0.15^2 \times 0.25 = 0.004418$ m³

Total cylinder volume $\qquad$ = Swept volume + clearance volume

$\qquad\qquad\qquad$ = 0.004418 + 0.0004 = 0.004818 m³

Volume at point of cut-off, $\quad V_3 = V_2 + \dfrac{5}{100} V_s$

$\qquad\qquad\qquad = 0.0004 + \dfrac{5}{100} \times 0.004418 = 0.000621$ m³

$\therefore \quad$ Cut-off ratio, $\qquad \rho = \dfrac{V_3}{V_2} = \dfrac{0.000621}{0.0004} = 1.55$

Compression ratio, $\qquad r = \dfrac{V_1}{V_2} = \dfrac{V_s + V_2}{V_2} = \dfrac{0.004418 + 0.0004}{0.0004} = 12.04$

Hence,
$$\eta_{\text{diesel}} = 1 - \frac{1}{\gamma(r)^{\gamma-1}}\left[\frac{\rho^\gamma - 1}{\rho - 1}\right] = 1 - \frac{1}{1.4 \times (12.04)^{1.4-1}}\left[\frac{(1.55)^{1.4} - 1}{1.55 - 1}\right]$$

$$= 1 - 0.264 \times 1.54 = \mathbf{0.593 \text{ or } 59.3\%}. \quad \textbf{(Ans.)}$$

Example 3.11. *Calculate the percentage loss in the ideal efficiency of a diesel engine with compression ratio 14 if the fuel cut-off is delayed from 5% to 8%.*

Solution. Let the clearance volume (V_2) be unity.

Then, compression ratio, $r = 14$

Now, when the fuel is cut-off at 5%, we have

$$\frac{\rho - 1}{r - 1} = \frac{5}{100} \quad \text{or} \quad \frac{\rho - 1}{14 - 1} = 0.05 \quad \text{or} \quad \rho - 1 = 13 \times 0.05 = 0.65$$

$$\therefore \quad \rho = 1.65$$

$$\eta_{\text{diesel}} = 1 - \frac{1}{\gamma(r)^{\gamma-1}}\left[\frac{\rho^\gamma - 1}{\rho - 1}\right] = 1 - \frac{1}{1.4 \times (14)^{1.4-1}}\left[\frac{(1.65)^{1.4} - 1}{1.65 - 1}\right]$$

$$= 1 - 0.248 \times 1.563 = 0.612 \quad \text{or} \quad 61.2\%$$

When the fuel is cut-off at 8%, we have

$$\frac{\rho - 1}{r - 1} = \frac{8}{100} \quad \text{or} \quad \frac{\rho - 1}{14 - 1} = \frac{8}{100} = 0.08$$

$$\therefore \quad \rho = 1 + 1.04 = 2.04$$

$$\eta_{\text{diesel}} = 1 - \frac{1}{\gamma(r)^{\gamma-1}}\left[\frac{\rho^\gamma - 1}{\rho - 1}\right] = 1 - \frac{1}{1.4 \times (14)^{1.4-1}}\left[\frac{(2.04)^{1.4} - 1}{2.04 - 1}\right]$$

$$= 1 - 0.248 \times 1.647 = \mathbf{0.591} \quad \text{or} \quad \mathbf{59.1\%}. \quad \textbf{(Ans.)}$$

Hence percentage loss in efficiency due to delay in fuel cut-off
$$= 61.2 - 59.1 = \mathbf{2.1\%}. \quad \textbf{(Ans.)}$$

Example 3.12. *The mean effective pressure of a Diesel cycle is 7.5 bar and compression ratio is 12.5. Find the percentage cut-off of the cycle if its initial pressure is 1 bar.*

Solution. Mean effective pressure, $p_m = 7.5$ bar

Compression ratio, $r = 12.5$

Initial pressure, $p_1 = 1$ bar

Refer Fig. 3.8.

The mean effective pressure is given by

$$p_m = \frac{p_1 r^\gamma [\gamma(\rho - 1) - r^{1-\gamma}(\rho^\gamma - 1)]}{(\gamma - 1)(r - 1)}$$

$$7.5 = \frac{1 \times (12.5)^{1.4}[1.4(\rho - 1) - (12.5)^{1-1.4}(\rho^{1.4} - 1)]}{(1.4 - 1)(12.5 - 1)}$$

$$7.5 = \frac{34.33[1.4\rho - 1.4 - 0.364\rho^{1.4} + 0.364]}{4.6}$$

GAS POWER CYCLES

$$7.5 = 7.46 (1.4\, \rho - 1.036 - 0.364\, \rho^{1.4})$$
$$1.005 = 1.4\, \rho - 1.036 - 0.364\, \rho^{1.4}$$
or
$$2.04 = 1.4\, \rho - 0.364\, \rho^{1.4} \quad \text{or} \quad 0.346\, \rho^{1.4} - 1.4\, \rho + 2.04 = 0$$

Solving by trial and error method, we get
$$\rho = 2.24$$

$$\therefore \quad \% \text{ cut-off} = \frac{\rho - 1}{r - 1} \times 100 = \frac{2.24 - 1}{12.5 - 1} \times 100 = \mathbf{10.78\%}. \quad \textbf{(Ans.)}$$

☞**Example 3.13.** *An engine with 200 mm cylinder diameter and 300 mm stroke works on theoretical Diesel cycle. The initial pressure and temperature of air used are 1 bar and 27°C. The cut-off is 8% of the stroke. Determine :*

(i) *Pressures and temperatures at all salient points.*
(ii) *Theoretical air standard efficiency.*
(iii) *Mean effective pressure.*
(iv) *Power of the engine if the working cycles per minute are 380.*

Assume that compression ratio is 15 and working fluid is air.

Consider all conditions to be ideal.

Solution. Refer to Fig. 3.10.

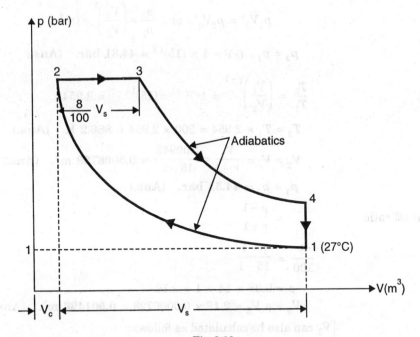

Fig. 3.10

Cylinder diameter, $D = 200$ mm or 0.2 m
Stroke length, $L = 300$ mm or 0.3 m
Initial pressure, $p_1 = 1.0$ bar
Initial temperature, $T_1 = 27 + 273 = 300$ K

Cut-off $= \dfrac{8}{100} V_s = 0.08 V_s$

(i) Pressures and temperatures at salient points :

Now, stroke volume, $V_s = \pi/4\, D^2 L = \pi/4 \times 0.2^2 \times 0.3 = 0.00942\ m^3$

$$V_1 = V_s + V_c = V_s + \dfrac{V_s}{r-1} \qquad \left[\because V_c = \dfrac{V_s}{r-1}\right]$$

$$= V_s\left(1 + \dfrac{1}{r-1}\right) = \dfrac{r}{r-1} \times V_s$$

i.e., $V_1 = \dfrac{15}{15-1} \times V_s = \dfrac{15}{14} \times 0.00942 = \mathbf{0.0101\ m^3}.$ **(Ans.)**

Mass of the air in the cylinder can be calculated by using the gas equation,

$$p_1 V_1 = mRT_1$$

$$m = \dfrac{p_1 V_1}{RT_1} = \dfrac{1 \times 10^5 \times 0.0101}{287 \times 300} = 0.0117\ kg/cycle$$

For the *adiabatic (or isentropic) process 1–2*

$$p_1 V_1^\gamma = p_2 V_2^\gamma \quad \text{or} \quad \dfrac{p_2}{p_1} = \left(\dfrac{V_1}{V_2}\right)^\gamma = (r)^\gamma$$

$\therefore \quad \boldsymbol{p_2} = p_1 \cdot (r)^\gamma = 1 \times (15)^{1.4} = \mathbf{44.31\ bar.}$ **(Ans.)**

Also, $\dfrac{T_2}{T_1} = \left(\dfrac{V_1}{V_2}\right)^{\gamma-1} = (r)^{\gamma-1} = (15)^{1.4-1} = 2.954$

$\therefore \quad \boldsymbol{T_2} = T_1 \times 2.954 = 300 \times 2.954 = \mathbf{886.2\ K.}$ **(Ans.)**

$$V_2 = V_c = \dfrac{V_s}{r-1} = \dfrac{0.00942}{15-1} = \mathbf{0.0006728\ m^3}.\ \textbf{(Ans.)}$$

$\boldsymbol{p_2} = p_3 = \mathbf{44.31\ bar.}$ **(Ans.)**

% cut-off ratio $= \dfrac{\rho - 1}{r - 1}$

$\dfrac{8}{100} = \dfrac{\rho - 1}{15 - 1}$

i.e., $\rho = 0.08 \times 14 + 1 = 2.12$

$\therefore \quad V_3 = \rho \cdot V_2 = 2.12 \times 0.0006728 = \mathbf{0.001426\ m^3}.$ **(Ans.)**

$$\left[\begin{array}{l} V_3 \text{ can also be calculated as follows :} \\ V_3 = 0.08 V_s + V_c = 0.08 \times 0.00942 + 0.0006728 = 0.001426\ m^3 \end{array}\right]$$

For the *constant pressure process 2–3*,

$$\dfrac{V_3}{T_3} = \dfrac{V_2}{T_2}$$

$\therefore \quad \boldsymbol{T_3} = T_2 \times \dfrac{V_3}{V_2} = 886.2 \times \dfrac{0.001426}{0.0006728} = \mathbf{1878.3\ K.}$ **(Ans.)**

GAS POWER CYCLES

For the *isentropic process 3-4*,

$$p_3 V_3^\gamma = p_4 V_4^\gamma$$

$$p_4 = p_3 \times \left(\frac{V_3}{V_4}\right)^\gamma = p_3 \times \frac{1}{(7.07)^{1.4}}$$

$$= \frac{44.31}{(7.07)^{1.4}} = 2.866 \text{ bar.} \quad \text{(Ans.)}$$

$$\left[\because \frac{V_4}{V_3} = \frac{V_4}{V_2} \times \frac{V_2}{V_3} = \frac{V_1}{V_2} \times \frac{V_2}{V_3}\right.$$

$$\left. = \frac{r}{\rho}, \because V_4 = V_1 = \frac{15}{2.12} = 7.07\right]$$

Also,
$$\frac{T_4}{T_3} = \left(\frac{V_3}{V_4}\right)^{\gamma-1} = \left(\frac{1}{7.07}\right)^{1.4-1} = 0.457$$

∴ $T_4 = T_3 \times 0.457 = 1878.3 \times 0.457 = $ **858.38 K.** (Ans.)

$V_4 = V_1 = $ **0.0101 m³.** (Ans.)

(ii) Theoretical air standard efficiency :

$$\eta_{\text{diesel}} = 1 - \frac{1}{\gamma(r)^{\gamma-1}} \left[\frac{\rho^\gamma - 1}{\rho - 1}\right] = 1 - \frac{1}{1.4 (15)^{1.4-1}} \left[\frac{(2.12)^{1.4} - 1}{2.12 - 1}\right]$$

$$= 1 - 0.2418 \times 1.663 = \textbf{0.598 or 59.8\%.} \quad \text{(Ans.)}$$

(iii) Mean effective pressure, p_m :

Mean effective pressure of Diesel cycle is given by

$$p_m = \frac{p_1(r)^\gamma [\gamma(\rho - 1) - r^{1-\gamma}(\rho^\gamma - 1)]}{(\gamma - 1)(r - 1)}$$

$$= \frac{1 \times (15)^{1.4}[1.4(2.12 - 1) - (15)^{1-1.4}(2.12^{1.4} - 1)]}{(1.4 - 1)(15 - 1)}$$

$$= \frac{44.31[1.568 - 0.338 \times 1.863]}{0.4 \times 14} = \textbf{7.424 bar.} \quad \text{(Ans.)}$$

(iv) Power of the engine, P :

Work done per cycle $= p_m V_s = \dfrac{7.424 \times 10^5 \times 0.00942}{10^3} = 6.99$ kJ/cycle

Work done per second = Work done per cycle × no. of cycles per second

$= 6.99 \times 380/60 = 44.27$ kJ/s = 44.27 kW

Hence *power of the engine* = **44.27 kW.** (Ans.)

HIGHLIGHTS

1. **A cycle** is defined as a repeated series of operations occurring in a certain order.
2. The efficiency of an engine using air as the working medium is known as an '**Air standard efficiency**'.
3. Relative efficiency, $\eta_{\text{relative}} = \dfrac{\text{Actual thermal efficiency}}{\text{Air standard efficiency}}$
4. **Carnot cycle** efficiency, $\eta_{\text{Carnot}} = \dfrac{T_1 - T_2}{T_1}$.

5. **Otto cycle** efficiency, $\eta_{Otto} = 1 - \dfrac{1}{(r)^{\gamma-1}}$.

   Mean effective pressure, $p_{m(Otto)} = \dfrac{p_1 r[(r^{\gamma-1}-1)(r_p-1)]}{(\gamma-1)(r-1)}$.

6. **Diesel cycle** efficiency, $\eta_{Diesel} = 1 - \dfrac{1}{\gamma(r)^{\gamma-1}}\left[\dfrac{\rho^{\gamma}-1}{\rho-1}\right]$

   Mean effective pressure, $p_{m(Diesel)} = \dfrac{p_1 r^{\gamma}[\gamma(\rho-1) - r^{1-\gamma}(\rho^{\gamma}-1)]}{(\gamma-1)(r-1)}$.

7. **Dual cycle** efficiency, $\eta_{Dual} = 1 - \dfrac{1}{(r)^{\gamma-1}}\left[\dfrac{(\beta.\rho^{\gamma}-1)}{(\beta-1) + \beta\gamma(\rho-1)}\right]$

   Mean effective pressure, $p_{m(Dual)} = \dfrac{p_1 r^{\gamma}[\beta(\rho-1) + (\beta-1) - r^{1-\gamma}(\beta\rho^{\gamma}-1)]}{(\gamma-1)(r-1)}$

8. **Atkinson cycle** efficiency, $\eta_{Atkinson} = 1 - \gamma \cdot \dfrac{(r-\alpha)}{r^{\gamma}-\alpha^{\gamma}}$

   where α = Compression ratio, r = Expansion ratio.

9. **Brayton cycle**, $\eta_{Brayton} = 1 - \dfrac{1}{(r_p)^{\frac{\gamma-1}{\gamma}}}$, where r_p = Pressure ratio.

OBJECTIVE TYPE QUESTIONS

Choose the Correct Answer:

1. The air standard Otto cycle comprises
   (a) two constant pressure processes and two constant volume processes
   (b) two constant pressure processes and two constant entropy processes
   (c) two constant volume processes and two constant entropy processes
   (d) none of the above.

2. The air standard efficiency of Otto cycle is given by
   (a) $\eta = 1 + \dfrac{1}{(r)^{\gamma+1}}$
   (b) $\eta = 1 - \dfrac{1}{(r)^{\gamma-1}}$
   (c) $\eta = 1 - \dfrac{1}{(r)^{\gamma+1}}$
   (d) $\eta = 2 - \dfrac{1}{(r)^{\gamma-1}}$.

3. The thermal efficiency of theoretical Otto cycle
   (a) increases with increase in compression ratio
   (b) increases with increase in isentropic index γ
   (c) does not depend upon the pressure ratio
   (d) follows all the above.

4. The work output of theoretical Otto cycle
   (a) increases with increase in compression ratio
   (b) increases with increase in pressure ratio
   (c) increases with increase in adiabatic index γ
   (d) follows all the above.

GAS POWER CYCLES   105

5. For same compression ratio
   (a) thermal efficiency of Otto cycle is greater than that of Diesel cycle
   (b) thermal efficiency of Otto cycle is less than that of Diesel cycle
   (c) thermal efficiency of Otto cycle is same as that for Diesel cycle
   (d) thermal efficiency of Otto cycle cannot be predicted.
6. In air standard Diesel cycle, at fixed compression ratio and fixed value of adiabatic index (γ)
   (a) thermal efficiency increases with increase in heat addition cut-off ratio
   (b) thermal efficiency decreases with increase in heat addition cut-off ratio
   (c) thermal efficiency remains same with increase in heat addition cut-off ratio
   (d) none of the above.

Answers

1. (b)   2. (b)   3. (d)   4. (d)   5. (a)   6. (b).

THEORETICAL QUESTIONS

1. What is a cycle ? What is the difference between an ideal and actual cycle ?
2. What is an air-standard efficiency ?
3. What is relative efficiency ?
4. Derive expressions of efficiency in the following cases :
   (i) Carnot cycle   (ii) Diesel cycle.
5. Explain "Air standard analysis" which has been adopted for I.C. engine cycles. State the assumptions made for air standard cycles.

UNSOLVED EXAMPLES

1. A Carnot engine working between 377°C and 37°C produces 120 kJ of work. Determine :
   (i) The heat added in kJ.   (ii) The entropy change during heat rejection process.
   (iii) The engine thermal efficiency.   [**Ans.** (i) 229.5 kJ ; (ii) 0.353 kJ/K ; (iii) 52.3%]
2. Find the thermal efficiency of a Carnot engine whose hot and cold bodies have temperatures of 154°C and 15°C respectively. [**Ans.** 32.55%]
3. Derive an expression for change in efficiency for a change in compression ratio. If the compression ratio is increased from 6 to 8, what will be the percentage increase in efficiency ? [**Ans.** 8%]
4. The efficiency of an Otto cycle is 50% and γ is 1.5. What is the compression ratio ? [**Ans.** 4]
5. An engine working on Otto cycle has a volume of 0.5 m³, pressure 1 bar and temperature 27°C at the commencement of compression stroke. At the end of compression stroke, the pressure is 10 bar. Heat added during the constant volume process is 200 kJ. Determine :
   (i) Percentage clearance   (ii) Air standard efficiency
   (iii) Mean effective pressure
   (iv) Ideal power developed by the engine if the engine runs at 400 r.p.m. so that there are 200 complete cycles per minutes. [**Ans.** (i) 23.76% ; (ii) 47.2% ; (iii) 2.37 bar ; (iv) 321 kW]
6. The compression ratio in an air-standard Otto cycle is 8. At the beginning of compression process, the pressure is 1 bar and the temperature is 300 K. The heat transfer to the air per cycle is 1900 kJ/kg of air. Calculate :
   (i) Thermal efficiency   (ii) The mean effective pressure.
                                                         [**Ans.** (i) 56.47% ; (ii) 14.24 bar]
7. An engine 200 mm bore and 300 mm stroke works on Otto cycle. The clearance volume is 0.0016 m³. The initial pressure and temperature are 1 bar and 60°C. If the maximum pressure is limited to 24 bar, find :

(i) The air-standard efficiency of the cycle (ii) The mean effective pressure for the cycle.
Assume ideal conditions. [**Ans.** (i) 54.08% ; (ii) 1.972 bar]

8. Calculate the air standard efficiency of a four stroke Otto cycle engine with the following data :
Piston diameter (bore) = 137 mm ; Length of stroke = 130 mm ;
Clearance volume 0.00028 m^3.
Express clearance as a percentage of swept volume. [**Ans.** 56.1% ; 14.6%]

9. In an ideal Diesel cycle, the temperatures at the beginning of compression, at the end of compression and at the end of the heat addition are 97°C, 789°C and 1839°C. Find the efficiency of the cycle.
[**Ans.** 59.6%]

10. An air-standard Diesel cycle has a compression ratio of 18, and the heat transferred to the working fluid per cycle is 1800 kJ/kg. At the beginning of the compression stroke, the pressure is 1 bar and the temperature is 300 K. Calculate : (i) Thermal efficiency, (ii) The mean effective pressure.
[**Ans.** (i) 61% ; (ii) 13.58 bar]

11. 1 kg of air is taken through a Diesel cycle. Initially the air is at 15°C and 1 ata. The compression ratio is 15 and the heat added is 1850 kJ. Calculate : (i) The ideal cycle efficiency, (ii) The mean effective pressure. [**Ans.** (i) 55.1% ; (ii) 13.4 bar]

12. What will be loss in the ideal efficiency of a Diesel engine with compression ratio 14 if the fuel cut-off is delayed from 6% to 9% ? [**Ans.** 2.1%]

13. The pressures on the compression curve of a diesel engine are at $\frac{1}{8}$ th stroke 1.4 bar and at $\frac{7}{8}$ th stroke 14 bar. Estimate the compression ratio. Calculate the air standard efficiency of the engine if the cut-off occurs at $\frac{1}{15}$ th of the stroke. [**Ans.** 18.54 ; 63.7%]

14. A compression ignition engine has a stroke 270 mm, and a cylinder diameter of 165 mm. The clearance volume is 0.000434 m^3 and the fuel ignition takes place at constant pressure for 4.5 per cent of the stroke. Find the efficiency of the engine assuming it works on the Diesel cycle. [**Ans.** 61.7%]

15. The following data belong to a Diesel cycle :
Compression ratio = 16 : 1 ; Heat added = 2500 kJ/kg ; Lowest pressure in the cycle = 1 bar ; Lowest temperature in the cycle = 27°C.
Determine :
(i) Thermal efficiency of the cycle. (ii) Mean effective pressure.
[**Ans.** (i) 45% ; (ii) 16.8 bar]

16. The compression ratio of an air-standard Dual cycle is 12 and the maximum pressure in the cycle is limited to 70 bar. The pressure and temperature of cycle at the beginning of compression process are 1 bar and 300 K. Calculate : (i) Thermal efficiency, (ii) Mean effective pressure.
Assume : cylinder bore = 250 mm, stroke length = 300 mm, c_p = 1.005, c_v = 0.718 and γ = 1.4.
[**Ans.** (i) 61.92% ; (ii) 9.847 bar]

17. The compression ratio of a Dual cycle is 10. The temperature and pressure at the beginning of the cycle are 1 bar and 27°C. The maximum pressure of the cycle is limited to 70 bar and heat supplied is limited to 675 kJ/kg of air. Find the thermal efficiency of the cycle. [**Ans.** 59.5%]

18. An air standard Dual cycle has a compression ratio of 16, and compression begins at 1 bar, 50°C. The maximum pressure is 70 bar. The heat transferred to air at constant pressure is equal to that at constant volume. Determine :
(i) The cycle efficiency. (ii) The mean effective pressure of the cycle.
Take : c_p = 1.005 kJ/kg K, c_v = 0.718 kJ/kg K. [**Ans.** (i) 66.5% ; (ii) 4.76 bar]

19. Compute the air standard efficiency of a Brayton cycle operating between a pressure of 1 bar and a final pressure of 12 bar. Take γ = 1.4. [**Ans.** 50.8%]

MODULE – 2

Chapters :

4. I.C. Engines

5. Refrigeration and Air-Conditioning

MODULE - 2

4

Internal Combustion Engines

4.1. Heat engines. 4.2. Development of I.C. engines. 4.3. Classification of I.C. engines. 4.4. Applications of I.C. engines. 4.5. Basic idea of I.C. engines. 4.6. Different parts of I.C. engines. 4.7. Terms connected with I.C. engines. 4.8. Working cycles. 4.9. Indicator diagram. 4.10. Four-stroke cycle engines. 4.11. Two-stroke cycle engines. 4.12. Comparison of four-stroke and two-stroke cycle engines. 4.13. Comparison of spark ignition (S.I.) and combustion ignition (C.I.) engines. 4.14. Comparison between a petrol engine and a diesel engine. 4.15. How to tell a two-stroke cycle engine from a four-stroke cycle engine. 4.16. Ignition system. 4.17. Cooling systems. 4.18. Lubrication systems—Highlights—Objective Type Questions—Theoretical Questions.

4.1. HEAT ENGINES

*Any type of engine or machine which derives heat energy from the combustion of fuel or any other source and converts this energy into mechanical work is termed as a **heat engine**.*

Heat engines may be classified into two main classes as follows :

1. External Combustion Engines.
2. Internal Combustion Engines.

1. External Combustion Engines (E.C. Engines)

In this case, *combustion of fuel takes place outside the cylinder* as in case of *steam engines* where the heat of combustion is employed to generate steam which is used to move a piston in a cylinder. Other examples of external combustion engines are *hot air engines, steam turbine* and *closed cycle gas turbine*. These engines are generally used for driving locomotives, ships, generation of electric power etc.

2. Internal Combustion Engines (I.C. Engines)

In this case, *combustion of the fuel with oxygen of the air occurs within the cylinder* of the engine. The internal combustion engines group includes engines employing mixtures of combustible gases and air, known as *gas engines*, those *using lighter liquid fuel* or spirit known as *petrol engines* and those using heavier liquid fuels, known as *oil compression ignition* or *diesel engines*.

The *external combustion engines* claim the following *advantages over internal combustion engines* :

1. Starting torque is generally high.
2. Because of external combustion of fuel, cheaper fuels can be used. Even solid fuels can be used advantageously.
3. Due to external combustion of fuel it is possible to have flexibility in arrangement.
4. These units are self-starting with the working fluid whereas in case of internal combustion engines, some additional equipment or device is used for starting the engines.

Reciprocating internal combustion engines offer the following *advantages over external combustion engines* :

1. Overall efficiency is high.
2. Greater mechanical simplicity.
3. Weight to power ratio is generally low.
4. Generally lower initial cost.
5. Easy starting from cold conditions.
6. These units are compact and thus require less space.

4.2. DEVELOPMENT OF I.C. ENGINES

Many experimental engines were constructed around 1878. The first really successful engine did not appear, however until 1879, when a German engineer Dr. Otto built his famous Otto gas engine. The operating cycle of this engines was based upon principles first laid down in 1860 by a French engineer named Bea de Rochas. The majority of modern I.C. engines operate according to these principles.

The development of the well known Diesel engine began about 1883 by Rudoff Diesel. Although this differs in many important respects from the Otto engine, the operating cycle of modern high speed Diesel engines is thermodynamically very similar to the Otto cycle.

4.3. CLASSIFICATION OF I.C. ENGINES

Internal combustion engines may be *classified* as given below :

1. **According to cycle of operation :**
    (*i*) Two-stroke cycle engines
    (*ii*) Four-stroke cycle engines.
2. **According to cycle of combustion :**
    (*i*) Otto cycle engine (combustion at constant volume)
    (*ii*) Diesel cycle engine (combustion at constant pressure)
    (*iii*) Dual-combustion or Semi-Diesel cycle engine (combustion partly at constant volume and partly at constant pressure).
3. **According to arrangement of cylinder :**
    (*i*) Horizontal engine   (*ii*) Vertical engine
    (*iii*) V-type engine   (*iv*) Radial engine etc.
4. **According to their uses :**
    (*i*) Stationary engine   (*ii*) Portable engine
    (*iii*) Marine engine   (*iv*) Automobile engine
    (*v*) Aero engine etc.
5. **According to the fuel employed and the method of fuel supply to the engine cylinder :**
    (*i*) Oil engine   (*ii*) Petrol engine
    (*iii*) Gas engine   (*iv*) Kerosene engine
    (*v*) Carburettor, hot bulb, solid injection and air injection engine.

INTERNAL COMBUSTION ENGINES

6. **According to the speed of the engine :**
   (i) Low speed engine   (ii) Medium speed engine
   (iii) High speed engine.

7. **According to method of ignition :**
   (i) Spark ignition (S.I.) engine   (ii) Compression ignition (C.I.) engine.

8. **According to method of cooling the cylinder :**
   (i) Air-cooled engine   (ii) Water-cooled engine.

9. **According to method of governing :**
   (i) Hit and miss governed engine   (ii) Quality governed engine
   (iii) Quantity governed engine.

10. **According to valve arrangement :**
    (i) Overhead valve engine   (ii) L-head type engine
    (iii) T-head type engine   (iv) F-head type engine.

11. **According to number of cylinders :**
    (i) Single cylinder engine   (ii) Multi-cylinder engine.

4.4. APPLICATIONS OF I.C. ENGINES

The I.C. engines are generally used for :
(i) Road vehicles (e.g., scooter, motorcycle, buses etc.)
(ii) Air craft
(iii) Locomotives
(iv) Construction in civil engineering equipment such as bull-dozer, scraper, power shovels etc.
(v) Pumping sets
(vi) Cinemas
(vii) Hospital
(viii) Several industrial applications.

Note. Prime movers in all *construction equipment* are invariable I.C. engines, unless of course, when drive is electric. Use of steam source for this equipment is almost absolete.

4.5. BASIC IDEA OF I.C. ENGINES

The basic idea of internal combustion engine is shown in Fig. 4.1. The cylinder which is closed at one end is filled with a mixture of fuel and air. As the crankshaft turns it pushes cylinder. The piston is forced up and compresses the mixture in the top of the cylinder. The mixture is set alight and, as it burns, it creates a gas pressure on the piston, forcing it down the cylinder. This motion is shown by arrow '1'. The piston pushes on the rod which pushes on the crank. The crank is given rotary (turning) motion as shown by the arrow '2'. The flywheel fitted on the end of the crankshaft stores energy and keeps the crank turning steadily.

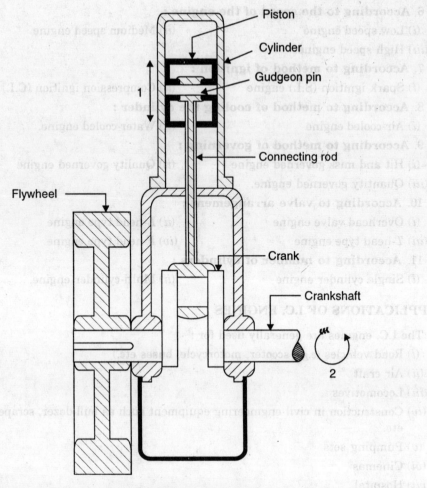

Fig. 4.1. Basic idea of I.C. engine.

4.6. DIFFERENT PARTS OF I.C. ENGINES

Here follows the detail of the various parts of an internal combustion engine.

A cross-section of an air-cooled I.C. engine with principal parts is shown in Fig. 4.2.

A. Parts common to both petrol and diesel engine :

1. Cylinder
2. Cylinder head
3. Piston
4. Piston rings
5. Gudgeon pin
6. Connecting rod
7. Crankshaft
8. Crank
9. Engine bearing
10. Crankcase
11. Flywheel
12. Governor
13. Valves and valve operating mechanism.

B. Parts for petrol engines only :

1. Spark plugs
2. Carburettor
3. Fuel pump.

INTERNAL COMBUSTION ENGINES

C. Parts for diesel engine only :
1. Fuel pump.
2. Injector.

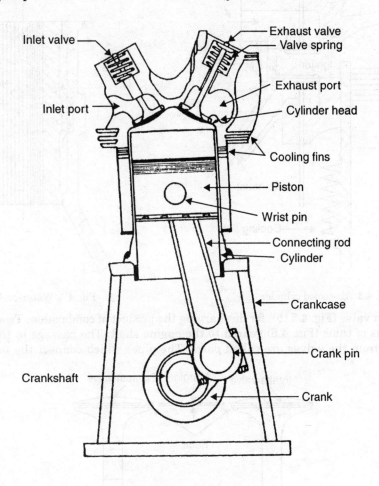

Fig. 4.2. Air-cooled I.C. engine.

A. Parts common to both petrol and diesel engines :

1. Cylinder

The cylinder contains gas under pressure and guides the piston. It is in direct contact with the products of combustion and it must be cooled. The ideal form consists of a plain cylindrical barrel in which the piston slides. The movement of the piston or stroke being in most cases, longer than the bore. This is known as the *"stroke bore ratio"*. The upper end consists of a combustion or clearance space in which the ignition and combustion of the charge takes place. In practice, it is necessary to depart from the ideal hemispherical shape in order to accommodate the valves, sparking plugs etc. and to control the combustion. Sections of an air-cooled cylinder and a water-cooled cylinder are shown in Figs. 4.3 and 4.4 respectively. *The cylinder is made of hard grade cast iron and is usually cast in one piece.*

2. Cylinder head

One end of the cylinder is closed by means of a *removable cylinder head* (Fig. 4.3) which usually contains the inlet or admission valve [Fig. 4.5 (a)] for admitting the mixture of air and

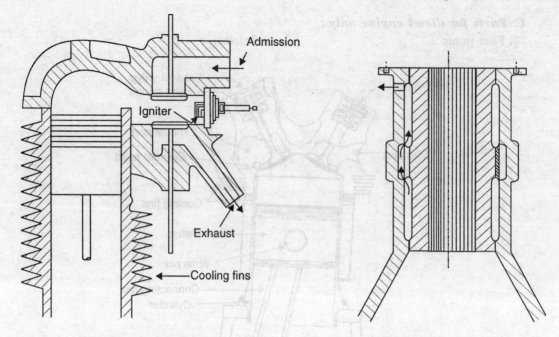

Fig. 4.3. Air-cooled cylinder.   Fig. 4.4. Water-cooled cylinder.

fuel and exhaust valve [Fig. 4.5 (b)] for discharging the product of combustion. Two valves are kept closed, by means of cams (Fig. 4.6) geared to the engine shaft. The passage in the cylinder head leading to and from the valves are called *ports*. The pipes which connect the inlet ports of the

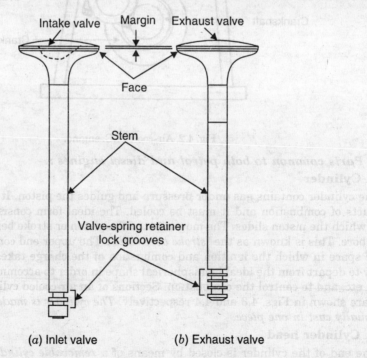

Fig. 4.5

INTERNAL COMBUSTION ENGINES

various cylinders to a common intake pipe for the engine is called the inlet *manifold*. If the exhaust ports are similarly connected to a common exhaust system, this system of piping is called *exhaust manifold*.

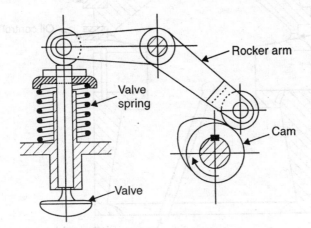

Fig. 4.6. Cam and rocker arm.

The main purpose of the cylinder head is to seal the working ends of the cylinders and not to permit entry and exit of gases on cover head valve engines. The inside cavity of head is called the *combustion chamber*, into which the mixture is compressed for firing. Its *shape controls the direction and rate of combustion*. Heads are drilled and tapped with correct thread to take the ignition spark plug. All the combustion chambers in an engine must be of same shape and size. The shape may be in part controlled by the piston shape.

The cylinder head is usually made of cast iron or aluminium.

3. Piston

A piston is fitted to each cylinder as a face to receive gas pressure and transmit the thrust to the connecting rod.

The piston must (*i*) give gas tight seal to the cylinder through bore, (*ii*) slide freely, (*iii*) be light and (*iv*) be strong. The thrust on the piston on the power stroke tries to tilt the piston as the connecting rod swings, sideways. The piston wall, called the *skirt* must be strong enough to stand upto this side thrust. *Pistons are made of cast iron or aluminium alloy for lightness.* Light alloy pistons expand more than cast iron one therefore they need large clearances to the bore, when cold, or special provision for expansion. Pistons may be solid skirt or split skirt. A section through a split skirt piston is shown in Fig. 4.7.

4. Piston rings

The piston must be a fairly loose fit in the cylinder. If it were a tight fit, it would expand as it got hot and might stick tight in the cylinder. If a piston sticks it could ruin the engine. On the other hand, if there is too much clearance between the piston and cylinder walls, much of the pressure from the burning gasoline vapour will leak past the piston. This means, that the push on the piston will be much less effective. It is the push on the piston that delivers the power from the engines.

To provide a good sealing fit between the piston and cylinder, pistons are equipped with piston rings, as shown in Fig. 4.8. The rings are usually made of cast iron of fine grain and high elasticity which is not affected by the working heat. Some rings are of alloy spring steel. They are split at one point so that they can be expanded and slipped over the end of the piston and into ring grooves which have been cut in the piston. When the piston is installed in the cylinder the rings

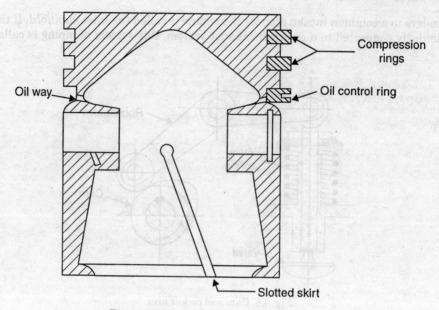

Fig. 4.7. Section through a split skirt piston.

are compressed into ring grooves which have been cut in the piston. When the piston is installed in the cylinder, the rings are compressed into the ring grooves so that the split ends come almost together. The rings fit tightly against the cylinder wall and against the sides of the ring grooves in the piston. Thus, they form a good seal between the piston and the cylinder wall. The rings can expand or contract as they heat and cool and still make a good deal. Thus they are free to slide up and down the cylinder wall.

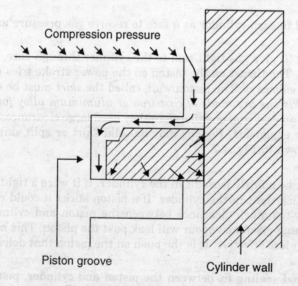

Fig. 4.8. Working of a piston ring.

Fig. 4.8 shows how the piston ring works to hold in the compression and combustion pressure. The arrows show the pressure above the piston passing through clearance between the piston and the cylinder wall. It presses down against the top and against the back of the piston rings as

INTERNAL COMBUSTION ENGINES

shown by the arrows. This pushes the piston ring firmly against the bottom of the piston ring groove. As a result there are good seals at both of these points. The higher the pressure in the combustion chamber, the better the seal.

Small two stroke cycle engines have two rings on the piston. Both are compression rings (Fig. 4.9). Two rings are used to divide up the job of holding the compression and combustion pressure. This produces better sealing with less ring pressure against the cylinder wall.

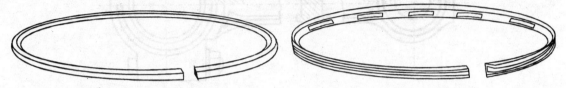

Fig. 4.9. Compression ring.   Fig. 4.10. Oil ring.

Four stroke cycle engines have an extra ring, called the oil control ring (Fig. 4.10). Four stroke cycle engines are so constructed that they get much more oil in the cylinder wall than do two stroke cycle engines. This additional oil must be scraped off to prevent it from getting up into the combustion chamber, where it would burn and cause trouble.

Refer Figs. 4.9 and 4.10 the compression rings have a rectilinear cross-section and oil rings are provided with a groove in the middle and with through holes spaced at certain interval from each other. The oil collected from the cylinder walls flows through these holes into the piston groove whence through the holes in the body of the piston and down its inner walls into the engine crankcase.

5. Gudgeon pin (or wrist pin or piston pin)

These are *hardened steel parallel spindles* fitted through the piston bosses and the small end bushes or eyes to allow the connecting rods to swivel. Gudgeon pins are a press fit in the piston bosses of light alloy pistons when cold. For removal or fitting, the piston should be dipped in hot water or hot oil, this expands the bosses and the pins can be removed or fitted freely without damage.

It is made hollow for lightness since it is a reciprocating part.

6. Connecting rod

Refer Fig. 4.11 (on next page). The connecting rod transmits the piston load to the crank, causing the latter to turn, thus converting the reciprocating motion of the piston into a rotary motion of the crankshaft. The lower or "big end" of the connecting rod turns on "crank pins".

The connecting rods are made of nickel, chrome and chrome vandium steels. For small engines the material may be aluminium.

7. Crank

The piston moves up and down in the cylinder. This up and down motion is called reciprocating motion. The piston moves in a straight line. The straight line motion must be changed to rotary, or turning motion, in most machines, before it can do any good. That is rotary motion is required to make wheels turn, a cutting blade spin or a pulley rotate. To change the reciprocating motion to rotary motion a crank and connecting rod are used. (Figs. 4.12 and 4.13). The connecting rod connects the piston to the crank.

Note. The crank end of the connecting rod is called rod "big end". The piston end of the connecting rod is called the rod "small end".

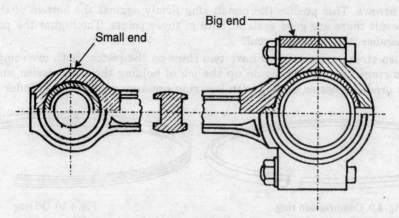

Fig. 4.11. Connecting rod.

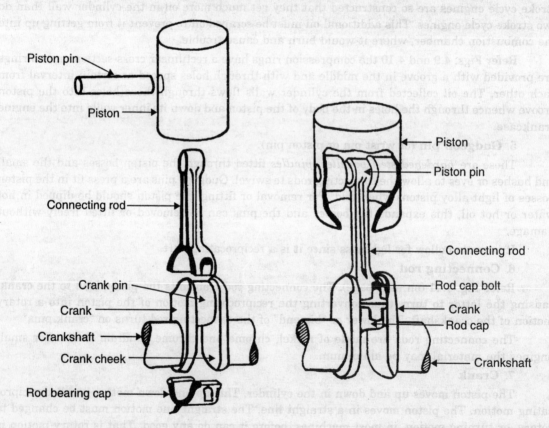

Fig. 4.12   Fig. 4.13

8. Crankshaft

The crank is part of the crankshaft. The crankshaft of an internal combustion engine receives via its cranks the efforts supplied by the pistons to the connecting rods. All the engines auxiliary mechanisms with mechanical transmission are geared in one way or the another to the crankshaft. *It is usually a steel forging, but some makers use special types of cast iron such as*

INTERNAL COMBUSTION ENGINES

spheroidal graphitic or nickel alloy castings which are cheaper to produce and have good service life. Refer Fig. 4.14. The crankshaft converts the reciprocating motion to rotary motion. The crankshaft mounts in bearings which, encircle the journals so it can rotate freely.

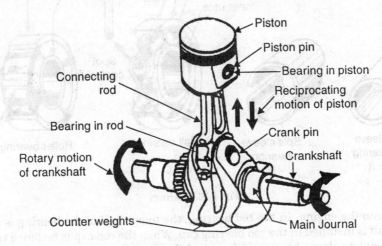

Fig. 4.14. Crankshaft and other parts.

The shape of the crankshaft *i.e.*, the mutual arrangement of the cranks depend on the number and arrangement of cylinders and the turning order of the engine. Fig. 4.15 shows a typical crankshaft layout for a four-cylinder engine.

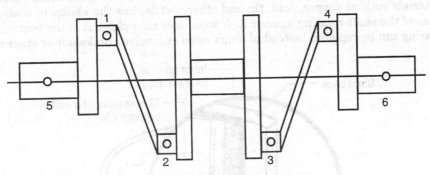

Fig. 4.15. Typical crankshaft layout.

9. Engine bearing

The crankshaft is supported by bearing. The connecting rod big end is attached to the crank pin on the crank of the crankshaft by a bearing. A piston pin at the rod small end is used to attach the rod to the piston. The piston pin rides in bearings. Everywhere there is rotary action in the engine, bearings are used to support the moving parts. The purpose of bearing is to reduce the friction and allow the parts to move easily. Bearings are lubricated with oil to make the relative motion easier.

Bearings used in engines are of two types : *sliding* or *rolling* (Fig. 4.16).

The sliding type of bearings are sometimes called bushings or sleeve bearings because they are in the shape of a sleeve that fits around the rotating journal or shaft. The sleeve-type connecting rod big end bearings usually called simply rod bearings and the crankshaft supporting bearings called the main bearings are of the split sleeve type. They must be split in order to permit

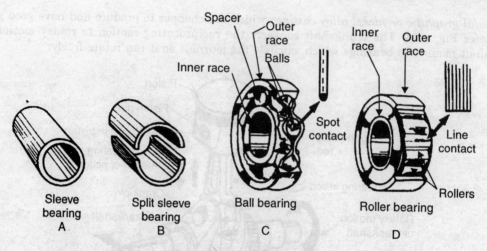

Fig. 4.16. Bearings.

their assembly into the engine. In the rod bearing, the upper half of the bearing is installed in the rod, the lower half is installed in the rod bearing cap. When the rod cap is fastened to the rod shown in Fig. 4.13 a complete sleeve bearing is formed. Likewise, the upper halves of the main bearings are assembled in the engine and then the main bearing caps, with the lower bearing halves are attached to the engine to complete the sleeve bearings supporting the crankshaft.

The typical bearing half is made of steel or bronze back to which a lining of relatively soft bearing material is applied. Refer Fig. 4.17. This relatively soft bearing material, which is made of several materials such as copper, lead, tin and other metals, has the ability to conform to slight irregularities of the shaft rotating against it. If wear does take place, it is the bearing that wears and the bearing can be replaced instead of much more expansive crankshaft or other engine part.

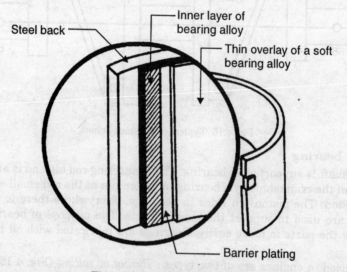

Fig. 4.17. Bearing half (details).

The rolling-type bearing uses balls or rollers between the stationary support and the rotating shaft. Refer Fig. 4.16. Since the balls or rollers provide rolling contact, the frictional resistance to movement is much less. In some roller bearing, the rollers are so small that they are hardly bigger than needles. These bearings are called *needle bearings*. Also some rollers bearings have the rollers set at an angle to the races, the rollers roll in are tapered. These bearings are called *tapered roller bearings*. Some ball and roller bearings are sealed with their lubricant already in place. Such bearings require no other lubrication. Other do require lubrication from the oil in the gasoline (two stroke cycle engines) or from the engine lubrication system (four stroke cycle engines).

The type of bearing selected by the designers of the engine depends on the design of the engine and the use to which the engine will be put. *Generally, sleeve bearings, being less expensive and satisfactory for most engine applications, are used. In fact sleeve bearings are used almost universally in automobile engines. But you will find some engines with ball and roller bearings to support the crankshaft and for the connecting rod and piston-pin bearings.*

10. Crankcase

The main body of the engine to which the cylinders are attached and which contains the crankshaft and crankshaft bearing is called *crankcase*. This member also holds other parts in alignment and resists the explosion and inertia forces. It also protects the parts from dirt etc. and serves as a part of lubricating system.

11. Flywheel

Refer Figs. 4.1 and 4.18. A flywheel (steel or cast iron disc) secured on the crankshaft performs the following *functions* :

(a) Brings the mechanism out of dead centres.

(b) Stores energy required to rotate the shaft during preparatory strokes.

(c) Makes crankshaft rotation more uniform.

(d) Facilitates the starting of the engine and overcoming of short time over loads as, for example, when the machine is started from rest.

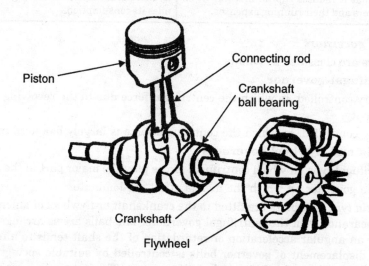

Fig. 4.18. Flywheel secured on crankshaft.

The weight of the flywheel depends upon the nature of variation of the pressure. The flywheel for a double acting steam engine is lighter than that of a single-acting one. Similarly, the flywheel for a two-stroke cycle engine is lighter than a flywheel used for a four-stroke cycle engine. *Lighter flywheels are used for multi-cylinder engines.*

12. Governor

A governor may be defined *as a device for regulating automatically output of a machine by regulating the supply of working fluid*. When the speed decreases due to increase in load the supply valve is opened by mechanism operated by the governor and the engine therefore speeds up again to its original speed. If the speed increases due to a decreases of load the governor mechanism closes the supply valve sufficiently to slow the engine to its original speed. *Thus the function of a governor is to control the fluctuations of engine speed due to changes of load.*

Comparison between a Flywheel and a Governor

	Flywheel	*Governor*
1.	It is provided on engines and fabricating machines *viz.*, rolling mills, punching machines ; shear machines, presses etc.	It is provided on primemovers such as engines and turbines.
2.	Its function is to store the available mechanical energy when it is in excess of the load requirement and to part with the same when the available energy is less than that required by the load.	Its function is to regulate the supply of driving fluid producing energy, according to the load requirement so that at different loads almost a constant speed is maintained.
3.	It works continuously from cycle to cycle.	It works intermittently *i.e.*, only when there is change in load.
4.	In engines it takes care of fluctuations of speed during thermodynamic cycle.	It takes care of fluctuations of speed due to variation of load over long range of working engines and turbines.
5.	In fabrication machines it is very economical to use it in that it reduces capital investment on primemovers and their running expenses.	But for governor, there would have been unnecessarily more consumption of driving fluid. Thus it economies its consumption.

Types of governors :

Governors are classified as follows :

1. Centrifugal governor

(*a*) Gravity controlled, in which the centrifugal force due to the revolving masses is largely balanced by gravity.

(*b*) Spring controlled, in which the centrifugal force is largely balanced by springs.

2. Inertia and flywheel governors

(*a*) Centrifugal type, in which centrifugal forces play the major part in the regulating action.

(*b*) Inertia governor, in which the inertia effect predominates.

The *inertia type* governors are fitted to the crankshaft or flywheel of an engine and so differ radically in appearance from the centrifugal governors. The balls are so arranged that the inertia force caused by an angular acceleration or retardation of the shaft tends to alter their positions. The amount of displacement of governor balls is controlled by suitable springs and through the governor mechanism, alters the fuel supply to the engine. The inertia governor is more sensitive than centrifugal but it becomes very difficult to balance the revolving parts. For this reason *centrifugal governors are more frequently used*. We shall discuss centrifugal governors only.

Important centrifugal governors are :

1. Watt governor   2. Porter governor
3. Proell governor   4. Hartnell governor.

INTERNAL COMBUSTION ENGINES

1. Watt governor

It is the primitive governor as used by Watt on some of his early steam engines. It is used for a very slow speed engine and this is why it has now become obsolete.

Refer Fig. 4.19. Two arms are hinged at the top of the spindle and two revolving balls are fitted on the other ends of the arms. One end of each of the links are hinged with the arms, while the other ends are hinged with the sleeve, which may slide over the spindle. The speed of the crankshaft is transmitted to the spindle through a pair of bevel gears by means of a suitable arrangement. So the rotation of the spindle of the governor causes the weights to move away from the centre due to the centrifugal force. This makes the sleeve to move in the upward direction. This movement of the sleeve is transmitted by the lever to the throttle valve which partially closes or opens the steam pipe and reduces or increases the supply of steam to the engine. So the engine speed may be adjusted to a normal limit.

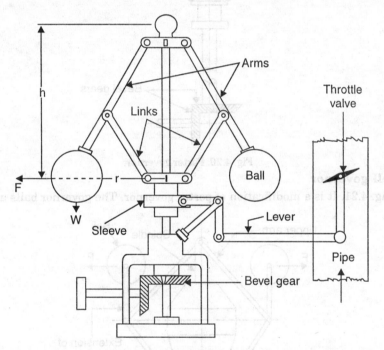

Fig. 4.19. Watt governor.

2. Porter governor

Fig. 4.20 shows diagrammatically a Porter governor where two or more masses called the governor balls rotate about the axis of the governor shaft which is driven through suitable gearing from the engine crankshaft. The governor balls are attached to the arms. The lower arms are attached to the *sleeve which acts as a central weight*. If the speed of the rotation of the balls increases owing to a decrease of load on the engine, the governor balls fly outwards and the sleeve moves upwards thus closing the fuel passage till the engine speed comes back to its designed speed. If the engine speed decreases owing to an increase of load, the governor balls fly inwards and the sleeve moves downwards thus opening the fuel passage more for oil till the engine speed comes back to its designed speed. The engine is said to be running at its designed speed when the outward inertia or centrifugal force is just balanced by the inward controlling force.

124 BASIC MECHANICAL ENGINEERING

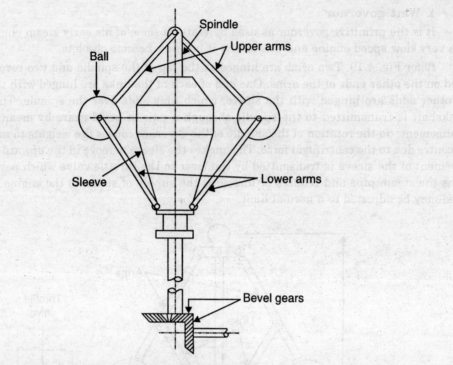

Fig. 4.20. Porter governor.

3. Proell governor

Refer Fig. 4.21. It is a modification of porter governor. The governor balls are carried on an

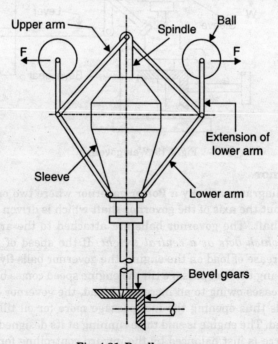

Fig. 4.21. Proell governor.

INTERNAL COMBUSTION ENGINES

extension of the lower arms. For given values of weight of the ball, weight of the sleeve and height of the governor, a Proell governor runs at a *lower speed* than a Porter governor. *In order to give the same equilibrium speed a ball of smaller mass may be used in Proell governor.*

4. Hartnell governor.

The Hartnell governor is a spring loaded governor in which the controlling force, to a great extent, is provided by the spring thrust.

Fig. 4.22 shows one of the types of Hartnell governors. It consists of casing fixed to the spindle. A compressed spring is placed inside the casing which presses against the top of the casing and on adjustable collars. The sleeve can move up and down on the vertical spindle depending upon the speed of the governor. Governor balls are carried on bell crank lever which are pivoted on the lower end of the casing. The balls will fly outwards or inwards as the speed of the governor shaft increases or decreases respectively.

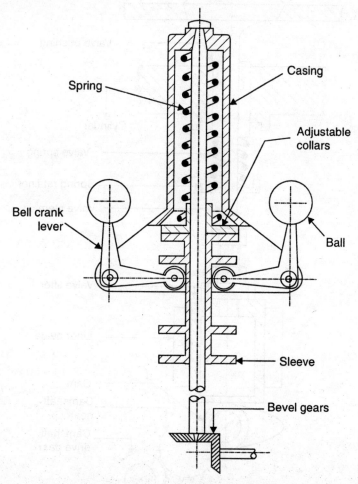

Fig. 4.22. Hartnell governor.

13. Valves and valve operating mechanism.

With few exceptions the inlet and exhaust of internal combustion engines are controlled by poppet valves. These valves are held to their seating by strong springs, and as the valves usually open inwards, the pressure in the cylinder helps to keep them closed. The valves are lifted from their seats and the ports opened either by cams having projecting portion designed to give the period of opening required or by eccentrics operating through link-work. Of these two methods the cam gear is more commonly used, but in either case it is necessary that the valve gear shaft of an engine should rotate but once from beginning to end of a complete cycle, however many strokes may be involved in the completion of that cycle. This is necessary to secure a continuous regulation of the valve gear as required. For this purpose the cams or eccentrics of four-stroke engines are mounted on shafts driven by gearing at half the speed of the crankshaft. The curves used for the acting faces of the cams depend on the speed of the engine and rapidity of valve opening desired.

Fig. 4.23 shows a valve gear for I.C. engine. It consists of poppet valve, the steam bushing or guide, valve spring, spring retainer, lifter or push rod, camshaft and half speed gear for a four-

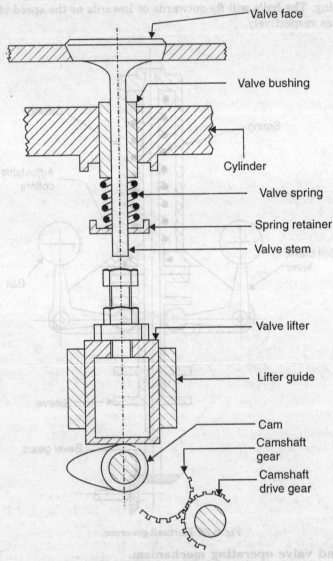

Fig. 4.23. Valve gear for I.C. engine.

stroke engine. The poppet valve, in spite of its shortcomings of noise and difficulties of cooling is commonly used due to its simplicity and capacity for effective sealing under all operating conditions.

INTERNAL COMBUSTION ENGINES

The valve is subjected to very heavy duty. It holds in combustion chamber and is exposed to high temperatures of burning gases. Exhaust valve itself may attain a high temperature while external cooling is not available. Special heat resisting alloys are therefore used in the construction of the exhaust valve and it may sometimes have a hollow construction filled with mineral salts to provide for heat dissipation. The salts become liquid when valve is working and transfer heat from the head to the stem from which it is carried through the stem guide to the cylinder block.

The timing of the valves *i.e.*, their opening and closing with respect to the travel of the piston is very important thing for efficient working of the engine. The drive of the camshaft is arranged through gears or chain and sprocket (called timing gear or timing chain). Any wearing of the gears or chain and sprocket would result in disturbing the precise timing of the valves. It is desirable, therefore, to avoid use of multiple gears of long chains in the camshaft drive.

Valve timing

Theoretically the valves open and close at top dead centre (T.D.C.) or at bottom dead centre (B.D.C.) but practically they do so sometime before or after the piston reaches the upper or lower limit of travel. There is a reason for this. Look at the inlet valve, for example. It normally opens several degrees of crankshaft-rotation before T.D.C. on the exhaust stroke. That is the intake valve begins to open before the exhaust stroke is finished. This gives the valve enough time to reach the fully open position before the intake stroke begins. Then, when the intake stroke starts, the intake valve is already wide open and air-fuel mixture can start to enter the cylinder, immediately. Likewise the intake valve remains open for quite a few degrees of crankshaft rotation after the piston has passed B.D.C. at the end of the intake stroke. This allows additional time for air-fuel mixture to continue to flow into the cylinder. The fact that the piston has already passed B.D.C. and is moving up or the compression stroke while the intake valve is still open does not effect the movement of air-fuel mixture into the cylinder. Actually air-fuel mixture is still flowing in as the intake valve starts to close.

This is due to the fact that air-fuel mixture has inertia. That is, it attempts to keep on flowing after it once starts through the carburettor and into the engine cylinder. The momentum of the mixture then keeps it flowing into the cylinder even though the piston has started up on the compression stroke. This packs more air-fuel mixture into the cylinder and results in a stronger power stroke. In other words, this improves *volumetric efficiency*.

For a some what similar reason, the exhaust valve opens well before the piston reaches B.D.C. on the power stroke. As the piston nears B.D.C., most of the push on the piston has ended and nothing is lost by opening the exhaust valve towards the end of the power stroke. This gives the exhaust gases additional time to start leaving the cylinder so that exhaust is well started by the time the piston passes B.D.C. and starts up on the exhaust stroke. The exhaust valve then starts opening for some degrees of crankshaft rotation after the piston has passed T.D.C. and intake stroke has started. This makes good use of momentum of exhaust gases. They are moving rapidly towards the exhaust port, and leaving the exhaust valve open for a few degrees after the intake stroke starts giving the exhaust gases some additional time to leave the cylinder. This allows more air-fuel mixture to enter on the intake stroke so that the stronger power stroke results. That is, it improves volumetric efficiency.

The actual timing of the valves varies with different four-stroke cycle engines, but the typical example for an engine is shown in Fig. 4.24. Note that the inlet valve opens 15° of crankshaft rotation before T.D.C. on the exhaust stroke and stays open until 50° of crankshaft rotation after B.D.C. on the compression stroke. The exhaust valve opens 50° before B.D.C. on the power stroke and stays open 15° after T.D.C. on the inlet stroke. This gives the two valves an overlap of 30° at the end of exhaust stroke and beginning of the *compression stroke*.

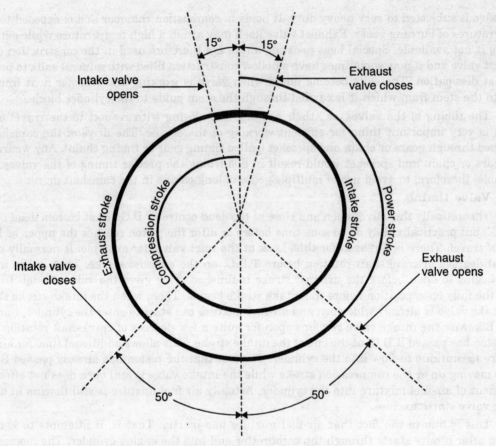

Fig. 4.24. Typical valve timing diagram.

B. Parts common to petrol engine only :
Spark plug :

The main function of a spark plug is to conduct the high potential from the ignition system into the combustion chamber. It provides the proper gap across which spark is produced by applying high voltage, to ignite the combustion chamber.

A *spark plug entails the following requirements* :
 (*i*) It must withstand peak pressures up to at least 55 bar.
 (*ii*) It must provide suitable insulation between two electrodes to prevent short circuiting.
 (*iii*) It must be capable of withstanding high temperatures to the tune of 2000°C to 2500°C over long periods of operation.
 (*iv*) It must offer maximum resistance to erosion burning away of the spark points irrespective of the nature of fuel used.
 (*v*) It must possess a high heat resistance so that the electrodes do not become sufficiently hot to cause the preignition of the charge within the engine cylinder.
 (*vi*) The insulating material must withstand satisfactorily the chemical reaction effects of the fuel and hot products of combustion.
 (*vii*) Gas tight joints between the insulator and metal parts are essential under all operating conditions.

Refer Fig. 4.25. The spark plug consists of a metal shell having two electrodes which are insulated from each other with an air gap. High tension current jumping from the supply electrode

INTERNAL COMBUSTION ENGINES

produces the necessary spark. Plugs are sometimes identified by the heat range or the relative temperature obtained during operation. The correct type of plug with correct width of gap between the electrodes are important factors. The spark plug gap can be easily checked by means of a feeler gauge and set as per manufacturer's specifications. It is most important that while adjusting the spark plug it is the outer earthed electrode *i.e.*, tip which is moved in or out gradually for proper setting of the gap. No bending force should be applied on the centre-electrode for adjusting the gap as this can cause crack and fracture of insulation and sender the plug absolutely useless.

Porcelain is commonly used as insulating material in spark plugs, as it is cheap and easy to manufacture. Mica can also be used as insulating material for spark plugs. Mica, however, cannot withstand high temperatures successfully.

Simple carburettor :

The function of a carburettor is to atomise and metre the liquid fuel and mix it with the air as it enters the induction system of the engine, maintaining under all conditions of operation fuel-air proportions appropriate to those conditions.

All modern carburettors are based upon Bernoulli's theorem,
$$C^2 = 2gh$$
where C is the velocity in metres/sec, g is the acceleration due to gravity in metre/sec^2 and h is the head causing the flow expressed in metres of height of a column of the fluid.

The equation of mass rate of flow is given by,
$$m = \rho A \sqrt{2gh}$$
where ρ is the density of the fluid and A is the cross-sectional area of fluid stream.

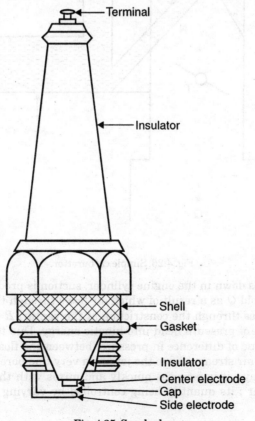

Fig. 4.25. Spark plug.

In Fig. 4.26 is shown simple carburettor. L is the float chamber for the storage of fuel. The fuel supplied under gravity action or by fuel pump enters the float chamber through the filter F. The arrangement is such that when the oil reaches a particular level the float valve M blocks the inlet passage and thus cuts off the fuel oil supply. On the fall of oil level, the float descends down, consequently intake passage opens and again the chamber is filled with oil. Then the float and the float valve maintains a constant fuel oil level in the float chamber. N is the jet from which the fuel is sprayed into the air stream as it enters the carburettor at the inlet S and passes through the throat or venturi R. The fuel level is slightly below the outlet of the jet when the carburettor is inoperative.

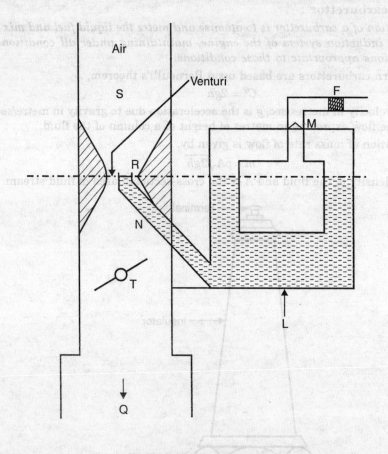

Fig. 4.26. Simple carburettor.

As the piston moves down in the engine cylinder, suction is produced in the cylinder as well as in the induction manifold Q as a result of which air flows through the carburettor. The velocity of air increases as it passes through the constriction at the venturi R and pressure decreases due to conversion of a portion of pressure head into kinetic energy. Due to decreased pressure at the venturi and hence by virtue of difference in pressure (between the float chamber and the venturi) the jet issues fuel oil into air stream. Since the jet has a very fine bore, the oil issuing from the jet is in the form of fine spray ; it vapourises quickly and mixes with the air. This air-fuel mixture enters the engine cylinder ; its quantity being controlled by varying the position of the throttle valve T.

INTERNAL COMBUSTION ENGINES

Limitations :

(*i*) Although theoretically the air-fuel ratio supplied by a simple (single jet) carburettor should remain constant as the throttle goes on opening, actually it provides increasingly richer mixture as the throttle is opened. This is because of the reason that the density of air tends to decrease as the rate of flow increases.

(*ii*) During idling, however, the nearly closed throttle causes a reduction in the mass of air flowing through the venturi. At such low rates of air flow, the pressure difference between the float chamber and the fuel discharge nozzle becomes very small. It is sufficient to cause fuel to flow through the jet.

(*iii*) Carburettor does not have arrangement for providing rich mixture during starting and warm up.

In order to *correct for faults* :

(*i*) number of compensating devices are used for (*ii*) an idling jet is used which helps in running the engine during idling. For (*iii*) choke arrangement is used.

Anti-fuel mixtures

- The theoretically correct mixture of air and petrol is 15 : 1 (approximately). Thus, the uniform supply of such mixture would result in burning without leaving excess of air of fuel. But it is difficult to get such a mixture in actual practice. When there is too little air, some of the fuel goes unburnt or simply changed to carbon. When there is too much air in mixture, it burns slowly and erratically and there is a loss of power.

There is however a *range of mixtures* between which combustion will take place.

— The *'lower limit'* is approximately 7 : 1 to 10 : 1. This mixture is *barely explosive*.

— The 'upper limit' is approximately 20 : 1. The mixture burns *irregularly*.

The above limits will also vary with the *characteristics* of *the fuel*, the *shape of the combustion space and the temperature and pressure in the combustion space.*

- *Mixture requirements of automotive engines* :

— For "*average cruising speeds*" the air-fuel ratio is approximately from 15 : 1 to 7 : 1.

— In order to obtain "*maximum power*" to be able to accelerate the engine quickly, a richer mixture of the ratio of about 12 : 1 is desirable. This is also called *maximum power ratio*. To start the engine *from cold* even a mixture richer than this but above the lower limit is obtained.

— For "*maximum economy*", *i.e.*, less fuel consumption for unit power, the mixture ratio may be approximately 16 : 1 to 17 : 1. The ratio, however, *entails loss of power.*

Hence, it is essential that the carburettor should be so designed that proper proportions of air and fuel are obtained to meet these varying operating conditions.

Fuel pump (for carburettor-petrol engine).

Refer Fig. 4.27. This type of pump is used in petrol engine for supply of fuel to the carburettor. Due to rotation of the crankshaft the cam pushes the lever in the upward direction. One end of the lever is hinged while the other end pulls the diaphragm rod with the *diaphragm*. So the diaphragm comes in the downward direction against the compression of the spring and thus a vacuum is produced in the pump chamber. This causes the fuel to enter into the pump chamber from the *glass bowl* through the *strainer* and the inlet valve, the impurities of the fuel ; if there is any, deposit at the bottom of the glass bowl. On the return stroke the spring pushes the diaphragm in the upward direction forcing the fuel from the pump chamber into the carburettor through the *outlet valve*.

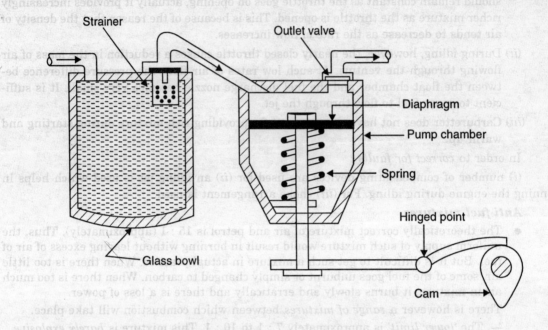

Fig. 4.27. Fuel pump for carburettor.

Parts for diesel engine only :
FUEL PUMP

Refer to Fig. 4.28. L is the plunger which is driven by a cam and tappet mechanism at the bottom (not shown in the figure) B is the barrel in which the plunger reciprocates. There is the rectangular vertical groove in the plunger which extends from top to another helical groove. V is the delivery valve which lifts off its seat under the liquid fuel pressure and against the spring force (S). The fuel pump is connected to fuel atomiser through the passage P, SP and Y are the spill and supply ports respectively. When the plunger is at its bottom stroke the ports SP and Y are uncovered (as shown in Fig. 4.28) and oil from low pressure pump (not shown) after being filtered is forced into the barrel. When the plunger moves up due to cam and tappet mechanism, a stage reaches when both the ports SP and Y are closed and with the further upward movement of the plunger the fuel gets compressed. The high pressure thus developed lifts the delivery valve off its seat and fuel flows to atomiser through the passage P. With further rise of the plunger, at a certain moment, the port SP is connected to the fuel in the upper part of the plunger through the rectangular vertical groove by the helical groove ; as a result of which a sudden drop in pressure occurs and the delivery valve falls back and occupies its seat against the spring force. The plunger is rotated by the rack R which is moved in or out by the governor. *By changing the angular position of the helical groove (by rotating the plunger) of the plunger relative to the supply port, the length of stroke during which the oil is delivered can be varied and thereby quantity of fuel delivered to the engine is also varied accordingly.*

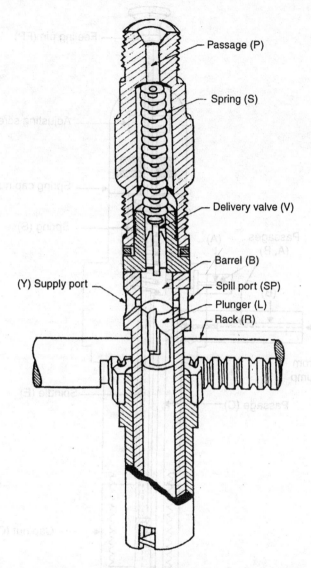

Fig. 4.28. Fuel pump.

Fuel atomiser or injector

Refer Fig. 4.29. It consists of a nozzle valve (*NV*) fitted in the nozzle body (*NB*). The nozzle valve is held on its seat by a spring '*S*' which exerts pressure through the spindle *E*. '*AS*' is the adjusting screw by which the nozzle valve lift can be adjusted. Usually the nozzle valve is set to lift at 135 to 170 bar pressure. *FP* is the feeling pin which indicates whether valve is working properly or not. The oil under pressure from the fuel pump enters the injector through the passages *B* and *C* and lifts the nozzle valve. The fuel travels down the nozzle *N* and injected into the engine cylinder in the form of fine sprays. When the pressure of the oil falls, the nozzle valve occupies its seat under the spring force and fuel supply is cut off. Any leakage of fuel accumulated above the valve is led to the fuel tank through the passage *A*. The leakage occurs when the nozzle valve is *worn out*.

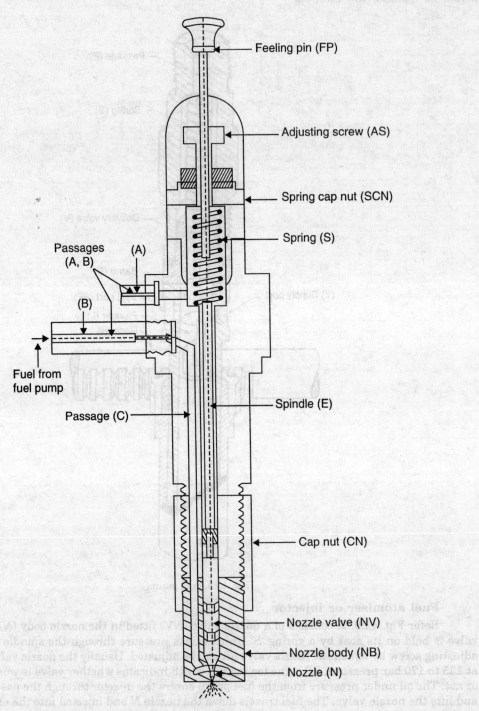

Fig. 4.29. Fuel atomiser or injector.

INTERNAL COMBUSTION ENGINES

List of engine parts, materials, method of manufacture and functions :

	Name of the part	Material	Function	Method of manufacture
1.	*Cylinder*	Hard grade cast iron	Contains gas under pressure and guides the piston.	Casting.
2.	*Cylinder head*	Cast iron or aluminium	Main function is to seal the working end of the cylinder and not to permit entry and exit of gases on overhead valve engines.	Casting, forging.
3.	*Piston*	Cast iron or aluminium alloy	It acts as a face to receive gas pressure and transmits the thrust to the connecting rod.	Casting, forging.
4.	*Piston rings*	Cast iron	Their main function is to provide a good sealing fit between the piston and cylinder.	Casting.
5.	*Gudgeon pin*	Hardened steel	It supports and allows the connecting rod to swivel.	Forging.
6.	*Connecting rod*	Alloy steel ; for small engines the material may be aluminium	It transmits the piston load to the crank, causing the latter to turn, thus converting the reciprocating motion of the piston into rotary motion of the crankshaft.	Forging.
7.	*Crankshaft*	In general the crankshaft is made from a high tensile forging, but special cast irons are sometimes used to produce a light weight crankshaft that does not require a lot of machining.	It converts the reciprocating motion of the piston into the rotary motion.	Forging.
8.	*Main bearings*	The typical bearing half is made of steel or bronze back to which a lining of relatively soft bearing material is applied.	The function of bearing is to reduce the friction and allow the parts to move easily.	Casting.
9.	*Flywheel*	Steel or cast iron.	In engines it takes care of fluctuations of speed during thermodynamic cycle.	Casting.
10.	*Inlet valve*	Silicon chrome steel with about 3% carbon.	Admits the air or mixture of air and fuel into engine cylinder.	Forging.
11.	*Exhaust valve*	Austenitic steel	Discharges the product of combustion.	Forging.

4.7. TERMS CONNECTED WITH I.C. ENGINES

Refer to Fig. 4.30.

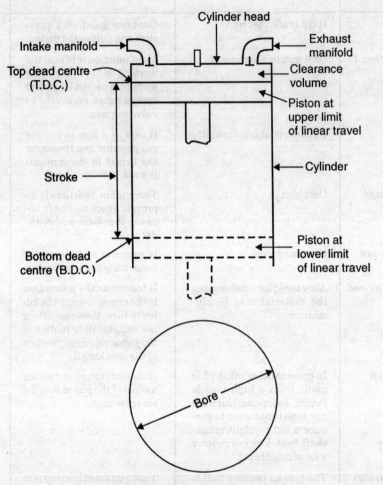

Fig. 4.30. Terms relating I.C. engines.

Bore. *The inside diameter of the cylinder is called **bore**.*

Stroke. As the piston reciprocates inside the engine cylinder, it has got limiting upper and lower positions beyond which it cannot move and reversal of motion takes place at these limiting positions.

*The linear distance along the cylinder axis between two limiting positions, is called **stroke**.*

Top Dead Centre (T.D.C.). *The top most position of the piston towards cover end side of the cylinder is called "top dead centre".* In case of horizontal engines, this is known as inner dead centre.

Bottom Dead Centre (B.D.C.). *The lowest position of the piston towards the crank end side of the cylinder is called "bottom dead centre".* In case of horizontal engines it is called outer dead centre.

Clearance volume. *The volume contained in the cylinder above the top of the piston, when the piston is at top dead centre, is called the **clearance volume**.*

INTERNAL COMBUSTION ENGINES

Swept volume. *The volume swept through by the piston in moving between top dead centre and bottom dead centre, is called **swept volume** or **piston displacement**.* Thus, when piston is at bottom dead centre, total volume = swept volume + clearance volume.

Compression ratio. *It is ratio of total cylinder volume to clearance volume.*

Refer Fig. 4.30. Compression ratio (r) is given by

$$r = \frac{V_s + V_c}{V_c}$$

where, V_s = Swept volume, and
V_c = Clearance volume.

The compression ratio varies from 5 : 1 to 11 : 1 (average value 7 : 1 to 9 : 1) in *S.I. engines* and from 12 : 1 to 24 : 1 (average value 15 : 1 to 18 : 1) in *C.I. engines*.

Piston speed. *The average speed of the piston is called **piston speed**.*

Piston speed = 2 LN

where, L = Length of the stroke, and
N = Speed of the engine in r.p.m.

4.8. WORKING CYCLES

An internal combustion engine can work on any one of the following cycles :

(*a*) Constant volume or Otto cycle

(*b*) Constant pressure or Diesel cycle

(*c*) Dual combustion cycle.

These may be either *four-stroke cycle or two-stroke cycle engines*.

(*a*) **Constant volume or Otto cycle.** The cycle is so called because heat is supplied at constant volume. Petrol, gas and light oil work on this cycle. In the case of a petrol engine the proper mixing of petrol and air takes place in the carburettor which is situated outside the engine cylinder. The proportionate mixture is drawn into the cylinder during the suction stroke. In a gas engine also, air and gas is mixed outside the engine cylinder and this mixture enters the cylinder during the suction stroke. In light oil engines the fuel is converted to vapours by a vapouriser which receives heat from the exhaust gases of the engine and their mixture flows towards engine cylinder during suction stroke.

(*b*) **Constant pressure or Diesel cycle.** In this cycle only air is drawn in the engine cylinder during the suction stroke, this air gets compressed during the compression stroke and its pressure and temperature increase by a considerable amount. Just before the end of the stroke a metered quantity of fuel under pressure adequately more than that developed in the engine cylinder is injected in the form of fine sprays by means of a fuel injector. Due to very high pressure and temperature of the air the fuel ignites and hot gases thus produced throw the piston downwards and work is obtained. *Heavy oil engines make use of this cycle.*

(*c*) **Dual combustion cycle.** This cycle is also called *semi-diesel cycle*. It is so named because heat is added *partly at constant volume and partly at constant pressure*. In this cycle only air is drawn in the engine cylinder during suction stroke. The air is then compressed in hot combustion chamber at the end of the cylinder during the compression stroke to a pressure of about 26 bar. The heat of compressed air together with heat of combustion chamber ignites the fuel. The fuel is injected into the cylinder just before the end of compression stroke where it ignites immediately. The fuel injection is continued until the point of cut-off is reached. The burning of fuel at first takes place at constant volume and continues to burn at constant pressure during the first part of expansion or working stroke. The field of application of this cycle is *heavy oil engines*.

4.9. INDICATOR DIAGRAM

An *"indicator diagram"* is *a graph between pressure and volume* ; the former being taken on vertical axis and the latter on the horizontal axis. This is obtained by an instrument known as *indicator*. The indicator diagrams are of two types : (*a*) Theoretical or hypothetical, (*b*) Actual. The theoretical or hypothetical indicator diagram is always longer in size as compared to the actual one, since in the former losses are neglected. *The ratio of the area of the actual indicator diagram to the theoretical one is called* **diagram factor**.

4.10. FOUR-STROKE CYCLE ENGINES

Here follows the description of the four-stroke otto and diesel cycle engines.

Otto engines. The Otto four-stroke cycle refers to its use in petrol engines, gas engines, light oil engines and heavy oil engines in which the mixture of air and fuel are drawn in the engine cylinder. Since ignition in these engines is due to a spark, therefore they are also called *spark ignition engines*.

The various strokes of a four-stroke (Otto) cycle engine are detailed below.
Refer Fig. 4.31.

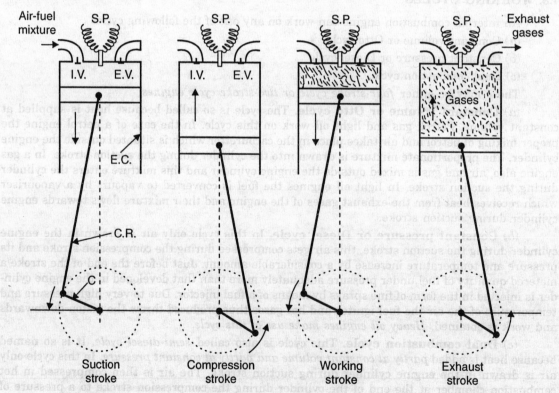

I.V = Intel valve, E.V. = Exhaust valve, E.C. = Engine cylinder, C.R. = Connecting rod
C = Crank, S.P. = Spark plug.

Fig. 4.31. Four-stroke Otto cycle engine.

INTERNAL COMBUSTION ENGINES

1. Suction stroke. During this stroke (also known as *induction stroke*) the piston moves from top dead centre (T.D.C.) to bottom dead centre (B.D.C.) ; the inlet valve opens and proportionate fuel-air mixture is sucked in the engine cylinder. This operation is represented by the line 5–1 (Fig. 4.32). The exhaust valve remains closed through out the stroke.

2. Compression stroke. In this stroke, the piston moves (1-2) towards (T.D.C.) and compresses the enclosed fuel-air mixture drawn in the engine cylinder during suction. The pressure of the mixture rises in the cylinder to a value of about 8 bar. Just before the end of this stroke the operating-plug initiates a spark which ignites the mixture and combustion takes place at constant volume (line 2-3) (Fig. 4.32). Both the inlet and exhaust valves remain closed during the stroke.

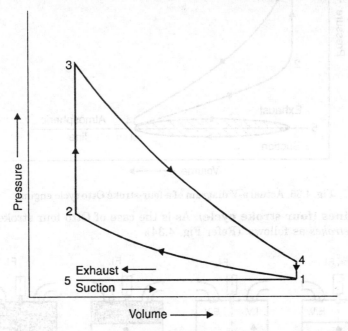

Fig. 4.32. Theoretical *p-V* diagram of a four-stroke Otto cycle engine.

3. Expansion or working stroke. When the mixture is ignited by the spark plug the hot gases are produced which drive or throw the piston from T.D.C. to B.D.C. and thus the work is obtained in this stroke. It is during this stroke when we get work from the engine ; the other three strokes namely suction, compression and exhaust being idle. *The flywheel mounted on the engine shaft stores energy during this stroke and supplies it during the idle strokes*. The expansion of the gases is shown by 3-4. (Fig. 4.32). Both the valves remain closed during the start of this stroke but when the piston just reaches the B.D.C. the exhaust valve opens.

4. Exhaust stroke. This is the last stroke of the cycle. Here the gases from which the work has been collected become useless after the completion of the expansion stroke and are made to escape through exhaust valve to the atmosphere. This removal of gas is accomplished during this stroke. The piston moves from B.D.C. to T.D.C. and the exhaust gases are driven out of the engine cylinder ; this is also called *scavenging*. This operation is represented by the line (1-5) (Fig. 4.32).

Fig. 4.33 shows the actual indicator diagram of four-stroke Otto cycle engine. It may be noted that line 5-1 is below the atmospheric pressure line. This is due to the fact that owing to restricted area of the inlet passages the entering fuel-air mixture cannot cope with the speed of the piston. The exhaust line 4-5 is slightly above the atmospheric pressure line. This is due to restricted exhaust passages which do not allow the exhaust gases to leave the engine-cylinder quickly.

The loop which has area 4-5-1 is called *negative loop* ; it gives the pumping loss due to admission of fuel air mixture and removal of exhaust gases. The area 1-2-3-4 is the total or gross negative work from the area 1-2-3-4 *i.e.*, gross work.

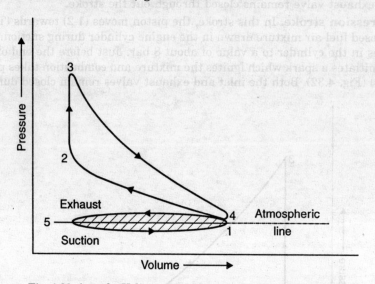

Fig. 4.33. Actual *p-V* diagram of a four-stroke Otto cycle engine.

Diesel engines (four-stroke cycle). As is the case of Otto four stroke ; this cycle too is completed in *four-strokes* as follows. (Refer Fig. 4.34).

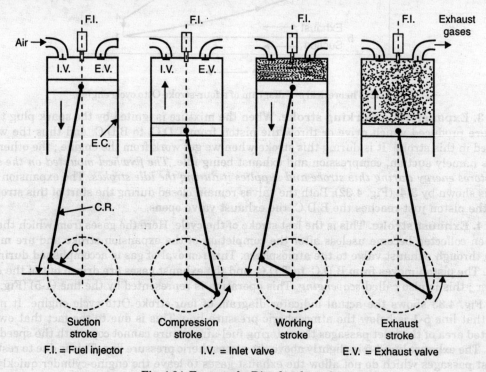

Fig. 4.34. Four-stroke Diesel cycle engine.

INTERNAL COMBUSTION ENGINES

1. Suction stroke. With the movement of the piston from T.D.C. to B.D.C. during this stroke, the inlet valve opens and the air at atmospheric pressure is drawn inside the engine cylinder ; the exhaust valve however remains closed. This operation is represented by the line 5-1 (Fig. 4.35).

2. Compression stroke. The air drawn at atmospheric pressure during the suction stroke is compressed to high pressure and temperature (to the value of 35 bar and 600°C respectively) as the piston moves from B.D.C. to T.D.C. This operation is represented by the line 1-2 (Fig. 4.35). Both the inlet and exhaust valves do not open during any part of this stroke.

3. Expansion or working stroke. As the piston starts moving from T.D.C. a matered quantity of fuel is injected into the hot compressed air in fine sprays by the fuel injector and it (fuel) starts burning at constant pressure shown by the line 2-3. At the point 3 fuel supply is cut off. The fuel is injected at the end of compression stroke but in actual practice the ignition of the fuel starts before the end of the compression stroke. The hot gases of the cylinder expand adiabatically to point 4, thus doing work on the piston. The expansion is shown by the line 3-4 (Fig. 4.35).

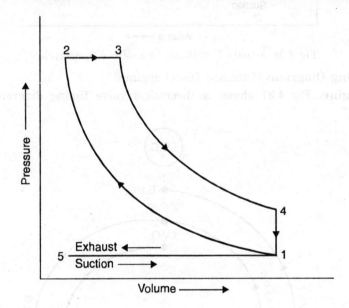

Fig. 4.35. Theoretical p-V diagram of a four-stroke Diesel cycle.

4. Exhaust stroke. The piston moves from the B.D.C. to T.D.C. and the exhaust gases escape to the atmosphere through the exhaust valve. When the piston reaches the T.D.C. the exhaust valve closes and the cycle is completed. This stroke is represented by the line 1-5 (Fig. 4.35).

Fig. 4.36 shows the actual indicator diagram for a four-stroke Diesel cycle engine. It may be noted that line 5-1 is below the atmospheric pressure line. This is due to the fact that owing to the restricted area of the inlet passages the entering air can't cope with the speed of the piston. The exhaust line 4-5 is slightly above the atmospheric line. This is because of the restricted exhaust passages which do not allow the exhaust gases to leave the engine cylinder quickly.

The loop of area 4-5-1 is called negative loop ; it gives the pumping loss due to admission of air and removal of exhaust gases. The area 1-2-3-4 is the total or gross work obtained from the piston and net work can be obtained by subtracting area 4-5-1 from area 1-2-3-4.

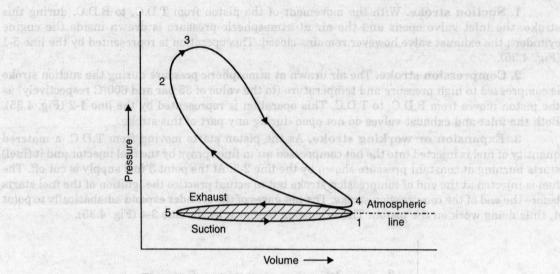

Fig. 4.36. Actual p-V diagram of four-stroke Diesel cycle.

Valve Timing Diagrams (Otto and Diesel engines)

1. Otto engine. Fig. 4.37 shows a theoretical valve timing diagram for *four stroke*

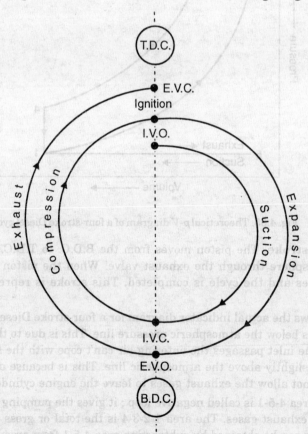

Fig. 4.37. Theoretical valve timing diagram (four-stroke Otto cycle engine).

INTERNAL COMBUSTION ENGINES

"Otto cycle" engines which is self-explanatory. In actual practice, it is difficult to open and close the valve instantaneously ; so as to get better performance of the engine the valve timings are modified. In Fig. 4.38 is shown an actual valve timing diagram. The inlet valve is opened 10° to 30° in advance of the T.D.C. position to enable the fresh charge to enter the cylinder and to help the burnt gases at the same time, to escape to the atmosphere. The suction of the mixture continues up to 30°–40° or even 60° after B.D.C. position. The inlet valve closes and the compression of the entrapped mixture starts.

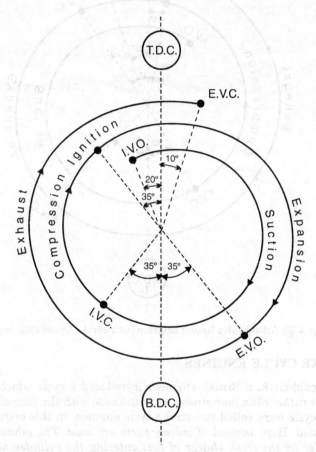

Fig. 4.38. Actual valve timing diagram (four-stroke Otto cycle engines).

The sparking plug produces a spark 30° to 40° before the T.D.C. position ; thus fuel gets more time to burn. The pressure becomes maximum nearly 10° past the T.D.C. position. The exhaust valve opens 30° to 60° before the B.D.C. position and the gases are driven out of the cylinder by piston during its upward movement. The exhaust valve closes when piston is nearly 10° past T.D.C. position.

2. Diesel engines. Fig. 4.39 shows the valve timing diagram of a *four-stroke "Diesel cycle" engine* (theoretical valve timing diagram, is however the same as Fig. 4.37). Inlet valve opens 10° to 25° in advance of T.D.C. position and closes 25° to 50° after the B.D.C. position. Exhaust valve opens 30° to 50° in advance of B.D.C. position and closes 10° to 15° after the T.D.C. position. The fuel injection takes place 5° to 10° before T.D.C. position and continues up to 15° to 25° near T.D.C. position.

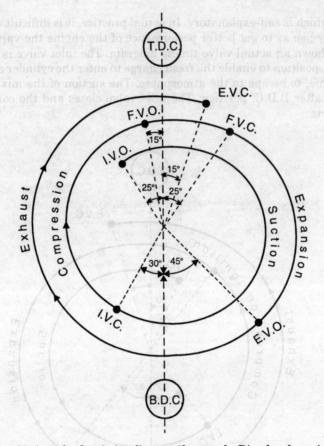

Fig. 4.39. Actual valve timing diagram (four-stroke Diesel cycle engines).

4.11. TWO-STROKE CYCLE ENGINES

In 1878, Dugald-clerk, a British engineer introduced a cycle which could be completed in two *strokes of piston rather than four strokes* as is the case with the four-stroke cycle engines. The engines using this cycle were called two-stroke cycle engines. In this engine suction and exhaust strokes are eliminated. Here *instead of valves, ports are used. The exhaust gases are driven out from engine cylinder by the fresh charge of fuel entering the cylinder nearly at the end of the working stroke.*

Fig. 4.40 shows a two-stroke petrol engine (used in scooters, motor cycles etc.). The cylinder L is connected to a closed crank chamber C.C. During the upward stroke of the piston M, the gases in L are compressed and at the same time fresh air and fuel (petrol) mixture enters the crank chamber through the valve V. When the piston moves downwards, V closes and the mixture in the crank chamber is compressed. Refer Fig. 4.40 (*i*), the piston is moving upwards and is compressing an explosive change which has previously been supplied to L. Ignition takes place at the end of the stroke. The piston then travels downwards due to expansion of the gases (Fig. 4.40 (*ii*)) and near the end of this stroke the piston uncovers the exhaust port (E.P.) and the burnt exhaust gases escape through this port (Fig. 4.40 (*iii*)). The transfer port (T.P.) then is uncovered immediately, and the compressed charge from the crank chamber flows into the cylinder and is deflected upwards by the *hump* provided on the head of the piston. It may be noted that the incoming air-petrol mixture helps the removal of gases from the engine-cylinder ; if, in case these exhaust gases do not

INTERNAL COMBUSTION ENGINES

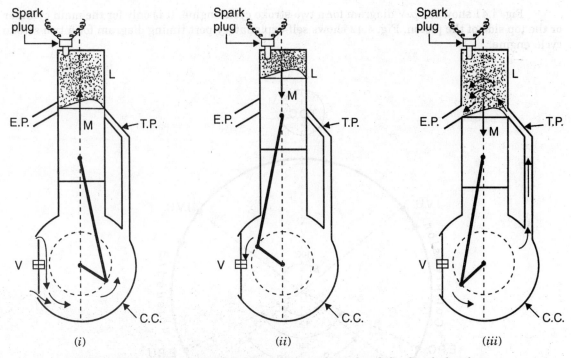

L = Cylinder, E.P. = Exhaust port, T.P. = Transfer port, V = valve, C.C = Crank chamber

Fig. 4.40. Two-stroke cycle engine.

leave the cylinder, the fresh charge gets diluted and efficiency of the engine will decrease. The piston then again starts moving from B.D.C. to T.D.C. and the charge gets compressed when E.P. (exhaust port) and T.P. are covered by the piston ; thus the cycle is repeated.

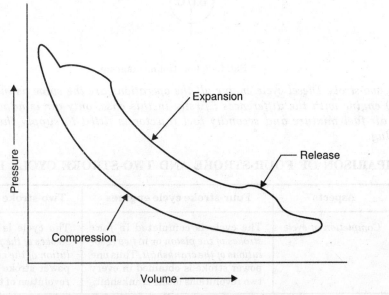

Fig. 4.41. *p-V* diagram for a two-stroke cycle engine.

Fig. 4.41 show the *p-V* diagram for a two-stroke cycle engine. It is only for the main cylinder or the top side of the piston. Fig. 4.42 shows self-explanatory port timing diagram for a two-stroke cycle engine.

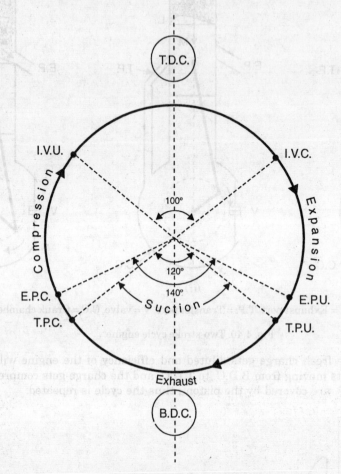

Fig. 4.42. Port timing diagram.

In a two-stroke Diesel cycle engine all the operations are the same as in the spark ignition (Otto cycle) engine with the differences ; firstly in this case, only air is admitted into cylinder instead of air-fuel mixture and secondly fuel injector is fitted to supply the fuel instead of a sparking plug.

4.12. COMPARISON OF FOUR-STROKE AND TWO-STROKE CYCLE ENGINES

S.No.	Aspects	Four-stroke cycle engines	Two-stroke cycle engines
1.	Completion of cycle	The cycle is completed in *four strokes of the piston* or in *two revolutions of the crankshaft*. Thus one power stroke is obtained in every two revolutions of the crankshaft.	The cycle is completed in *two strokes of the piston* or in *one revolution of the crankshaft*. Thus one power stroke is obtained in each revolution of the crankshaft.

S.No.	Aspects	Four-stroke cycle engines	Two-stroke cycle engines
2.	*Flywheel required-heavier or lighter*	Because of (i) turning-movement is not so uniform and hence *heavier* flywheel is needed.	More uniform turning movement and hence *lighter* flywheel is needed.
3.	*Power produced for same size of engine*	Again because of one power stroke for two revolutions, power produced for same size of engine is *small* or for the same power the engine is heavy and bulky.	Because of one power stroke for one revolution, power produced for same size of engine in *more* (theoretically twice, actually about 1.8 times) or for the same power the engine is light and compact.
4.	*Cooling and lubrication requirements*	Because of one power stroke in two revolutions *lesser* cooling and lubrication requirements. Lesser rate of wear and tear.	Because of one power stroke in one revolution *greater* cooling and lubrication requirement. Great rate of wear and tear.
5.	*Value and valve mechanism*	The four-stroke engine *contains* valve and valve mechanism.	Two-stroke engines have *no* valves but only ports (some two-stroke engines are fitted with conventional exhaust valves).
6.	*Initial cost*	Because of the heavy weight and complication of valve mechanism, *higher* is the initial cost.	Because of light weight and simplicity due to absence of valve mechanism, *cheaper* in initial cost.
7.	*Volumetric efficiency*	Volumetric efficiency *more* due to more time of induction.	Volumetric efficiency *less* due to lesser time for induction.
8.	*Thermal and part-load efficiencies*	Thermal efficiency higher, part load efficiency better than two stroke cycle engine.	Thermal efficiency lower, part load efficiency lesser than four stroke cycle engine.
9.	*Applications*	Used where efficiency is important; in *cars, buses, trucks tractors, industrial engines, aeroplane, power generators* etc.	In two-stroke petrol engine some fuel is exhausted during scavenging. Used where (a) *low cost*, and (b) *compactness and light weight important*. Two-stroke (air-cooled) petrol engines used in very small sizes only, *lawn movers, scooters motor cycles* (lubricating oil mixed with petrol). Two-stroke diesel engines used in *very large sizes* more than 60 cm bore, for *ship propulsion* because of low weight and compactness.

4.13. COMPARISON OF SPARK IGNITION (S.I.) AND COMBUSTION IGNITION (C.I.) ENGINES

S.No.	Aspects	S.I. engines	C.I. engines
1.	*Thermodynamic cycle*	Otto cycle	Diesel cycle ...... For slow speed engines Dual cycle ...... For high speed engines
2.	*Fuel used*	Petrol	Diesel.

S.No.	Aspects	S.I. engines	C.I. engines
3.	Air-fuel ratio	10 : 1 to 20 : 1	18 : 1 to 100 : 1.
4.	Compression ratio	Upto 11 ; Average value 7 to 9 ; Upper limit of compression ratio fixed by *anti-knock quality of fuel*.	12 to 24 ; Average value 15 to 18 ; Upper limit of compression ratio is limited by *thermal and mechanical stresses*.
5.	Combustion	Spark ignition	Compression ignition.
6.	Fuel supply	By carburettor ...... cheap method	By injection ...... explosive method.
7.	Operating pressure (i) Compression pressure (ii) Maximum pressure	7 bar to 15 bar 45 bar to 60 bar	30 bar to 50 bar 60 bar to 120 bar.
8.	Operating speed	High speed : 2000 to 6000 r.p.m.	Low speed : 400 r.p.m. Medium speed : 400 to 1200 r.p.m. High speed : 1200 to 3500 r.p.m.
9.	Control of power	Quantity governing ...... by throttle	Quality governing ...... by rack.
10.	Calorific value	44 MJ/kg	42 MJ/kg.
11.	Cost of running	High	Low.
12.	Maintenance cost	Minor maintenance required	Major overhaul required but less frequently.
13.	Supercharging	Limited by *detonation*. Used only in *aircraft engines*.	Limited by *blower power and mechanical and thermal stresses*. Widely used.
14.	Two-stroke operation	*Less suitable*, fuel loss in scavenging. But small two-stroke engines are used in mopeds, scooters and motorcycles due to their *simplicity and low cost*.	No fuel loss in scavenging. *More suitable*.
15.	High powers	No	Yes.
16.	Uses	Mopeds, scooters, motorcycles, simple engine passenger cars, air-crafts etc.	Buses, trucks locomotives, tractors, earth moving machinery and stationary generating plants.

4.14. COMPARISON BETWEEN A PETROL ENGINE AND A DIESEL ENGINE

S.No.	Petrol engine	Diesel engine
1.	Air-petrol mixture is sucked in the engine cylinder during suction stroke.	Only air is sucked during suction stroke.
2.	Spark plug is used.	Employs an injector.
3.	Power is produced by spark ignition.	Power is produced by compression ignition.
4.	Thermal efficiency up to 25%.	Thermal efficiency up to 40%.

INTERNAL COMBUSTION ENGINES 149

S.No.	Petrol engine	Diesel engine
5.	Occupies less space.	Occupies more space.
6.	More running cost.	Less running cost.
7.	Light in weight.	Heavy in weight.
8.	Fuel (Petrol) costlier.	Fuel (Diesel) cheaper.
9.	Petrol being volatile is dangerous.	Diesel is non-dangerous as it is non-volatile.
10.	Pre-ignition possible.	Pre-ignition not possible.
11.	Works on Otto cycle.	Works on Diesel cycle.
12.	Less dependable.	More dependable.
13.	Used in *cars* and *motor cycles*.	Used in heavy duty vehicles like *trucks*, *buses* and *heavy machinery*.

4.15. HOW TO TELL A TWO-STROKE CYCLE ENGINE FROM A FOUR-STROKE CYCLE ENGINE ?

S.No.	Distinguishing features	Four-stroke cycle engine	Two-stroke cycle engine
1.	*Oil sump and oil-filter plug*	It has an oil sump and oil-filter plug.	It does not have oil sump and oil-filter plug.
2.	*Oil drains etc.*	It requires oil drains and refills periodically, just an automobile do.	In this type of engine, the oil is added to the gasoline so that a mixture of gasoline and oil passes through the carburettor and enters the crankcase with the air.
3.	*Location of muffler (exhaust silencer)*	It is installed at the head end of the cylinder at the exhaust valve location.	It is installed towards the middle of the cylinder, at the exhaust port location.
4.	Name plate	If the name plate mentions the type of oil and the crankcase capacity, or similar data, it is a four-stroke cycle engine.	If the name plate tells to mix oil with the gasoline, it is a two stroke cycle engine.

4.16. IGNITION SYSTEM
(Petrol Engines)

The operator of a spark ignition engine expects the ignition system to five thousands of consecutive cycles in cylinders full of fuel-air mixture without a *"miss"* although the manifod pressure may vary from 0.35 to 2.7 bar, the fuel-air ratio may vary from 0.06 to 0.12 and the r.p.m. from 400 to 5000. In addition the ignition must occur at the proper crank angle so that the time losses are held at a minimum. In a view of this, the *requirements of ignition-systems* may be put down as given below :

 1. A source of electric energy must be there.
 2. A means for stepping up the voltage from the source to the very high potential required to produce a high tension arc across the spark plug gap that ignites the combustible mixture.

3. A means for timing and distributing the high voltage i.e., supply of high voltage to each spark plug at the exact instant is required in every cycle in each cylinder.

4. Adjustment of spark advance with speed and load.

The two basic ignition systems in current use are :

(a) Battery or coil ignition system.

(b) Magneto ignition system.

Both the systems have been preferred to considerable extent and these systems fulfil the requirements for satisfactory operation.

(a) **Battery or coil-ignition system**

Most of the modern spark ignition engines use battery ignition system. This system consists of the following components :

1. Battery (6 or 12 volts)
2. Ignition switch
3. Induction coil
4. Circuit/contact breaker
5. Condenser
6. Distributor.

Refer to Fig. 4.43.

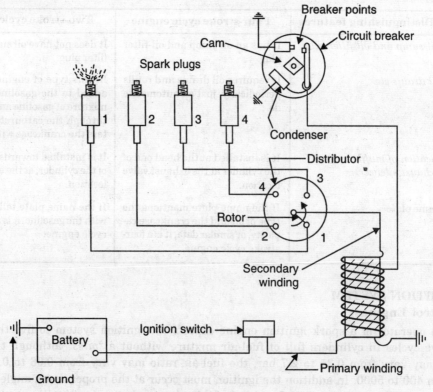

Fig. 4.43. Battery or coil ignition system.

One terminal of the battery is ground to the frame of the engine, and other is connected through the ignition switch to one primary terminal of the ignition coil (consisting of a comparatively few turns of thick wire wound round an iron core). The other primary terminal is connected to one end of the contact points of the circuit breaker and through closed points to ground. The primary circuit of the ignition coil thus gets completed when contact points of the circuit breaker

INTERNAL COMBUSTION ENGINES

are together and switch is closed. The secondary terminal of the coil is connected to the central contact of the distributor and hence to distributor rotor. The secondary circuit consists of secondary winding (consisting of a large number of turns of fine wire) of the coil, distributor and four spark plugs. The contact breaker is driven by a cam whose speed is *half* the engine speed (for *four-stroke engines*) and breaks the primary circuit one for each cylinder during one complete cycle of the engine.

The breaker points are held on contact by a spring except when forced apart by lobs of the cam.

To start with, the ignition switch is made on and the engine is cranked *i.e.,* turned by hand when the contacts touch, the current flows from battery through the switch, primary winding of the induction coil to circuit breaker points and the circuit is completed through the ground. A condenser connected across the terminals of the contact breaker points prevent the sparking at these points. The rotating cam breaks open the contacts immediately and breaking of this primary circuit brings about a change of magnetic field ; due to which a very high voltage to the tune of 8000 to 12000 V is produced across the secondary terminals. (The number of turns in the secondary winding may be 50 to 100 times than in primary winding). Due to high voltage the spark jumps across the gap in the spark plug and air-fuel mixture is ignited in the cylinder.

On account of its combined cheapness, convenience of maintenance, attention and general suitability, it has been adopted universally on automobiles.

(b) Magneto ignition system

The magneto ignition system is similar in principle to the battery system except that the magnetic field in the core of the primary and secondary windings is produced by a rotating permanent magnet (Fig. 4.44). As the magnet turns, the field is produced from a positive maximum to a negative maximum and back again. As this magnetic field falls from a positive maximum value, a voltage and current are induced in the primary winding. The primary current produces a magnetic field of its own which keeps the total magnetic field surrounding the primary and secondary winding approximately constant. When the permanent magnet has turned for enough so that its contribution to the total field is strongly negative, the breaker points are opened and the magnetic field about the secondary winding suddenly goes from a high positive value to a high negative value. This induces a high voltage in the secondary winding which is led to the proper spark plug by the distributor.

The *magneto is an efficient, reliable, self contained unit which is often preferred for aircraft engines because storage batteries are heavy and troublesome.* Special starting means are required, however, as the magneto will not furnish enough voltage for ignition at low speeds. Variation in ignition timing is more difficult with the magneto, since the breaker point must be opened when the rotating magnets are in the most favourable position. It is possible to change the engine crank angle at which the mangeto points open without disturbing the relationship between point opening and magnet position by designing the attachment pad so that the entire magneto body may be rotated a few degrees about its own shaft. Obviously this method is not as satisfactory as rotating a timer cam-plate.

Advantages of battery system :

1. It offers better sparks at low speeds, starting and for cranking purposes.
2. The initial cost of the system is low.
3. It is a reliable system and periodical maintenance required is negligible except for battery.
4. Items requiring attention can be easily located in more accessible position than those of magnetos.
5. The high speed engine drive is usually simpler than magneto drive.

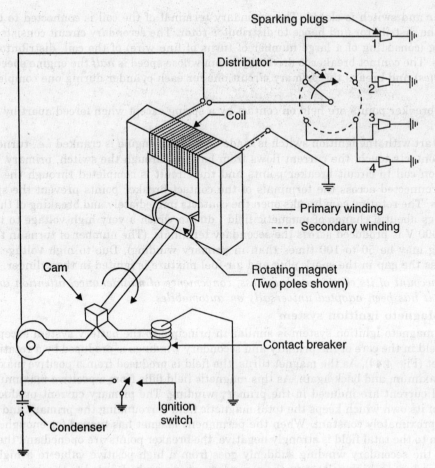

Fig. 4.44. Magneto ignition system.

6. Adjustment of spark timing has no deteriorental effect over the complete ignition timing range.

Disadvantages :
1. With the increasing speed, sparking voltage drops.
2. Battery, the only unreliable component of the system needs regular attention. In case battery runs down, the engine cannot be started as induction coil fails to operate.
3. Because of battery, bulk of the system is high.

Advantages of magneto system :
1. The system is more reliable as there is no battery or connecting cable.
2. The system is more suitable for medium and very high speed engines.
3. With the use of cobalt steel and nickel-aluminium magnet metals very light and compact units can be made which require very little room.
4. With recent development this system has become fairly reliable.

Disadvantages :
1. At low speeds and during cranking the voltage is very low. This has been overcome by suitable modifications in the circuit.

INTERNAL COMBUSTION ENGINES

2. Adjustment of the spark timing *i.e.*, advance or retard, has deterimental effect upon the spark voltage or energy.

3. The powerful sparks at high engine speeds cause burning of the electrodes.

Firing order :

In case of multi-cylinder engines the order in which spark inside engine cylinders must occur is decided on the consideration of balancing of forces. Firing orders for various engines are given below :

No. of cylinders	Firing order
Two	1, 2
Three	1, 3, 2
Four	1, 4, 3, 2 or 1, 3, 4, 2
Six	1, 5, 3, 6, 2, 4 or 1, 4, 2, 6, 3, 5
Eight	1, 6, 2, 5, 8, 3, 7, 4 or 1, 8, 7, 3, 6, 5, 4, 2

Eight (Vee) : Either of the following alternatives.

(A)   1L, 1R, 4L, 4R, 2R, 3L, 3R, 2L

(B)   1L, 4R, 2R, 2L, 3R, 3L, 4L, 1R

(C)   1L, 3L, 2R, 4R, 3R, 2L, 4L, 1R

(D)   1L, 4R, 4L, 2L, 3R, 3L, 2R, 1R

(E)   1L, 3L, 3R, 2L, 2R, 1R, 4L, 4R

L and R indicate cylinder on left and right hand side respectively. The firing order for a four-stroke engine with its cylinder numbered consecutively from 1 to n will be 1, 3, 5, 7, ... to n for one revolution of the crankshaft and 2, 4, 6, 8, ... to $(n-1)$ for the next revolution.

4.17. COOLING SYSTEMS

In an I.C. engine, the temperature of the gases inside the engine cylinder may very from 35°C or less to as high as 2750°C during the cycle. If an engine is allowed to run without external cooling, the cylinder walls, cylinder and pistons will tend to assume the average temperature of the gases to which they are exposed, which may be of the order of 1000 to 1500°C. Obviously at such high temperature ; the metals will loose their characteristics and piston will expand considerably and sieze the liner. Of course theoretically thermal efficiency of the engine will improve without cooling but actually the engine will sieze to run. If the cylinder wall temperature is allowed to rise above a certain limit, about 65°C, the lubricating oil will begin to evaporate rapidly and both cylinder and piston may be damaged. Also high temperature may cause excessive stress in some parts rendering them useless for further operation. In view of this, part of the heat generated inside the engine cylinder is allowed to be carried away by the cooling system. *Thus cooling system is provided on an engine for the following* **reasons** :

1. The even expansion of piston in the cylinder may result in seizure of the piston.

2. High temperatures reduce strength of piston and cylinder liner.

3. Overheated cylinder may lead to preignition of the charge, in case of spark ignition engine.

4. Physical and chemical changes may occur in lubricating oil which may cause sticking of piston rings and excessive wear of cylinder.

Almost 25 to 35 per cent of total heat supplied in the fuel is removed by the cooling medium. Heat carried away by lubricating oil and heat lost by radiation amounts 3 to 5 per cent of total heat supplied.

There are mainly two methods of cooling I.C. engine :

(a) Air cooling.

(b) Liquid cooling.

(a) **Air cooling**

In this method, heat is carried away by the air flowing over and around the engine cylinder. It is used in scooters, motorcycles etc. Here fins are cast on the cylinder head and cylinder barrel which provide additional conductive and radiating surface. (Fig. 4.45). The fins are arranged in such a way that they are at right angles to the cylinder axis.

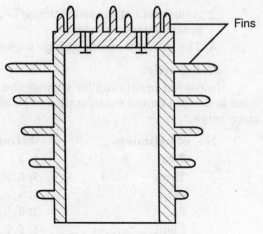

Fig. 4.45. Air cooling.

Advantages :

1. The design of the engine becomes simpler as no water jackets are required. The cylinder can be of identical dimensions and individually detachable and therefore cheaper to renew in case of accident etc.
2. Absence of cooling pipes, radiator etc. makes the cooling system simpler.
3. No danger of coolant leakage etc.
4. The engine is not subjected to freezing troubles etc. usually encountered in case of water-cooled engine.
5. The weight per B.H.P. of the air-cooled engine is less than that of water-cooled engine.
6. In this case engine is rather a self contained unit as it requires no external components *e.g.* radiator, headers, tank etc.
7. Installation of air-cooled engines is easier.

Disadvantages :

1. Their movement is noisy.
2. Non-uniform cooling.
3. The output of air cooled engine is less than that of a liquid cooled engine.
4. Maintenance is not easy.
5. Smaller useful compression ratio.

(b) **Liquid cooling**

In this method of cooling engines, the cylinder walls and heads are provided with jackets through which the cooling liquid can circulate. The heat is transferred from cylinder walls to the liquid by convection and conduction. The liquid becomes heated in its passage through the jackets and is itself cooled by means of an air-cooled radiator system. The heat-from liquid in turn is transferred to air.

Various methods are used for circulating the water around the cylinder and cylinder head. These are :

1. Thermo-syphon cooling
2. Forced or pump cooling

INTERNAL COMBUSTION ENGINES

3. Cooling with thermostatic regulator
4. Pressurised water cooling
5. Evaporative cooling.

1. Thermo-syphon cooling

The basis of this type of cooling is the fact that water becomes light on heating. Fig. 4.46 shows the thermo-syphon cooling arrangement. The top of radiator is connected to the top of water

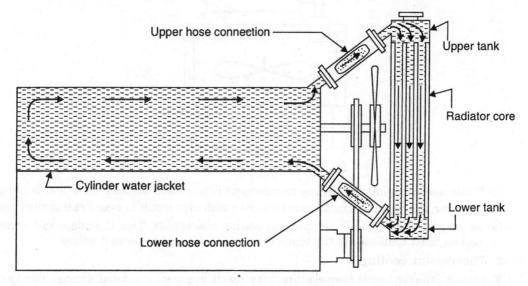

Fig. 4.46. Thermo-syphon cooling.

jacket by a pipe and bottom of the radiator to the bottom of the water jacket. Water travels down the radiator across which air is passed to cool it. The air flow can take place due to vehicle motion or a fan can be provided for the purpose.

This system has the advantage that it is quite simple and automatic and is without any water pump unless there is leak, there is nothing to get out of order.

The major shortcoming of this system is that cooling depends only on the temperature and is independent of the engine speed. The rate of circulation is slow and insufficient. The circulation of water starts only after the engine has become hot enough to cause thermo-syphon action. This system requires that the radiator be above the engine for gravity flow of water to engine.

Thermo-syphon system is not widely used at present.

2. Forced or pump system

Refer Fig. 4.47. In this system, a pump is used to cause positive circulation of water in the water jacket. Usually the pump is belt driven from the engine.

The advantage of forced system is that *cooling is ensured* under all conditions of operation.

This system entails the following *demerits* :

(i) The cooling is independent of temperature. This may, under certain circumstances result in over cooling the engine.

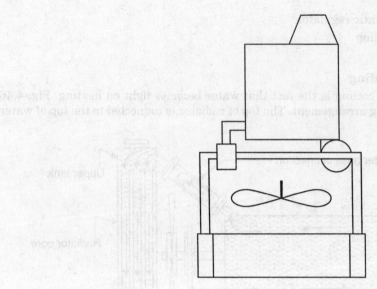

Fig. 4.47. Forced or pump system.

(*ii*) While moving uphill the cooling requirement is increased because more fuel is burned. However, the coolant circulation is reduced which may result in over-heating the engine.

(*iii*) As soon as the engine is stopped the cooling also ceases. This is undesirable because cooling must continue till the temperatures are reduced to normal values.

3. Thermostat cooling

Too lower cylinder barrel temperature, may result in severe corrosion damage due to condensation of acids on the barrel wall. To avoid such a situation it is customary to use a thermostat (a temperature controlling device) to stop flow of coolant below a pre-set cylinder barrel temperature. Most modern cooling system employ a thermostatic device which prevents the water in the engine jacket from circulating through the radiator for cooling until its temperature has reached to a value suitable for efficient engine operation.

Fig. 4.48 shows a systematic diagram of a thermostatically controlled cooling system. Also shown is a typical thermostat (Fig. 4.49). It consists of bellows which are made of thin copper tubes, partially filled with a volatile liquid like ether or methyl alcohol. The volatile liquid changes into vapour at the correct working temperature, thus creating enough pressure to expand the bellows. The temperature at which the thermostat operates is set by the manufacturers and cannot be altered. The movement of the bellows opens the main valve in the ratio of temperature rise, increasing or restricting the flow of water from engine to the radiator. Hence when the normal temperature of the engine has been reached the valve opens and circulation of water commences. When the unit is closed the gas condenses and so the pressure falls. The bellows collapse and the thermostat seats on its seat and circulation around thermostat stops. When the thermostat valve is not open and the engine is running the water being pumped rises in pressure and causes the pressure relief valve to open. Thus the water completes its circulation through the by-pass as shown in Figs. 4.48 and 4.49. Now when the temperature of water around the engine-cylinder rises upto a certain limit, it causes the thermostat valve to open. The pressure of water being pumped falls and pressure relief valve closes. So the flow of cooling water in the normal circuit commences through the radiator. This accelerates the rise of temperature of the cylinder walls and water and more power is developed in a few moments of the starting of the engine.

INTERNAL COMBUSTION ENGINES

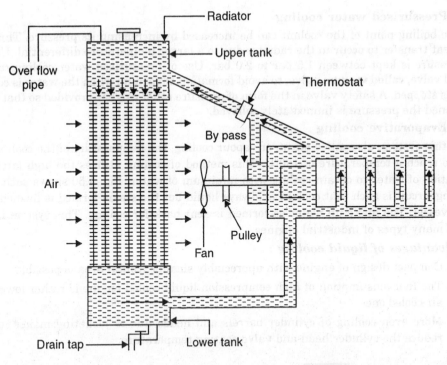

Fig. 4.48. Thermostatically controlled cooling system.

Another method of warming up the radiator water upto the normal temperature is by utilising the shutter in the radiator in order to restrict the incoming air through the radiator till the engine warms up. Thereafter, the shutter is opened gradually so that the desired rate of cooling is achieved.

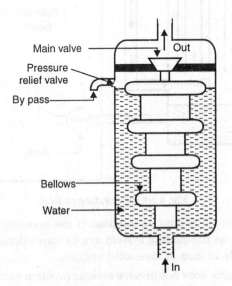

Fig. 4.49. Typical thermostat.

4. Pressurised water cooling

The boiling point of the coolant can be increased by increasing its pressure. This allows a greater heat transfer to occur in the radiator due to a larger temperature differential. Usually the water pressure is kept between 1.5 bar to 2.0 bar. Use of pressurised water cooling requires an additional valve, called vacuum valve, to avoid formation of vacuum when the water is cooled after engine has stopped. A safety valve in the form of pressure relief valve is provided so that whenever cap is opened the pressure is immediately relieved.

5. Evaporative cooling

In this system, also called steam or vapour cooling, the temperature of the cooling water is allowed to reach a temperature of 100°C. This method of cooling utilises the high latent heat of vapourisation of water to obtain cooling with minimum of water. Fig. 4.50 shows such a system. The cooling circuit is such that coolant is always liquid but the steam formed is flashed off in the separate vessel. The make up water so formed is sent back for cooling. This system is used for cooling of many types of industrial engines.

Advantages of liquid cooling :

1. Compact design of engine with appreciably smaller frontal area is possible.
2. The fuel consumption of high compression liquid-cooled engine is rather lower than for air-cooled one.
3. More even cooling of cylinder barrels and heads due to jacketing makes it easier to reduce the cylinder head and valve seating temperature.

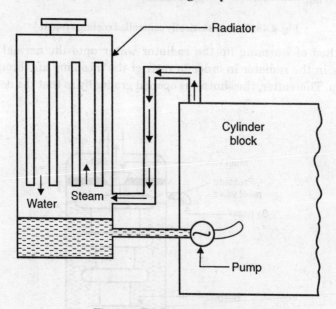

Fig. 4.50. Evaporating cooling.

4. In case of water-cooled engine installation is not necessary at the front of mobile vehicles, air crafts etc. as the cooling system can be conveniently located wherever required. This is not possible in case of air-cooled engines.
5. The size of the engine does not involve serious problem as far as design of cooling system is concerned. In case of air-cooled engines particularly in high horse power range difficulty is encountered in circulation of required quantity of air for cooling purposes.

Disadvantages :
1. This is dependent system in which supply of water for circulation in the jacket is required.
2. Power absorbed by the pump for water circulation is considerably higher than that for cooling fans.
3. In the event of failure of cooling system serious damage may be caused to the engine.
4. Cost of system is considerably high.
5. System requires considerable attention for the maintenance of various parts of system.

4.18. LUBRICATION SYSTEMS

Lubrication is the admittance of oil between two surfaces having relative motion. The purpose of lubrication may be one or more of the following :
1. Reduce friction and wear between the parts having relative motion.
2. Cool the surfaces by carrying away heat generated due to friction.
3. Seal a space adjoining the surfaces such as piston rings and cylinder liner.
4. Clean the surface by carrying away the carbon and metal particles caused by wear.
5. Absorb shock between bearings and other parts and consequently reduce noise.

Properties of lubricants :

The chief qualities to be considered in selecting oil for lubrication are :

1. Viscosity	2. Flash point
3. Fire point	4. Cloud point
5. Pour point	6. Oiliness
7. Corrosion	8. Emulsification
9. Physical stability	10. Chemical stability
11. Neutralisation number	12. Adhesiveness
13. Film strength	14. Specific gravity.

1. **Viscosity.** *It is the ability of the oil to resist internal deformation due to mechanical stresses and hence it is a measure of the ability of the oil film to carry a load.* A more viscous oil can carry a greater load, but it will offer greater friction to sliding movement of the one bearing surface over the other. Viscosity varies with the temperature and hence if a surface to be lubricated is normally at high temperature it should be supplied with oil of a higher viscosity than would be suitable for, say journal bearings.

2. **Flash point.** *It is defined as the lowest temperature at which the lubricating oil will flash when a small flame is passed across its surface.* The flash point of the oil should be sufficiently high so as to avoid flashing of oil vapours at the temperatures occurring in common use. High flash point oils are needed in air compressors.

3. **Fire point.** *It is the lowest temperature at which the oil burns continuously.* The fire point also must be high in a lubricating oil, so that oil does not burn in service.

4. **Cloud point.** When subject to low temperatures the oil changes from liquid stage to a plastic or solid state. In some cases the oil starts solidifying which makes it to appear cloudy. The temperature at which this takes place is called the *cloud point*.

5. **Pour point.** *Pour point is the lowest temperature at which the lubricating oil will pour.* It is an indication of its ability to move at low temperatures. This property must be considered because of its effect on starting an engine in cold weather and on free circulation of oil through exterior feed pipes when pressure is not applied.

6. Oiliness. *This is the property which enables oil to spread over and adhere to the surface of the bearing.* It is most important in boundary lubrication.

7. Corrosion. A lubricant should not corrode the working parts and it must retain its properties even in the presence of foreign matter and additives.

8. Emulsification. A lubricating oil, when mixed with water is emulsified and loses its lubricating property. The emulsification number is an index of the tendency of an oil to emulsify with water.

9. Physical stability. A lubricating oil must be stable physically at the lowest and highest temperatures between which the oil is to be used. At the lowest temperature there should not be any separation of solids, and at the highest temperature it should not vapourize beyond a certain limit.

10. Chemical stability. A lubricating oil should also be stable chemically. There should not be any tendency for oxide formation.

11. Neutralisation number. An oil may contain certain impurities that are not removed during refining. The neutralisation number test is a simple procedure to determine acidity or alkalinity of an oil. It is the weight in milligrams of potassium hydroxide required to neutralise the acid content of one gram of oil.

12. Adhesiveness. *It is the property of lubricating oil due to which the oil particles stick with the metal surfaces.*

13. Film strength. *It is the property of a lubricating oil due to which the oil retains a thin film between the two surfaces even at high speed and load.* The film does not break and the two surfaces do not come in direct contact. Adhesiveness and film strength cause the lubricant to enter the metal pores and cling to the surfaces of the bearings and journals keeping them wet when the journals are at rest and presenting metal to contact until the film of lubricant is built up.

14. Specific gravity. It is a measure of density of oil. It is an indication regarding the grade of lubricant by comparing one lubricant with other. It is determined by a hydrometre which flats in the oil, and the gravity is read on the scale of the hydrometer at the surface of the oil.

Main parts of an engine to be lubricated :

The main parts of an engine which need lubrication are as given under :

(*i*) Main crankshaft bearings. (*ii*) Big-end bearings.
(*iii*) Small-end or gudgeon pin bearings. (*iv*) Piston rings and cylinder walls.
(*v*) Timing gears. (*vi*) Camshaft and camshaft bearings.
(*vii*) Valve mechanism. (*viii*) Valve guides, valve tappets and rocker arms.

Various lubrication systems used for I.C. engines may be classified as :

1. Wet sump lubrication system.
2. Dry sump lubrication system.
3. Mist lubrication system.

1. Wet Sump Lubrication System

These systems employ a large capacity oil sump at the base of crank chamber, from which the oil is drawn by a low pressure oil pump and delivered to various parts. Oil there gradually returns back to the sump after serving the purpose.

(*a*) **Splash system.** Refer Fig. 4.51. This system is used on some *small four-stroke stationary engines*. In this case the caps on the big ends bearings of connecting rods are provided with scoops which, when the connecting rod is in the lowest position, just dip into oil troughs and thus direct the oil through holes in the caps to the big end bearings. Due to splash of oil it reaches the lower portion of the cylinder walls, crankshaft and other parts requiring lubrication. Surplus oil

INTERNAL COMBUSTION ENGINES

eventually flows back to the oil sump. Oil level in the troughs is maintained by means of a oil pump which takes oil from sump, through a filter.

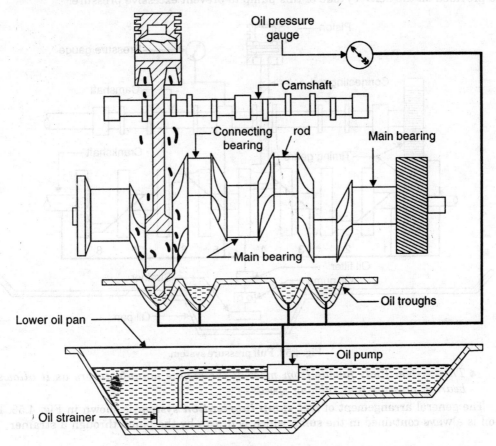

Fig. 4.51. Splash system.

Splash system is suitable for low and medium speed engines having moderate bearing load pressures. For high performance engines, which normally operate at high bearing pressures and rubbing speeds this system does not serve the purpose.

(b) **Semi-pressure system**. This method is a combination of splash and pressure systems. It incorporates the advantages of both. In this case main supply of oil is located in the base of crank chamber. Oil is drawn from the lower portion of the sump through a filter and is delivered by means of a gear pump at pressure of about 1 bar to the main bearings. The big end bearings are lubricated by means of a spray through nozzles. Thus oil also lubricates the cams, crankshaft bearings, cylinder walls and timing gears. An oil pressure gauge is provided to indicate satisfactory oil supply.

The system is less costly to install as compared to pressure system. It enables higher bearing loads and engine speeds to be employed as compared to splash system.

(c) **Full pressure system**. In this system, oil from oil sump is pumped under pressure to the various parts requiring lubrication. Refer Fig. 4.52. The oil is drawn from the sump through filter and pumped by means of a gear pump. Oil is delivered by the pressure pump at pressure ranging from 1.5 to 4 bar. The oil under pressure is supplied to main bearings of crankshaft and camshaft. Holes drilled through the main crankshafts bearing journals, communicate oil to the big end bearings and also small end bearings through hole drilled in connecting rods. A pressure gauge is provided to confirm the circulation of oil to the various parts. A pressure regulating valve is also provided on the delivery side of this pump to prevent excessive pressure.

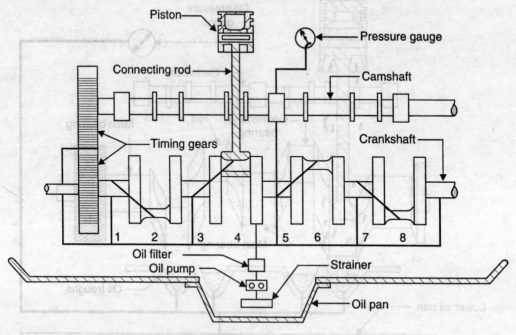

Fig. 4.52. Full pressure system.

- *This system finds favour from most of the engine manufacturers as it allows high bearing pressure and rubbing speeds.*

The general arrangement of **wet sump lubrication system** is shown in Fig. 4.53. In this case oil is always contained in the sump which is drawn by the pump through a strainer.

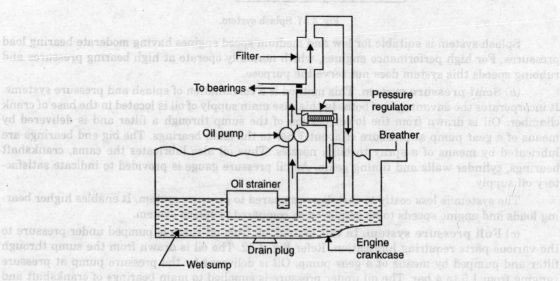

Fig. 4.53. Wet sump lubrication system.

INTERNAL COMBUSTION ENGINES

2. Dry Sump Lubrication System

Refer Fig. 4.54. In this system, the oil from the sump is carried to a separate storage tank outside the engine cylinder block. The oil from sump is pumped by means of a sump pump through filters to the storage tank. Oil from storage tank is pumped *to the engine cylinder* through oil cooler. Oil pressure may vary from 3 to 8 bar. *Dry sump lubrication system is generally adopted for high capacity engines.*

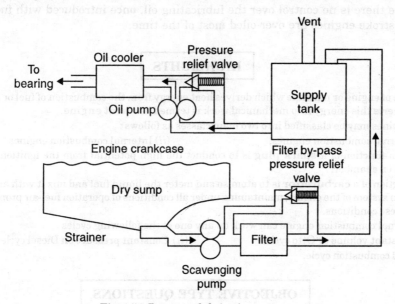

Fig. 4.54. Dry sump lubrication system.

3. Mist Lubrication System

This system is used for two-stroke cycle engines. Most of these engines are crank charged, *i.e.*, they employ crankcase compression and thus, are not suitable for crankcase lubrication. These engines are lubricated by adding 2 to 3 per cent lubricating oil in the fuel tank. The oil and fuel mixture is induced through the carburettor. The gasoline is vaporised ; and the oil in the form of mist, goes via crankcase into the cylinder. The oil which impinges on the crankcase walls lubricates the main and connecting rod bearings, and rest of the oil which passes on the cylinder during charging and scavenging periods, lubricates the piston, piston rings and the cylinder.

Advantages :
1. System is simple.
2. Low cost (because no oil pump, filter etc. are required.)

Disadvantages :
1. A portion of the lubricating oil invariably burns in combustion chamber. This bearing oil when burned, and still worse, when partially burned in combustion chamber leads to heavy exhaust emissions and formation of heavy deposit on piston crown, ring grooves and exhaust port which interferes with the efficient engine operation.
2. One of the main functions of lubricating oil is the protection of anti-friction bearings etc. against corrosion. Since the oil comes in close contact with acidic vapours produced during the combustion process, it rapidly loses its anti-corrosion properties resulting in corrosion damage of bearings.

3. For effective lubrication oil and fuel must be thoroughly mixed. This requires either separate mixing prior to use or use of some additive to give the oil good mixing characteristics.
4. Due to higher exhaust temperature and less efficient scavenging the crankcase oil is diluted. In addition some lubricating oil burns in combustion chamber. This results in 5 to 15 per cent higher lubricant consumption for two-stroke engine of similar size.
5. Since there is no control over the lubricating oil, once introduced with fuel, most of the two-stroke engines are over-oiled most of the time.

HIGHLIGHTS

1. Any type of engine or machine which derives heat energy from the combustion of fuel or any other source and converts this energy into mechanical work is termed as a **heat engine**.
2. Heat engines may be classified into two main classes as follows :
   (i) External combustion engines, (ii) Internal combustion engines.
3. The main function of a **spark plug** is to conduct the high potential from the ignition system into the combustion chamber.
4. The function of a **carburettor** is to atomise and meter the liquid fuel and mix it with air as it enters the induction system of the engine, maintaining under all conditions of operation fuel-air proportions appropriate to these conditions.
5. An internal combustion engine can work on any one of the following cycles :
   (i) Constant volume or Otto cycle (ii) Constant pressure on Diesel cycle
   (iii) Dual combustion cycle.

OBJECTIVE TYPE QUESTIONS

Choose the Correct Answer :

1. In a four-stroke cycle engine, the four operations namely suction, compression, expansion and exhaust are completed in the number of revolutions of crankshaft equal to
   (a) four (b) three
   (c) two (d) one.
2. In a two-stroke cycle engine, the operations namely suction, compression, expansion and exhaust are completed in the number of revolutions of crankshaft equal to
   (a) four (b) three
   (c) two (d) one.
3. In a four-stroke cycle S.I. engine the camshaft runs
   (a) at the same speed as crankshaft (b) at half the speed of crankshaft
   (c) at twice the speed of crankshaft (d) at any speed irrespective of crankshaft speed.
4. The following is an S.I. engine
   (a) Diesel engine (b) Petrol engine
   (c) Gas engine (d) none of the above.
5. The following is C.I. engine
   (a) Diesel engine (b) Petrol engine
   (c) Gas engine (d) none of the above.
6. In a four-stroke cycle petrol engine, during suction stroke
   (a) only air is sucked in (b) only petrol is sucked in
   (c) mixture of petrol and air is sucked in (d) none of the above.

INTERNAL COMBUSTION ENGINES

7. In a four-stroke cycle diesel engine, during suction stroke
   (a) only air is sucked in
   (b) only fuel is sucked in
   (c) mixture of fuel and air is sucked in
   (d) none of the above.
8. The two-stroke cycle engine has
   (a) one suction valve and one exhaust valve operated by one cam
   (b) one suction valve and one exhaust valve operated by two cams
   (c) only ports covered and uncovered by piston to effect charging and exhausting
   (d) none of the above.
9. For same output, same speed and same compression ratio the thermal efficiency of a two-stroke cycle petrol engine as compared to that for four-stroke cycle petrol engine is
   (a) more
   (b) less
   (c) same as long as compression ratio is same
   (d) same as long as output is same.
10. The ratio of brake power to indicated power of an I.C. engine is called
    (a) mechanical efficiency
    (b) thermal efficiency
    (c) volumetric efficiency
    (d) relative efficiency.
11. The specific fuel consumption of a diesel engine as compared to that for petrol engines is
    (a) lower
    (b) higher
    (c) same for same output
    (d) none of the above.
12. The thermal efficiency of petrol engine as compared to diesel engine is
    (a) lower
    (b) higher
    (c) same for same power output
    (d) same for same speed.
13. Compression ratio of petrol engines is in the range of
    (a) 2 to 3
    (b) 7 to 10
    (c) 16 to 20
    (d) none of the above.
14. Compression ratio of diesel engines may have a range
    (a) 8 to 10
    (b) 10 to 15
    (c) 16 to 20
    (d) none of the above.
15. The thermal efficiency of good I.C. engine at the rated load is in the range of
    (a) 80 to 90%
    (b) 60 to 70%
    (c) 30 to 35%
    (d) 10 to 20%.
16. In case of S.I. engine, to have best thermal efficiency the fuel-air mixture ratio should be
    (a) lean
    (b) rich
    (c) may be lean or rich
    (d) chemically correct.
17. The fuel-air ratio, for maximum power of S.I. engines, should be
    (a) lean
    (b) rich
    (c) may be lean or rich
    (d) chemically correct.
18. In case of petrol engine, at starting
    (a) rich fuel-air ratio is needed
    (b) weak fuel-air ratio is needed
    (c) chemically correct fuel-air ratio is needed
    (d) any fuel-air ratio will do.
19. Carburettor is used for
    (a) S.I. engines
    (b) Gas engines
    (c) C.I. engines
    (d) none of the above.
20. Fuel injector is used in
    (a) S.I. engines
    (b) Gas engines
    (c) C.I. engines
    (d) none of the above.
21. Very high speed engines are generally
    (a) Gas engines
    (b) S.I. engines
    (c) C.I. engines
    (d) Steam engines.

22. In S.I. engine, to develop high voltage for spark plug
    (a) battery is installed
    (b) distributor is installed
    (c) carburettor is installed
    (d) ignition coil is installed.
23. In S.I. engine, to obtain required firing order
    (a) battery is installed
    (b) distributor is installed
    (c) carburettor is installed
    (d) ignition coil is installed.
24. For petrol engines, the method of governing employed is
    (a) quantity governing
    (b) quality governing
    (c) hit and miss governing
    (d) none of the above.
25. For diesel engines, the method of governing employed is
    (a) quantity governing
    (b) quality governing
    (c) hit and miss governing
    (d) none of the above.
26. Voltage developed to strike spark in the spark plug is in the range
    (a) 6 to 12 volts
    (b) 1000 to 2000 volts
    (c) 20000 to 25000 volts
    (d) none of the above.
27. In a 4-cylinder petrol engine the standard firing order is
    (a) 1-2-3-4
    (b) 1-4-3-2
    (c) 1-3-2-4
    (d) 1-3-4-2.
28. The torque developed by the engine is maximum
    (a) at minimum speed of engine
    (b) at maximum speed of engine
    (c) at maximum volumetric efficiency speed of engine
    (d) at maximum power speed of engine.
29. Iso-octane content in a fuel for S.I. engines
    (a) retards auto-ignition
    (b) accelerates auto-ignition
    (c) does not affect auto-ignition
    (d) none of the above.
30. Normal heptane content in fuel for S.I. engines
    (a) retards auto-ignition
    (b) accelerates auto-ignition
    (c) does not affect auto-ignition
    (d) none of the above.
31. The knocking in S.I. engines increases with
    (a) increase in inlet air temperature
    (b) increase in compression ratio
    (c) increase in cooling water temperature
    (d) all of the above.
32. The knocking in S.I. engines gets reduced
    (a) by increasing the compression ratio
    (b) by retarding the spark advance
    (c) by increasing inlet air temperature
    (d) by increasing the cooling water temperature.
33. Increasing the compression ratio in S.I. engines
    (a) increases the tendency for knocking
    (b) decreases tendency for knocking
    (c) does not affect knocking
    (d) none of the above.
34. The knocking tendency in petrol engines will increase when
    (a) speed is decreased
    (b) speed is increased
    (c) fuel-air ratio is made rich
    (d) fuel-air ratio is made lean.
35. The ignition quality of fuels for S.I. engines is determined by
    (a) cetane number rating
    (b) octane number rating
    (c) calorific value rating
    (d) volatility of the fuel.
36. Petrol commercially available in India for Indian passenger cars has octane number in the range
    (a) 40 to 50
    (b) 60 to 70
    (c) 80 to 85
    (d) 95 to 100.
37. Octane number of the fuel used commercially for diesel engine in India is in the range
    (a) 80 to 90
    (b) 60 to 80
    (c) 60 to 70
    (d) 40 to 45.

INTERNAL COMBUSTION ENGINES

38. The knocking tendency in C.I. engines increases with
   (a) decrease of compression ratio
   (b) increase of compression ratio
   (c) increasing the temperature of inlet air
   (d) increasing cooling water temperature.

39. Desirable characteristic of combustion chamber for S.I. engines to avoid knock is
   (a) small bore
   (b) short ratio of flame path to bore
   (c) absence of hot surfaces in the end region of gas
   (d) all of the above.

Answers

1. (c)	**2.** (d)	**3.** (b)	**4.** (b)	**5.** (a)	**6.** (c)	**7.** (a)
8. (c)	**9.** (b)	**10.** (a)	**11.** (a)	**12.** (a)	**13.** (b)	**14.** (c)
15. (c)	**16.** (a)	**17.** (b)	**18.** (a)	**19.** (a)	**20.** (c)	**21.** (b)
22. (d)	**23.** (b)	**24.** (a)	**25.** (a)	**26.** (c)	**27.** (d)	**28.** (c)
29. (a)	**30.** (b)	**31.** (d)	**32.** (b)	**33.** (b)	**34.** (a)	**35.** (b)
36. (c)	**37.** (d)	**38.** (a)	**39.** (d).			

THEORETICAL QUESTIONS

1. Name the two general classes of combustion engines and state how do they basically differ in principle ?
2. Discuss the relative advantages and disadvantages of internal combustion and external combustion engines.
3. What are the two basic types of internal combustion engines ? What are the fundamental differences between the two ?
4. What is the function of a governor ? Enumerate the types of governors and discuss with a neat sketch the Porter governor.
5. Differentiate between a flywheel and a governor.
6. (a) State the function of a carburettor in a petrol engine.
   (b) Describe a simple carburettor with a neat sketch and also state its limitations.
7. Explain with neat sketches the construction and working of the following :
   (i) Fuel pump,   (ii) Injector.
8. Explain the following terms as applied to I.C. engines :
   Bore, stroke, T.D.C., B.D.C., clearance volume, swept volume, compression ratio and piston speed.
9. Explain with suitable sketches the working of a four-stroke Otto engine.
10. Discuss the difference between ideal and actual valve timing diagrams of a petrol engine.
11. In what respects four-stroke diesel cycle (compression ignition) engine differs from four-stroke cycle spark ignition engine ?
12. Discuss the difference between theoretical and actual valve timing diagrams of a diesel engine.
13. What promotes the development of two-stroke engines ? What are the two main types of two-stoke engines.
14. Describe with a suitable sketch the two-stroke cycle spark ignition (SI) engine. How its indicator diagram differs from that of four-stroke cycle engine ?
15. Compare the relative advantages and disadvantages of four-stroke and two-stroke cycle engines.

5
Refrigeration and Air-Conditioning

5.1. Fundamentals of refrigeration—Introduction—Elements of refrigeration systems—Refrigeration systems—Co-efficient of performance (C.O.P.)—Standard rating of a refrigeration machine. 5.2. Simple vapour compression system—Introduction—Simple vapour compression cycle—Functions of parts of a simple vapour compression system. 5.3. Domestic refrigerator. 5.4. Vapour absorption system—Introduction—Simple vapour absorption system—Practical vapour absorption system—Comparison between vapour compression and vapour absorption systems. 5.5. Refrigerants—Classification of refrigerants—Desirable properties of an ideal refrigerant—Properties and uses of commonly used refrigerants. 5.6. Air-conditioning—Introduction—Air-conditioning systems—Applications of air conditioning—Highlights—Objective Type Questions—Theoretical Questions.

5.1. FUNDAMENTALS OF REFRIGERATION (August 2000, 2001)

5.1.1. Introduction

Refrigeration is the science of producing and maintaining temperatures below that of the surrounding atmosphere. This means the removing of heat from a substance to be cooled. Heat always passes downhill, from a warm body to a cooler one, until both bodies are at the same temperature. Maintaining perishables at their required temperatures is done by refrigeration. Not only perishables but to-day many human work spaces in offices and factory buildings are air-conditioned and a refrigeration unit is the heart of the system.

Before the advent of mechanical refrigeration water was kept cool by storing it in semi-porous jugs so that the water could seep through and evaporate. The evaporation carried away heat and cooled the water. This system was used by the Egyptians and by Indians in the Southwest. Natural ice from lakes and rivers was often cut during winter and stored in caves, straw-lined pits, and later in sawdust-insulated buildings to be used as required. The Romans carried pack trains of snow from Alps to Rome for cooling the Emperor's drinks. Though these methods of cooling all make use of natural phenomena, they were used to maintain a lower temperature in a space or product and may properly be called refrigeration.

In simple, *refrigeration means the cooling of or removal of heat from a system*. The equipment employed to maintain the system at a low temperature is termed as *refrigerating system* and the system which is kept at lower temperature is called *refrigerated system*. Refrigeration is generally produced in one of the following three ways :

  (i) by *melting of a solid*
  (ii) by *sublimation of a solid* and
  (iii) by *evaporation of a liquid.*

Most of the commercial refrigeration is produced by the evaporation of a liquid called *refrigerant*. **Mechanical refrigeration** depends upon the evaporation of liquid refrigerant and its circuit includes the equipments naming **evaporator, compressor, condenser** and **expansion valve.** It is used for preservation of food, manufacture of ice, solid carbon dioxide and control of air temperature and humidity in the air conditioning system.

REFRIGERATION AND AIR-CONDITIONING

Important refrigeration applications:
1. Ice making
2. Transportation of foods above and below freezing
3. Industrial air conditioning
4. Comfort air conditioning
5. Chemical and related industries
6. Medical and surgical aids
7. Processing food products and beverages
8. Oil refining and synthetic rubber manufacturing
9. Manufacturing and treatment of metals
10. Freezing food products
11. Miscellaneous applications:
    (i) Extremely low temperatures
    (ii) Building construction etc.

5.1.2. Elements of Refrigeration Systems

All refrigeration systems must include at least four basic units as given below:

(i) *A low temperature thermal "sink" to which heat will flow from the space to be cooled.*
(ii) *Means of extracting energy from the sink, raising the temperature level of this energy, and delivering it to a heat receiver.*
(iii) *A receiver to which heat will be transferred from the high temperature high-pressure refrigerant.*
(iv) *Means of reducing of pressure and temperature of the refrigerant as it returns from the receiver to the "sink".*

5.1.3. Refrigeration Systems

The various refrigeration systems may be enumerated as below:
1. Ice refrigeration
2. Air refrigeration system
3. Vapour compression refrigeration system
4. Vapour absorption refrigeration system
5. Special refrigeration systems:
   (i) Adsorption refrigeration system   (ii) Cascade refrigeration system
   (iii) Mixed refrigeration system   (iv) Vortex tube refrigeration system
   (v) Thermoelectric refrigeration   (vi) Steam jet refrigeration system.

5.1.4. Co-efficient of Performance (C.O.P.)

The performance of a refrigeration system is expressed by a term known as the *co-efficient of performance*, which is defined as the *ratio of heat absorbed by the refrigerant while passing through the evaporator to the work input required to compress the refrigerant in the compressor*; in short it is the *ratio between heat extracted and work done* (in heat units).

If, R_n = net refrigerating effect

W = work expanded in by the machine during the same interval of time

Then   C.O.P. = $\dfrac{R_n}{W}$

and $\quad$ Relative C.O.P. $= \dfrac{\text{actual C.O.P.}}{\text{theoretical C.O.P.}}$

where $\quad$ Actual C.O.P. = ratio of R_n and W *actually measured during a test*
and, Theoretical C.O.P. = ratio of *theoretical values* of R_n and W *obtained by applying laws of thermodynamics to the refrigeration cycle*.

5.1.5. Standard Rating of a Refrigeration Machine

The rating of a refrigeration machine is obtained by refrigerating effect or amount of heat extracted in a given time from a body. The rating of the refrigeration machine is given by a unit of refrigeration known as *standard commercial tonne of refrigeration* which is defined as the refrigerating effect produced by the melting of 1 tonne of ice from and at 0°C in 24 hours. Since the latent heat of fusion of ice is 336 kJ/kg, the refrigerating effect of 336 × 1000 kJ in 24 hours is rated as one *tonne*, i.e.,

$$1 \text{ tonne of refrigeration TR} = \dfrac{336 \times 1000}{24} = 14000 \text{ kJ/h.}$$

5.2. SIMPLE VAPOUR COMPRESSION SYSTEM $\hfill$ (March 1999, 2001)

5.2.1. Introduction

Out of all refrigeration systems, the vapour compression system is the most important system from the view point of *commercial* and *domestic utility*. It is the most practical form of refrigeration. In this system the *working fluid is a vapour*. It readily evaporates and condenses or changes alternately between the vapour and liquid phases without leaving the refrigerating plant. During evaporation, it absorbs heat from the cold body. This heat is used as its latent heat for converting it from the liquid to vapour. In condensing or cooling or liquifying, it rejects heat to external body, thus creating a cooling effect in the working fluid. This refrigeration system thus acts as a latent heat pump since it pumps its latent heat from the cold body or brine and rejects it or delivers it to the external hot body or cooling medium. The principle upon which the vapour compression system works apply to all the vapours for which tables of Thermodynamic properties are available.

5.2.2. Simple Vapour Compression Cycle

In a simple vapour compression system fundamental processes are completed in one cycle. These are:

1. Compression $\quad$ 2. Condensation $\quad$ 3. Expansion $\quad$ 4. Vapourisation.

The flow diagram of such a cycle is shown in Fig. 5.1.

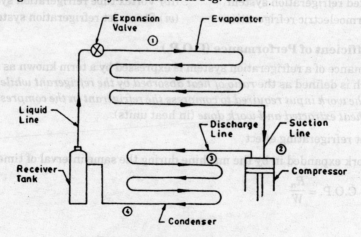

Fig. 5.1. Vapour compression system

The *vapour at low temperature and pressure* (state '2') enters the *"compressor"* where it is compressed isentropically and subsequently its temperature and pressure increase considerably (state '3'). This vapour after leaving the compressor enters the *"condenser"* where it is condensed into *high pressure liquid* (state '4') and is collected in a *"receiver tank"*. From receiver tank it passes through the *"expansion valve"*, here it is *throttled down to a lower pressure* and has a low temperature (state '1'). After finding its way through expansion "valve" it finally passes on to *"evaporator"* where it *extracts heat from the surroundings* or *circulating fluid being refrigerated and vapourises to low pressure vapour (state '2')*.

Merits and demerits of vapour compression system over Air refrigeration system :

Merits :

1. C.O.P. is quite high as the working of the cycle is very near to that of reversed Carnot cycle.

2. When used on ground level the running cost of vapour-compression refrigeration system is only 1/5th of air refrigeration system.

3. For the same refrigerating effect the size of the evaporator is smaller.

4. The required temperature of the evaporator can be achieved simply by adjusting the throttle valve of the same unit.

Demerits :

1. Initial cost is high.

2. The major disadvantages are *inflammability, leakage of vapours* and *toxity*. These have been overcome to a great extent by improvement in design.

5.2.3. Functions of Parts of a Simple Vapour Compression System

Here follows the brief description of various parts of a simple vapour compression system shown in Fig. 5.1.

1. Compressor. The function of a compressor is to remove the *vapour* from the evaporator, and to *raise its temperature and pressure to a point such that it* (vapour) *can be condensed with available condensing media*.

2. Discharge line (or **hot gas line**). A hot gas or discharge line *delivers the high-pressure, high-temperature vapour from the discharge of the compressor to the condenser*.

3. Condenser. The function of a condenser is to *provide a heat transfer surface through which heat passes from the hot refrigerant vapour to the condensing medium*.

4. Receiver tank. A receiver tank is used to provide *storage for a condensed liquid* so that a constant supply of liquid is available to the evaporator as required.

5. Liquid line. A liquid line carries the liquid refrigerant from the receiver tank to the refrigerant flow control.

6. Expansion valve (refrigerant flow control). Its function is to *meter the proper amount of refrigerant to the evaporator and to reduce the pressure of liquid entering the evaporator so that liquid will vapourize in the evaporator at the desired low temperature and take out sufficient amount of heat*.

7. Evaporator. An evaporator *provides a heat transfer surface through which heat can pass from the refrigerated space into the vapourizing refrigerant*.

8. Suction line. The suction line *conveys the low pressure vapour from the evaporator to the suction inlet of the compressor*.

5.3. DOMESTIC REFRIGERATOR

Refrigerators, these days, are becoming the common item for *house hold use, vendor's shop, hotels, motels, offices, laboratories, hospitals, chemists* and *druggists shops, studios etc*. They are manufactured in different size to meet the needs of various groups of people. They are usually rated with *internal gross volume* and the *freezer volume*. The freezer space is meant to preserve perishable products at a temperature much below 0°C such as fish, meat, chicken etc. and to produce ice and ice cream as well. The refrigerators in India are available in different sizes of various makes, *i.e.*, 90, 100, 140, 160, 200, 250, 380 litres of gross volume. The freezers are usually provided at top portion of the refrigerator space occupying around one-tenth to one-third of the refrigerator volume. In some refrigerators, freezers are provided at the bottom.

Construction :

A domestic refrigerator consists of the following two main parts :
1. The refrigeration system (containing compressor, condenser, capillary tube and evaporator).
2. The insulated cabinet.

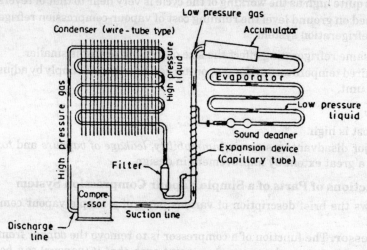

Fig. 5.2. Domestic refrigerator.

Fig. 5.2 shows a flow diagram of a typical refrigeration system used in a domestic refrigerator. A simple domestic refrigerator consists of a *hermetic compressor placed in the cabinet base. The condenser is installed at the back and the evaporator is placed inside the cabinet at the top.*

Working :

The working of the refrigerator is as follows :

- The low pressure and low temperature refrigerant vapour (*usually* R12) is drawn through the suction line to the compressor. The **accumulator** provided between the **suction line** and the evaporator collects liquid refrigerant coming out of the evaporator due to incomplete evaporation, if any, prevents it from entering the compressor. The **compressor** then compresses the refrigerant vapour to a high pressure and high temperature. The compressed vapour flows through the *discharge line* into condenser (vertical natural draft, wire-tube type).

- In the **condenser** the vapour refrigerant at high pressure and at high temperature is condensed to the liquid refrigerant at high pressure and low temperature.

- The high pressure liquid refrigerant then flows through the **filter** and then enters the **capillary tube** (expansion device). The capillary tube is attached to the suction line as shown in Fig. 5.2. The warm refrigerant passing through the capillary tube gives some of its heat to cold suction line vapour. This increases the heat absorbing quality of the liquid refrigerant slightly and increases the superheat of vapour entering the compressor.

The capillary tube expands the liquid refrigerant at high pressure to the liquid refrigerant at low pressure so that a measured quantity of liquid refrigerant is passed into the **evaporator.**

- In the **evaporator** the liquid refrigerant gets evaporated by absorbing heat from the container/articles placed in the evaporative chamber and is sucked back into the compressor and the cycle is repeated.

5.4. VAPOUR ABSORPTION SYSTEM   (August 2000, 2001)

5.4.1. Introduction

In a *vapour absorption system the refrigerant is absorbed on leaving the evaporator, the absorbing medium being a solid or liquid.* In order that the sequence of events should be continuous it is necessary for the refrigerant to be separated from the absorbent and subsequently condensed before being returned to the evaporator. The separation is accomplished by the application of direct heat in a *'generator'*. The solubility of the refrigerant and absorbent must be suitable and the plant which uses *ammonia as the refrigerant and water as absorbent* will be described.

5.4.2. Simple Vapour Absorption System

Refer to Fig. 5.3 for a simple absorption system. The solubility of ammonia in water at low temperatures and pressures is higher than it is at higher temperatures and pressures. The ammonia vapour leaving the evaporator at point 2 is readily absorbed in the low temperature hot solution in the absorber. This process is accompanied by the rejection of heat. The ammonia in water solution is pumped to the higher pressure and is heated in the generator. *Due to reduced solubility of ammonia in water at the higher pressure and temperature, the vapour is removed from the solution.* The vapour then passes to the condenser and the weakened ammonia in water solution is returned to the absorber.

In this system the *work done on compression is less than in vapour compression cycle* (since pumping a liquid requires much less work than compressing a vapour between the same pressures) but a heat input to the generator is required. The heat may be supplied by any convenient form *e.g.,* steam or gas heating.

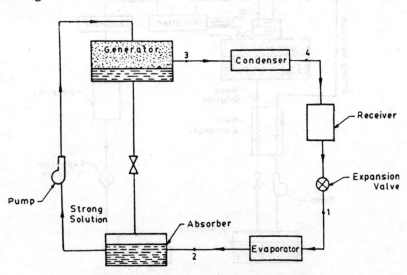

Fig. 5.3. (a) Simple vapour absorption system.

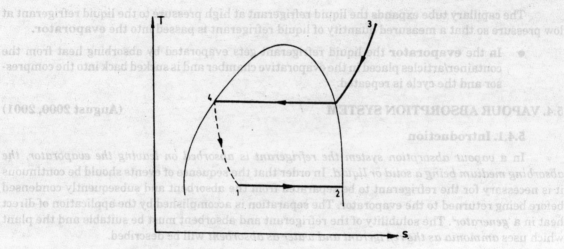

Fig. 5.3 (b) Simple vapour absorption system—T-s diagram.

5.4.3. Practical Vapour Absorption System

Refer Fig. 5.4. Although a simple vapour absorption system can provide refrigeration *yet its operating efficiency is low*. The following *accessories* are fitted to make the system more practical and improve the performance and working of the plant.

    1. Heat exchanger.   2. Analyser.   3. Rectifier.

    1. **Heat exchanger.** A heat exchanger is located between the generator and the absorber. The strong solution which is pumped from the absorber to the generator must be heated ; and the weak solution from the generator to the absorber must be cooled. This is accomplished by a heat exchanger and consequently *cost of heating the generator and cost of cooling the absorber are reduced.*

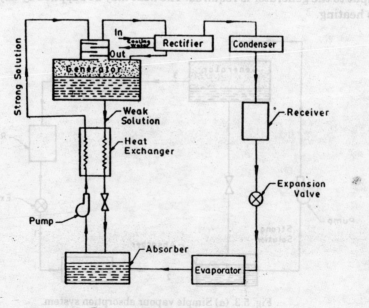

Fig. 5.4

REFRIGERATION AND AIR-CONDITIONING 175

2. **Analyser.** An analyser consists of a series of trays mounted above the generator. Its *main function is to remove partly some of the unwanted water particles associated with ammonia vapour going to condenser.* If these water vapours are permitted to enter condenser they may enter the expansion valve and freeze ; as a result the pipe line may get choked.

3. **Rectifier.** A rectifier is a water cooled heat exchanger *which condenses water vapour and some ammonia and sends back to the generator.* Thus *final reduction or elemination of the percentage of water vapour takes place in a rectifier.*

The co-efficient of performance (C.O.P.) of this system is given by :

$$\text{C.O.P.} = \frac{\text{heat extracted from the evaporator}}{\text{heat supplied in the generator} + \text{work done by the liquid pump}}.$$

5.4.4. Comparison between Vapour Compression and Vapour Absorption Systems

S. No.	Particulars	Vapour compression system	Vapour absorption system
1.	*Type of energy supplied*	Mechanical—a high grade energy	Mainly heat—a low grade energy
2.	*Energy supply*	Low	High
3.	*Wear and tear*	More	Less
4.	*Performance at part loads*	Poor	System not affected by variations of loads.
5.	*Suitability*	Used where high grade mechanical energy is available	Can also be used at remote places as it can work even with a simple kerosene lamp (of course in small capacities)
6.	*Charging of refrigerant*	Simple	Difficult
7.	*Leakage of refrigerant*	More chances	No chance as there is no compressor or any reciprocating component to cause leakage.
8.	*Damage*	Liquid traces in suction line may damage the compressor	Liquid traces of refrigerant present in piping at the exit of evaporator constitute no danger.

5.5. REFRIGERANTS

A *'refrigerant' is defined as any substance that absorbs heat through expansion or vapourisation* and loses it through condensation in a refrigeration system. The term 'refrigerant' in the broadest sense is also applied to such *secondary cooling mediums* as cold water or brine, solutions. Usually refrigerants include only those working medium which pass through the cycle of *evaporation, recovery, compression, condensation* and *liquification.* These substances absorb heat at one place at low temperature level and reject the same at some other place having higher temperature and pressure. *The rejection of heat takes place at the cost of some mechanical work.* Thus circulating cold mediums and a cooling mediums (such as ice and solid carbon dioxide) are not primary refrigerants. In the early days only four refrigerants, *Air, ammonia* (NH_3), *Carbon dioxide* (CO_2), *Sulphur dioxide* (SO_2), possessing chemical, physical and thermodynamic properties permitting their efficient application and service in the practical design of refrigeration equipment were used. All the refrigerants change from liquid state to vapour state during the process.

5.5.1. Classification of Refrigerants

The refrigerants are classified as follows :
1. Primary refrigerants.
2. Secondary refrigerants.

1. **Primary refrigerants** are those working mediums or heat carriers which *directly* take part in the refrigeration system and cool the substance by the absorption of latent heat *e.g., Ammonia, Carbon dioxide, Sulphur dioxide, Methyl chloride, Methylene chloride, Ethyl chloride and Freon group* etc.

2. **Secondary refrigerants** are those circulating substances which are first cooled with the help of the primary refrigerants and are then employed for cooling purposes, *e.g., ice, solid carbon dioxide* etc. These refrigerants cool substances by absorption of their sensible heat.

The primary refrigerants are grouped as follows :

(*i*) **Halocarbon compounds.** In 1928, Charles Kettening and Dr. Thomas Mighey invented and developed this group of refrigerant. In this group are included refrigerants which contain one or more of three halogens, chlorine and bromine and they are sold in the market under the names as *Freon, Genetron, Isotron*, and *Areton*. Since the refrigerants belonging to this group have outstanding merits over the other group's refrigerants, therefore they find wide field of application in domestic, commercial and industrial purposes.

The list of the halocarbon-refrigerants commonly used is given below :

R10 — Carbon tetrachloride (CCl_4)
R11 — Trichloro-monofluoro methane (CCl_3F)
R12 — Dichloro-difluoro methane (CCl_2F_2)
R13 — Mono-bromotrifluoro methane ($CBrF_3$)
R21 — Dichloro monofluoro methane ($CHCl_2F$)
R22 — Mono chloro difluoro methane ($CHClF_2$)
R30 — Methylene-chloride (CH_2Cl_2)
R40 — Methyle chloride (CH_3Cl)
R41 — Methyle fluoride (CH_3F)
R100 — Ethyl chloride (C_2H_5Cl)
R113 — Trichloro trifluoroethane ($C_2F_3Cl_3$)
R114 — Tetra-fluoro dichloroethane ($Cl_2F_4Cl_2$)
R152 — Difluoro-ethane ($C_2H_6F_2$)

(*ii*) **Azeotropes.** The refrigerants belonging to this group consists of mixtures of different substances. These substances cannot be separated into components by distillation. They possess fixed thermodynamic properties and do not undergo any separation with changes in temperature and pressure. An azeotrope behaves like a simple substance.

Example. R-500. It contains 73.8% of (R-12) and 26.2% of (R-152).

(*iii*) **Hydrocarbons.** Most of the refrigerants of this group are organic compounds. Several hydrocarbons are used successfully in commercial and industrial installation. Most of them possess satisfactory thermodynamic properties but are highly inflammable. Some of the important refrigerants of this group are :

R50 — Methane (CH_4)
R170 — Ethane (C_2H_6)
R290 — Propane (C_2H_8)
R600 — Butane (C_4H_{10})
R601 — Isobentane [$CH(CH_3)_3$]

REFRIGERATION AND AIR-CONDITIONING

(iv) **Inorganic compounds.** Before the introduction of hydrocarbon group these refrigerants were most commonly used for all purposes.

The important refrigerants of this group are :

- R717 — Ammonia (NH_3)
- R718 — Water (H_2O)
- R729 — Air (mixture of O_2, N_2, CO_2 etc.)
- R744 — Carbon dioxide (CO_2)
- R764 — Sulphur dioxide (SO_2)

(v) **Unsaturated organic compound.** The refrigerants belonging to this group possess ethylene or propylene as their constituents. They are :

- R1120 — Trichloroethylene ($C_3H_4Cl_3$)
- R1130 — Dichloroethylene ($C_2H_4Cl_2$)
- R1150 — Ethylene (C_3H_6)
- R1270 — Propylene.

5.5.2. Desirable Properties of an Ideal Refrigerant (March 1999)

An ideal refrigerant should possess the following properties :

1. Thermodynamic properties :
 (i) Low boiling point
 (ii) Low freezing point
 (iii) Positive pressures (but not very high) in condenser and evaporator.
 (iv) High saturation temperature
 (v) High latent heat of vapourisation.

2. Chemical Properties :
 (i) Non-toxicity
 (ii) Non-flammable and non-explosive
 (iii) Non-corrosiveness
 (iv) Chemical stability in reacting
 (v) No effect on the quality of stored (food and other) products like flowers, with other materials i.e., furs and fabrics.
 (vi) Non-irritating and odourless.

3. Physical Properties :
 (i) Low specific volume of vapour
 (ii) Low specific heat
 (iii) High thermal conductivity
 (iv) Low viscosity
 (v) High electrical insulation.

4. Other Properties :
 (i) Ease of leakage location
 (ii) Availability and low cost
 (iii) Ease of handling
 (iv) High C.O.P.
 (v) Low power consumption per tonne of refrigeration.
 (vi) Low pressure ratio and pressure difference.

Some important properties (mentioned above) are discussed below :

Freezing point. As the refrigerant must operate in the cycle above its freezing point, it is evident that the same for the refrigerant *must be lower than system temperatures*. It is found that except in the case of water for which the freezing point is 0°C, other refrigerants have reasonably low values. Water, therefore, can be used only in air-conditioning applications which are above 0°C.

Condenser and evaporator pressures. The evaporating pressure should be as near atmospheric as possible. If it is *too low*, it would result in *a large volume of the suction vapour*. If it is *too high, overall high pressures including condenser pressure would result necessitating stronger equipment and consequently higher cost. A positive pressure is required in order to eliminate the possibility of the entry of air and moisture into the system. The normal boiling point of the refrigerant should, therefore, be lower than the refrigerant temperature.*

Critical temperature and pressure. Generally, for high C.O.P. the *critical temperature should be very high so that the condenser temperature line on p-h diagram is far removed from the critical point.* This ensures reasonable refrigerating effect as it is very small with the state of liquid before expansion near the critical point.

The *critical pressure should be low so as to give low condensing pressure.*

Latent heat of vapourisation. It should be *as large as possible to reduce the weight of the refrigerant to be circulated in the system. This reduces initial cost of the refrigerant. The size of the system will also be small and hence low initial cost.*

Toxicity. Taking into consideration comparative hazard to life due to gases and vapours underwriters Laboratories have divided the compounds into *six groups*. Group six contains compounds with a very low degree of toxicity. It includes R_{12}, R_{114}, R_{13}, R_1, etc. Group one, at the other end of the scale, includes the most toxic substances such as SO_2.

Ammonia is not used in comfort air-conditioning and in domestic refrigeration because of inflammability and toxicity.

Inflammability. Hydrocarbons (*e.g.* methane, ethane etc.) are highly explosive and inflammable. Fluorocarbons are neither explosive nor inflammable. *Ammonia is explosive* in a mixture with air in concentration of 16 to 25% by volume of ammonia.

Volume of suction vapour. The *size of the compressor depends on the volume of suction vapour per unit* (say per tonne) *of refrigeration. Reciprocating compressors are used with refrigerants with high pressures and small volumes of the suction vapour. Centrifugal or turbo compressors are used with refrigerants with low pressures and large volumes of the suction vapour.* A high volume flow rate for a given capacity is required for centrifugal compressors to permit flow passages of sufficient width to minimise drag and obtain high efficiency.

Thermal conductivity. *For a high heat transfer co-efficient a high thermal conductivity is desirable.* R_{22} has better heat transfer characteristics than R_{12} ; R_{21} is still better, R_{13} has poor heat transfer characteristics.

Viscosity. For *a high heat transfer co-efficient a low viscosity is desirable.*

Leak tendency. The *refrigerants should have low leak tendency.* The greatest drawback of *fluorocarbons* is the fact that they are *odourless*. This, at times, results in a complete loss of costly gas from leaks without being detected. An ammonia leak can be very easily detected by pungent odour.

Refrigerant cost. The cost factor is only relevant to the extent of the price of the initial charge of the refrigerant which is very small compared to the total cost of the plant and its installation. The cost of losses due to leakage is also important. In small-capacity units requiring only a small charge of the refrigerant, the cost of refrigerant is immaterial.

The *cheapest refrigerant is Ammonia. R_{12} is slightly cheaper than R_{22}. R_{12}* and R_{22} have replaced ammonia in the dairy and frozen food industry (and even in cold storages) *because of the tendency of ammonia to attack some food products.*

Co-efficient of performance and power per tonne. Practically all common refrigerants have approximately same C.O.P. and power requirement.

Table 5.1. Gives the values of C.O.P. for some important refrigerants.

Table 5.1. C.O.P. of some important refrigerant

S. No.	Refrigerant	C.O.P.
1.	Carnot value	5.74
2.	R_{11}	5.09
3.	R_{113}	4.92
4.	Ammonia	4.76
5.	R_{12}	4.70
6.	R_{22}	4.66
7.	R_{114}	4.49
	CO_2	2.56

Action with oil. No chemical reaction between refrigerant and lubricating oil of the compressor should take place. Miscibility of the oil is quite important as some oil should be carried out of the compressor crank case with the hot refrigerant vapour to lubricate the pistons and discharge valves properly.

Reaction with materials of construction. While selecting a material to contain the refrigerant this material should be given a due consideration. Some metals are attacked by the refrigerants ; *e.g., ammonia reacts with copper, brass or other cuprous alloys in the presence of water,* therefore in *ammonia systems the common metals used are iron and steel. Freon group* does not react with steel, copper, brass, zinc, tin and aluminium but is corrosive to Magnesium and Aluminium having magnesium more than 2%. Freon group refrigerants tend to dissolve natural rubber in packing and gaskets but synthetic rubber such as neoprene are entirely suitable. The hydrogenerated hydrocarbons may react with zinc but not with copper, aluminium, iron and steel.

5.5.3. Properties and uses of Commonly used Refrigerants

1. Air

Properties :

(*i*) No cost involved ; easily available.

(*ii*) Completely non-toxic.

(*iii*) Completely safe.

(*iv*) The C.O.P. of air cycle operating between temperatures of 80°C and – 15°C is 1.67.

Uses :

(*i*) Air was one of the earliest refrigerants and was widely used even as late as World War I wherever a completely non-toxic medium was needed.

(*ii*) Because of low C.O.P., it is used only where *operating efficiency is secondary* as in *air craft refrigeration.*

2. Ammonia (NH_3)

Properties :

(*i*) It is highly toxic and flammable.

(ii) It has the excellent thermal properties.
(iii) It has the *highest refrigerating effect per kg of refrigerant*.
(iv) Low volumetric displacement.
(v) Low cost.
(vi) Low weight of liquid circulated per tonne of refrigeration.
(vii) High efficiency.
(viii) The evaporator and condenser pressures are 3.5 bar abs. and 13 bar abs. (app.) respectively at standard conditions of – 15°C and 30°C.

Uses :
(i) It is widely used in large industrial and commercial reciprocating compression systems where high toxicity is secondary.
It is extensively used in *ice plants, packing plants, large cold storages and skating rinks* etc.
(ii) It is widely used as the refrigerant in *absorption systems*.

Note :
(i) Ammonia *should never be used with copper, brass and other copper alloys ; iron and steel should be used in ammonia systems instead*.
(ii) In ammonia systems, to *detect the leakage a sulphur candle is used which gives off a dense white smoke when ammonia vapour is present*.

3. Sulphur dioxide (SO_2)

Properties :
(i) It is a colourless gas or liquid.
(ii) It is extremely toxic and has a pungent irritating odour.
(iii) It is non-explosive and non-flammable.
(iv) It has a liquid specific gravity of 1.36.
(v) Works at low pressures.
(vi) Possesses small latent heat of vapourisation.

Uses :
It finds little use these days. However its use was made in small machines in early days.
Note. The leakage of sulphur dioxide may be detected by bringing aqueous ammonia near the leak, this gives off a white smoke.

4. Carbon dioxide (CO_2)

Properties :
(i) It is a colourless and odourless gas, and is heavier than air.
(ii) It has liquid specific gravity of 1.56.
(iii) It is non-toxic and non-flammable.
(iv) It is non-explosive and non-corrosive.
(v) It has *extremely high operating pressures*.
(vi) It gives *very low refrigerating effect*.

Uses :
This refrigerant has received only limited use because of the high power requirements per tonne of refrigeration and the high operating pressures. In *former years* it was selected for *marine refrigeration*, for *theater air-conditioning systems*, and for *hotel and institutional refrigeration* instead of ammonia because it is non-toxic.

REFRIGERATION AND AIR-CONDITIONING

At the present-time its use is limited primarily to the *manufacture of dry ice* (solid carbon dioxide).

Note. The *leak detection of CO_2* is done by *soap solution*.

5. Methyl chloride (CH_3Cl)
Properties:

(i) It is a colourless liquid with a faint, sweet, non-irritating odour.
(ii) It has liquid specific gravity of 1.002 at atmospheric pressure.
(iii) It is neither flammable nor toxic.

Uses:

It has been used in the past in both domestic and commercial applications. It should never be used with Aluminium.

6. R11 (Trichloro monofluoro methane)
Properties:

(i) It is composed of one carbon, three chlorine and one fluorine atoms (or parts by weight) and is *non-corrosive, non-toxic* and *non-flammable*.
(ii) It dissolves natural rubber.
(iii) It has a boiling point of – 24°C.
(iv) It mixes completely with mineral lubricating oil under all conditions.

Uses:

It is employed for 50 tonnes capacity and over in small office buildings and factories. A centrifugal compressor is used in the plants employing this refrigerant.

Its leakage is detected by a halide torch.

7. R12 (Dichloro-difluoro methane) or Freon-12
Properties:

(i) It is *non-toxic, non-flammable*, and *non-explosive*, therefore it is *most suitable refrigerant*.
(ii) It is fully oil miscible therefore it simplifies the problem of oil return.
(iii) The operating pressures of R12 in evaporator and condenser under *standard tonne of refrigeration* are 1.9 bar abs. and 7.6 bar abs. (app.).
(iv) Its latent heat at – 15°C is 161.6 kJ/kg.
(v) C.O.P. = 4.61.
(vi) It does not break even under the extreme operating conditions.
(vii) It condenses at moderate pressure and under atmospheric conditions.

Uses:

1. It is suitable for high, medium and low temperature applications.
2. It is used for domestic applications.
3. It is *excellent electric insulator therefore it is universally used in sealed type compressors*.

8. R22 (Monochloro-difluro methane) or Freon-22

R22 refrigerant is superior to R12 in many respects. It has the following properties and uses:

Properties:

(i) The compressor displacement per tonne of refrigeration with R22 is 60% less than the compressor displacement with R12 as refrigerant.

(ii) R22 is miscible with oil at condenser temperature but tries to separate at evaporator temperature when the system is used for very low temperature applications (– 90°C). Oil separators must be incorporated to return the oil from the evaporator when the system is used for such low temperature applications.

(iii) The pressures in the evaporator and condenser at standard tonne of refrigeration are 2.9 bar abs. and 11.9 bar abs. (app.).

(iv) The latent heat at – 15°C is low and is 89 kJ/kg.

The *major disadvantage of R22 compared with R12 is the high discharge temperature which requires water cooling of the compressor head and cylinder.*

Uses :

R22 is universally used in commercial and industrial low temperature systems.

5.6. AIR-CONDITIONING

5.6.1. Introduction

"Air conditioning" is the simultaneous control of temperature, humidity, motion and purity of the atmosphere in confined space. Thus the important factors which are involved in a complete air conditioning installation are (i) Temperature control ; (ii) Humidity control ; (iii) Air movement and circulation and (iv) Air filtering, cleaning and purification. Complete air conditioning provides simultaneous control of these factors for both summer and winter. In addition to comfort phases of air conditioning many industries have found that air conditioning of their plants has made possible *more complete control of manufacturing processes and material and improves the quality of the finished products.*

The development of the central heating plant was an early step towards modern air conditioning. Another step was *development of automatic control* for *regulating the heating plant and providing* the proper humidity. The third step was the *development of automatic* refrigeration devices which could be employed for summer cooling and dehumidifying the air.

5.6.2. Air-Conditioning Systems

5.6.2.1. Introduction

An air-conditioning system is defined as an assembly of different parts of the system used to produce a specified condition of air within a required space or building.

The basic elements of air-conditioning systems (of whatever form) are :

1. **Fans :** for moving air.
2. **Filters :** for cleaning air, either fresh, recirculated or both.
3. **Refrigerating plant :** connected to heat exchange surface, such as finned coils or chilled water sprays.
4. **Means for warming :** the air, such as hot water or steam heated coils or electrical elements.
5. **Means for humidification ; and or dehumidification.**
6. **Control system :** to regulate automatically the amount of cooling or warming.

5.6.2.2. Air-conditioning cycle

Refer Fig. 5.5. An air-conditioning cycle comprises the following steps :

— The fan forces air into duct work which is connected to the openings in the room. These openings are commonly called *outlets* or *terminals*.

— The duct work directs the air to the room through the outlets.

REFRIGERATION AND AIR-CONDITIONING

— The air enters the room and either heats or cools as required. Dust particles from the room enter the air stream and are carried along with it.

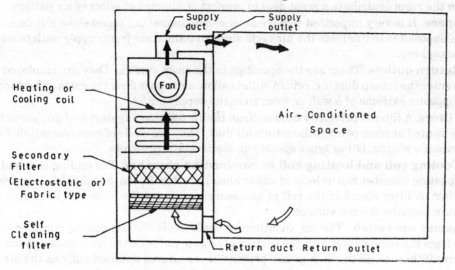

Fig. 5.5. Air-conditioning cycle.

— Air then flows from the room through a second outlet (sometimes called the return outlet) and enters the return duct work, where dust particles are removed by a *filter*.

— After the air is cleaned, it is either heated or cooled depending upon the condition in the room. If cool air is required, the air is passed over the surface of a *cooling coil* ; if warm air required, the air is passed through a *combustion chamber* or over the surface of a *heating coil*.

— Finally the air flows back to the fan, and the cycle is completed.

The main parts of the equipment in the air-conditioning cycle are :

(*i*) Fan (*ii*) Supply ducts
(*iii*) Supply outlets (*iv*) Space to be conditioned
(*v*) Return outlets (*vi*) Return ducts
(*vii*) Filter and (*viii*) Heating chamber or cooling coil.

1. Fan. The primary function of a fan is move air *to and from the room*. The air that a fan moves in air conditioning system is made up of :

(*i*) all outdoor air

(*ii*) all indoor room air (called recirculated air)

(*iii*) a combination of outdoor and indoor air.

The fans *pulls* air from outdoors or from the room but in most systems it *pulls* air from both sources at the same time. The amount of air supplied by the fan must be *regulated* since drafts in the room cause discomfort, and poor air movement slows the heat rejection processes. This can be achieved by (*i*) chosing a fan that can deliver the correct amount of air, and (*ii*) by controlling the speed of the fan so that air stream in the room provides good circulation but does not cause drafts. Of course, the fan is only one of the pieces of equipment that contributes body comfort.

2. Supply duct. The function of a supply duct is to direct the air from fan to the room. In order that air may flow freely it should be as short as possible and have minimum number of turns.

3. Supply outlets. The function of supply outlets is to distribute the air evenly in a room. These outlets may (*i*) fan the air, (*ii*) direct air in a jet stream and (*iii*) may do a combination of both.

Since supply outlets can either fan or jet the air stream, therefore they are able to exert some control on the direction of air delivered by the fan. This direction control plus the location and the number of outlets in the room contribute a great deal to comfort or discomfort effect of air pattern.

4. **Space.** It is very important to have an enclosed space (*i.e.*, room) since if it does not exist it would be impossible to complete the air cycle since contained air from supply outlets would flow into the atmosphere.

5. **Return outlets.** These are the openings in the room surface. They are employed to allow room air to enter the return duct (*i.e.* return outlets allow air to pass *from* the room). They are usually located at opposite extreme of a wall or room from the supply outlet.

6. **Filters.** A filter is primarily used to *clean the air by removing dust and dirt particles*. They are usually located at some point in the return air duct. They are made of many materials from *spun glass to composite plastic*. Other types operate on electrostatic principle.

7. **Cooling coil and heating coil or combustion chamber.** The cooling coil, and heating coil or combustion chamber can be located either ahead or after the fan, *but should always be located after the filter*. A filter ahead of the coil is necessary to prevent the excessive dirt, dust and dirt particles from covering the coil surface.

Summer operation. The air conditioning cycle *cools the air* during summer operation. Return air from the room passes over the surface of *cooling coil*, and the air is cooled to the required temperature. If there is too much moisture present, it is removed automatically as the air is cooled by the coil.

Winter operation. The air conditioning cycle *adds heat to the air* during winter operation. This is achieved by passing the return air from the room over the surface of a *heating coil* etc.

5.6.2.3. Air Conditioning Systems

The air conditioning systems are mainly classified as :

1. Central systems 2. Zoned systems
3. Unitary systems 4. Unitary central systems.

Another method of classification of air conditioning system is as follows :

1. Single-air systems 2. Dual-air systems
3. Primary-air systems 4. Unit systems
5. Panel systems.

1. Central system. This type of system is suitable for air conditioning large spaces such as *theatres, cinemas restaurants, exhibition halls, or big factory spaces where no sub-division exists*. The central systems, are generally employed for the loads above 25 TR and 2500 m^3/min of conditioned air. The unitary systems can be more economically employed for low capacity (below 25 TR) units.

In central system, the equipment such as fans, coils, filters and their encasement are designed for assembly in the field. A central system serves different rooms, requires individual control of each room. The condenser, compressor, dampers, heating, cooling and humidifying coils and fan are located at one place say basement. The conditioned air is carried to the different rooms by means of supply ducts and returned back to the control plant through return ducts. Part of the supply air to the rooms may be exhausted outdoors. Outdoor air enters from a intake which should be situated on that side of the building least exposed to solar heat. It should not be close to the ground or to dust collected roof. The air after passing through damper passes through filters. The filters may be of a mechanical cleaned type, replaceable-cell type or may be electrostatic. The cleaned air then passes to the conditioning equipment in the following order : Tempering (or preheater) coil, cooling coil, humidifier (Air washer), heating coil and finally fan. Fig. 5.6, shows a schematic diagram of complete (year round) air conditioning system.

REFRIGERATION AND AIR-CONDITIONING

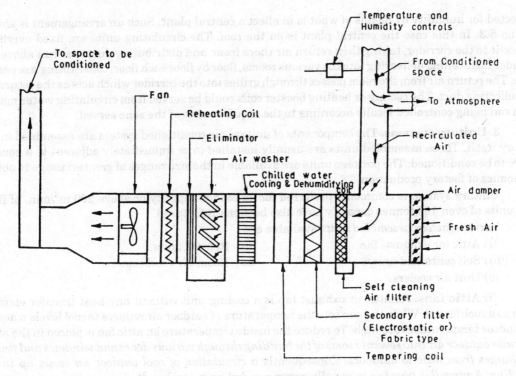

Fig. 5.6. Central system.

Advantages of central system :

1. Low investment cost as compared to total cost of separate unit.

2. Space occupied is unimportant as compared to a room unit conditioner which must be placed in the room.

3. Better accessibility for maintenance.

4. The running cost is less per unit of refrigeration.

5. Noise and vibration troubles are less to the people in air conditioned places as the air conditioning plant is far away from the air-conditioned places.

6. The exhaust air can be returned and partly reused with obvious saving in heating and refrigeration.

Disadvantages of central system :

The disadvantage of the central system is that it results in *large size ducts which are costly and occupy large space.*

2. Zoned system. When in a building several rooms or floors are to be served, it is necessary to consider means by which the varying heat gains in the different departments can be dealt with. Some rooms may have some, and others not ; some may be crowded and others empty ; again some may contain heat-producing equipment. Variations in requirements of this kind are the most common case with which air conditioning has to deal, and for this central system is unsuitable. *Zoned system is one approach.*

In zoned system the building is divided into zones such that as nearly as possible conditions may be expected to exist. Each zone is provided with its own local recirculating fan and booster cooler or heater, and this unit receives fresh air supply conditioned to some average temperature and corrected for humidity by means of what is in effect a central plant. Such an arrangement is shown in Fig. 5.3. In this case the central plant is on the roof. The circulating units are fixed overhead adjacent to the corridor, taking then return air there from, and distributing ducts are run above the corridor false ceiling delivering into the various rooms, floor by floor each floor, constituting a separate zone. The return air from the room passes through grilles into the corridor which acts as the returning air collecting duct. The cooling or heating booster coils could be served from circulating water mains, each coil being controlled locally according to the requirements of the zone served.

3. Unitary systems. The components of unitary air conditioned system are assembled in the factory itself. These assembled units are usually installed in or immediately adjacent to a zone or space to be conditioned. The package units are available in the size ranges of greatest usage to obtain economics of factory production.

Unitary system is commonly preferred for 15 tonnes capacity or above 200 m^3/min. of flow. The units of even 100 tonnes capacity have also been manufactured.

Various *factory assembled units* available are :

(*i*) Attic (or exhaust) fan. (*ii*) Remote units.
(*iii*) Self contained units. (*iv*) Room air conditioners.
(*v*) Unit air coolers.

(*i*) **Attic fans.** An attic or exhaust fan is a cooling unit without any heat transfer element such as a cooling coil. When the sun sets the temperature of outdoor air reduces to cool levels whereas the indoor temperatures are high. To reduce the inside temperature an attic fan is placed in the attic. It *draws outdoor air into several rooms of the building through various doors and windows and finally discharges from attic to outdoors.* Consequently *a circulation of cool outdoor air is set up in the building.* A *propeller type fan* is usually recommended as it can handle large volume of air at low pressure efficiently.

(*ii*) **Remote units.** A system in which air handling unit is separated from the condensing unit is called a *remote system*. The conditioning or air handling unit is called *remote unit*. It consists of a fan (either propeller or centrifugal type) with its driving motor, cooling coil, heating coil, filters, drip pan, louvers etc. with or without the duct connections at the outlet. Remote units are available in capacities ranging from 2 to 100 TR. These units are available for floor mounting or for suspension from ceiling. Some remote units have air-washing and coil wetting features.

(*iii*) **Self contained units.** In a self contained unit the condensing unit and other functional elements (such as coil and fans) are encased in the same cabinet. Fresh air can be introduced if required. The discharge from the casing may be free pressure type (*i.e.*, with or without duct work). Proper means should be adopted to cool the compressor.

(*iv*) **Room air conditioners.** Fig. 5.8 shows a unit air conditioner for mounting in a window or wall bracket. Unit air conditioners of small size generally have the condenser of the refrigerator air cooled but in large sizes the condenser may be water-cooled, in which case piping connections are required. Apart from this the only services needed are an electric supply and a connection to drain to conduct away any moisture condensed out of the atmosphere during dehumidification. Compressors in most units are *hermetic* and *therefore quiet in running.*

Units of considerable size are suitable for industrial applications, in which case ducting may be connected for distribution.

(*v*) **Unit air coolers.** A unit air cooler is a special application of remote units. It primarily reduces the temperature in insulated and sealed storage rooms. The rating of these coolers is the basic rating expressed in kJ per hour per C-degree temperature differential between the refrigerant and the air. A defrosting coil is necessary when the temperature is below 2°C. It may be mounted on the floor or wall or suspended from the ceiling.

REFRIGERATION AND AIR-CONDITIONING

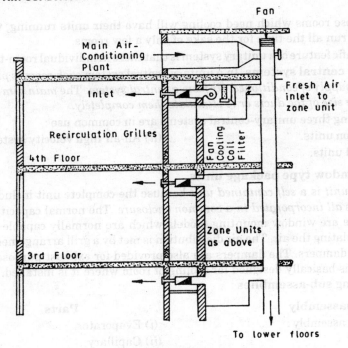

Fig. 5.7. Zoned system.

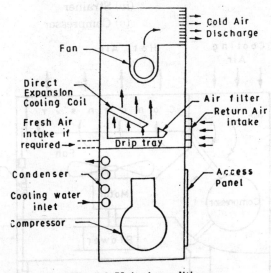

Fig. 5.8. Unit air conditioner.

Advantages of unitary system :

The unitary system commands following advantages over the central system.

1. There is saving in the installation and assembly labour charges.

2. Zoning and duct work eliminated.

3. In unitary system exact requirement of each separate room is met whereas in central system the individual needs of separate rooms cannot the met.

4. Failure of the unit puts off conditioning in only one room whereas the failure of the central plant off-sets all the rooms to be served.

188   BASIC MECHANICAL ENGINEERING

5. Only those rooms which need cooling will have their units running, whereas the central plant will have to run all the time for the sake of only a few rooms.

6. The specific feature of a unitary system is that there in individual room-temperature control.

4. Unitary central systems. In a unitary-central system *each room is provided with a room unit which gets a supply of conditioned air from a central system.* The *main aim of such systems it to either decrease the size of the ducts or to eliminate them completely.*

The following three unitary-central systems are in common use :

  (*i*) Induction units.   (*ii*) All-air high velocity systems.

  (*iii*) Fan coil units.

5.6.2.4. Window type package units   (April 2000)

A *package unit* is a *self contained unit* because the complete unit including evaporator and condensing unit is *all incorporated in a common enclosure*. The normal capacity of such a unit is 1 and 1.5 TR. There are window mounting models which are normally capable of cooling, heating, cleaning and circulating the air. The air distribution is met by a grill arrangement which also allows fresh air through dampers. The dampers are also provided for exhaust purposes. A **window type air-conditioner** is basically designed for cooling of room where it is installed. The entire systems consists of following **sub-assemblies** :

Subassembly	Parts
1. System assembly :	(*i*) Evaporator
	(*ii*) Capillary
	(*iii*) Condenser
	(*iv*) Strainer
	(*v*) Compressor

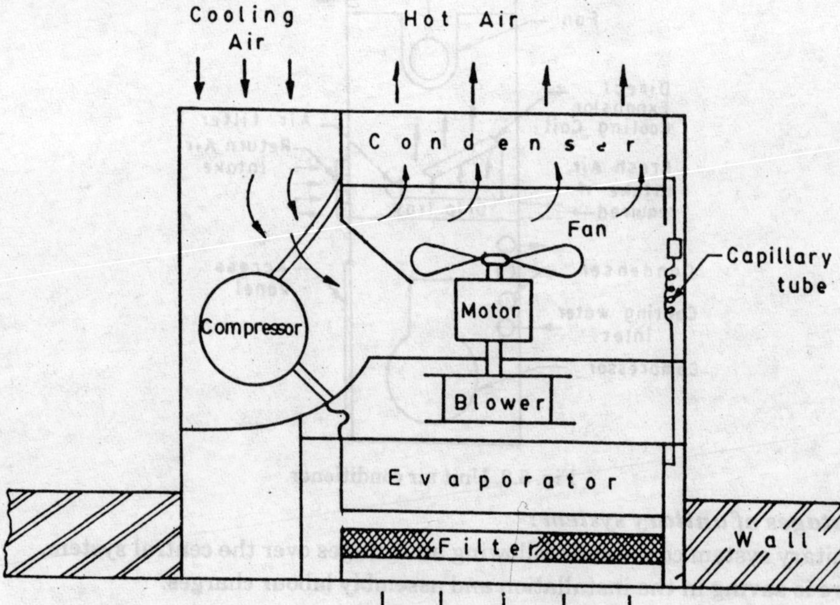

Fig. 5.9. Window type air conditioner.

2. Motor, fan and blower assembly
    (*i*) Fan
    (*ii*) Blower motor
    (*iii*) Motor mounting brackets
3. Cabinet and grill assembly
    (*i*) Cabinet
    (*ii*) Grill
4. Switch board panel assembly
    (*i*) Selector switch
    (*ii*) Relay
    (*iii*) Thermostat
    (*iv*) Fan motor capacitor
    (*v*) Running capacitor for compressor motor and starting capacitor for compressor motor.

Fig. 5.9 shows a schematic diagram of a **window type air-conditioner**.

5.6.3. Applications of Air-Conditioning

1. Industrial applications :
    (*i*) Food industry.
    (*ii*) Photographic industry.
    (*iii*) Textile industry.
    (*iv*) Printing industry.
    (*v*) Machine tool industry.
2. Theatres air conditioning.
3. Departmental-store air conditioning.
4. Transport air conditioning :
    (*i*) Automobile air-conditioning.
    (*ii*) Train air-conditioning.
    (*iii*) Air craft air-conditioning.
    (*iv*) Ship air-conditioning.
5. Air conditioning for television centre.
6. Air conditioning for computer centre.
7. Air conditioning of automatic telephone exchange building.
8. Museum air conditioning.
9. Hospital air-conditioning.

HIGHLIGHTS

1. Refrigeration is the science of producing and maintaining temperatures below that of the surrounding atmosphere.
2. Refrigeration is generally produced in one of the following three ways :
    (*i*) By melting a solid ;
    (*ii*) By sublimation of a solid and
    (*iii*) By evaporation of a liquid.
3. Co-efficient of performance (C.O.P.) is defined as the ratio of heat absorbed by the refrigerant while passing through the evaporator to the work input required to compress the refrigerant in the compressor ; in short it is the ratio between heat extracted and work done (in heat units).
4. Relative C.O.P. = $\dfrac{\text{actual C.O.P.}}{\text{theoretical C.O.P.}}$.
5. 1 tonne of refrigeration = 14000 kJ/h.
6. The main characteristic feature of air refrigeration system is that throughout the cycle the refrigerant remains in *gaseous state*.
    The air refrigeration system may be of two types :
    (*i*) Closed system and
    (*ii*) Open system.
7. Co-efficient of performance of a 'refrigerator' working on a reversal Carnot cycle
$$= \dfrac{T_2}{T_1 - T_2}$$

For a Carnot cycle 'heat pump'   C.O.P. = $\dfrac{T_1}{T_1 - T_2}$

8. In a simple vapour compression cycle the following processes are completed :
   (i) Compression   (ii) Condensation
   (iii) Expansion   (iv) Vapourisation.
9. The various parts of a simple vapour compression cycle are : Compressor, Discharge line (or hot gas line), Condenser, Receiver tank, Liquid line, Expansion valve, Evaporator and Suction line.
10. If the vapour is not superheated after compression, the operation is called 'Wet compression' and if the vapour is superheated at the end of compression, it is known as 'Dry compression'. Dry compression, in actual practice is always preferred as it gives higher volumetric efficiency and mechanical efficiency and there are less chances of compressor damage.
11. p-h chart gives directly the changes in enthalpy and pressure during a process for thermodynamic analysis.
12. When suction pressure is decreased, the refrigerating effect is decreased and work required is increased. The net effect is to reduce the refrigerating capacity of the system and the C.O.P.
13. The overall effect of superheating is to give a low value of C.O.P.
14. 'Subcooling' results in increase of C.O.P. provided that no further energy has to be spent to obtain the extra cold coolant required.
15. The refrigerating system should always be designed to operate at the highest possible vapourising temperature and lowest possible condensing temperature, of course, keeping in view the requirements of the application.
16. 'Volumetric efficiency' is defined as the ratio of the actual volume of gas drawn into the compressor (at evaporator temperature and pressure) on each stroke to the piston displacement. If the effect of clearance alone is considered, the resulting expression may be termed 'clearance volumetric efficiency'. The expression used for grouping into one constant all the factors affecting efficiency may be termed 'total volumetric efficiency'.
17. An air conditioning system is defined as an assembly of different parts of the system used to produce specified condition of air within a required space or building.
18. The basic elements of an air-conditioning system are : Fans, filters, refrigerating plant, means for warming the air, means for humidification and/or dehumidification ; and control system.
19. The main parts of the equipment in the air conditioning cycle are :
    Fan, supply ducts, supply outlets, space to be conditioned, return outlets, return ducts, filter and heating chamber or cooling coil.
20. The air conditioning systems are mainly classified as follows :
    (i) Central systems   (ii) Zoned systems
    (iii) Unitary systems   (iv) Unitary-central systems.
21. Air conditioning equipment primarily include package units and central units. Package unit may be of window or console type.
22. A package units is a self contained unit because the complete unit including evaporator and condensing unit is all incorporated in a common enclosure.
23. Central units are available in two types : horizontal and vertical depending upon position of filter and the drain pan.
24. Air conditioning components : Filters, fans, Air washer, radiator, convector etc.
25. Air conditioning controls may be manual, automatic and semi-automatic.
26. Some important control devices are : Thermostats, automatic humidity control, air movement control, automatic temperature control, limit switches and time clocks.
27. Air distribution system comprises of :
    (i) Air distribution system   (ii) Duct system
    (iii) Fan.
28. Draft is defined as any localized feeling of coolness or warmth of any portion of the body due to both air movement and air temperature with humidity and radiation considered constant.

29. Outlets may be classified as follows :
    (i) Grille outlets (ii) Slot diffuser outlets
    (iii) Ceiling diffuser outlets (iv) Perforated ceiling panels.
30. The supply ducts may be arranged in the following ways :
    (i) Loop perimeter duct system (ii) Radial perimeter duct system
    (iii) Extended plenum duct system.
31. Air distribution systems may be divided into three types :
    (i) Ejector system (ii) Downward system
    (iii) Upward system.

OBJECTIVE TYPE QUESTIONS

Fill in the blanks :

1. ...... means the cooling of or removal of heat from a system.
2. Most of the commercial refrigeration is produced by the evaporation of a liquid ......
3. ...... is the ratio between the heat extracted and the work done.
4. ...... = $\dfrac{\text{actual C.O.P.}}{\text{theoretical C.O.P.}}$
5. The C.O.P. for Carnot refrigerator is equal to ......
6. The C.O.P. for a Carnot heat pump is equal to .....
7. The C.O.P. for a Carnot refrigerator is ...... than that of Carnot heat pump.
8. The C.O.P. of an air refrigeration system is ...... than a vapour compression system.
9. In a refrigeration system the heat rejected at higher temperature = ... + ...
10. Out of all the refrigeration systems, the ...... system is the most important system from the stand point of commercial and domestic utility.
11. The function of a ...... is to remove the vapour from the evaporator and to raise its temperature and pressure to a point such that it (vapour) can be condensed with normally available condensing media.
12. The function of a ...... is to provide a heat transfer surface through which a heat passes from the hot refrigerant vapour to the condensing medium.
13. The function of ...... is to meter the proper amount of refrigerant to the evaporator and to reduce the pressure of liquid entering the evaporator so that liquid will vapourize in the evaporator at the desired low temperature.
14. ...... provides a heat transfer surface through which heat can pass from the refrigerated space or product into the vapourizing refrigerant.
15. If the vapour is not superheated after compression, the operation is called .....
16. If the vapour is superheated at the end of compression, the operation is called ......
17. When the suction pressure decreases the refrigerating effect and C.O.P. are ......
18. ...... results in increase of C.O.P. provided that no further energy has to be spent to obtain the extra cold coolant required.
19. ...... efficiency is defined as the ratio of actual volume of gas drawn into the compressor (at evaporator temperature and pressure) on each stroke to the piston displacement.
20. The simultaneous control of temperature, humidity, motion and purity of the atmosphere in a confined space is called ......
21. The art of measuring the moisture content of air is termed ......
22. ...... is the temperature of air as registered by an ordinary thermometer.
23. The difference between the dry bulb and wet bulb temperature is called ......
24. ...... is the ratio of the partial pressure of water vapour in the mixture to the saturated partial pressure at the dry bulb temperature, expressed as percentage.
25. The heat that does not affect the temperature but changes the state of a substance when added to or subtracted from it is called ......

26. The ratio of mass of water vapour associated with unit mass of water vapour associated with saturated unit mass of dry air is called ......
27. Air undergoes sensible ...... whenever it passes over a surface that is at a temperature less than the dry bulb temperature of the air but greater than dew point temperature.
28. The ratio of the room sensible heat to the sum of room sensible heat and room latent heat is called ......
29. ...... is the ratio of total sensible heat to the grand total heat that the cooling or the conditioning apparatus should handle.
30. An assembly of different parts of the system used to produce specified condition of air with a required space or building is called ......
31. ...... are used for moving air.
32. ...... are employed for cleaning air.
33. The function of a ...... is to direct the air from fan to the room.
34. ...... distribute the air evenly in a room.
35. ...... is a cooling unit without any heat transfer element such as a cooling coil.
36. A system in which air handling unit is separated from condensing unit is called a ...... system.
37. ...... unit uses the principle of induction as a means of recirculation.
38. In ...... system zoning and duct work is eliminated.
39. In ...... system each room is provided with a room unit which gets a supply of conditioned air from a central system.
40. An opening through which air is supplied to the treated space is called ......
41. ...... is a grille provided with a damper.
42. ...... is an outlet grille designed to guide the direction of the air.

Answers

1. Refrigeration
2. Refrigerant
3. C.O.P.
4. Relative C.O.P.
5. $\dfrac{T_2}{T_1 - T_2}$
6. $\dfrac{T_1}{T_1 - T_2}$
7. Less
8. Less
9. Refrigeration effect + work done
10. Vapour compression
11. Compressor
12. Condenser
13. Expansion valve
14. Evaporator
15. Wet compression
16. Dry compression
17. Reduced
18. Subcooling
19. Volumetric
20. Air conditioning
21. Psychrometry
22. DBT
23. Wet bulb depression
24. Relative humidity
25. Latent heat
26. Degree of saturation
27. Cooling
28. RSHF
29. GSHF
30. Air conditioning system
31. Fans
32. Filters
33. Supply duct
34. Supply outlets
35. Attic fan
36. Remote
37. Induction
38. Unitary
39. Unitary-central system
40. Outlet
41. Register
42. Diffuser.

THEORETICAL QUESTIONS

1. Define the following :
    (i) Refrigeration
    (ii) Refrigerating system
    (iii) Refrigerated system.
2. Enumerate different ways of producing refrigeration.
3. Enumerate important refrigeration applications.
4. State elements of refrigeration systems.
5. Enumerate systems of refrigeration.

6. Define the following:
    (i) Actual C.O.P.
    (ii) Theoretical C.O.P.
    (iii) Relative C.O.P.
7. What is a standard rating of a refrigeration machine?
8. Describe a simple vapour compression cycle giving clearly its flow diagram.
9. State the functions of the following parts of a simple vapour compression system:
    (i) Compressor,
    (ii) Condenser,
    (iii) Expansion valve and
    (iv) Evaporator.
10. Show the vapour compression cycle on 'Temperature-Entropy' (T-s) diagram for the following cases:
    (i) When the vapour is dry and saturated at the end of compression.
    (ii) When the vapour is superheated after compression.
    (iii) When the vapour is wet after compression.
11. What is the difference between 'Wet compression' and 'Dry compression'?
12. Write a short note on 'Pressure Enthalpy (p-h) chart'.
13. Show the simple vapour compression cycle on a p-h chart.
14. Discuss the effect of the following on the performance of a vapour compression system:
    (i) Effect of suction pressure
    (ii) Effect of delivery pressure
    (iii) Effect of superheating
    (iv) Effect of subcooling of liquid
    (v) Effect of suction temperature and condenser temperature.
15. Show with the help of diagrams, the difference between theoretical and actual vapour compression cycles.
16. Define the terms 'Volumetric efficiency' and 'Clearance volumetric efficiency'.
17. Derive an expression for 'Clearance volumetric efficiency'.
18. Explain briefly the term 'Total volumetric efficiency'.
19. Explain briefly simple vapour absorption system.
20. Give the comparison between a vapour compression system and a vapour absorption system.
21. List the desirable properties of a good refrigerant.
22. Define 'Air-conditioning'.
23. Define an 'air conditioning system'. Name its basic elements.
24. Explain with a neat sketch an 'air conditioning cycle'.
25. Enumerate the main parts of the equipment in the conditioning cycle.
26. How are air conditioning systems classified?
27. Explain with neat diagram the working of central system of air conditioning.
28. Write a short note on zone system of air conditioning.
29. Define a 'Unitary system'. Where is it commonly preferred? Explain a room air conditioner with a neat sketch.
30. State the advantages of central system over unitary system of air conditioning.

REFRIGERATION AND AIR CONDITIONING

5. Define the following:
   (i) Actual C.O.P.   (ii) Theoretical C.O.P.
   (iii) Relative C.O.P.
6. What is a standard rating of a refrigeration machine ?
7. Describe a simple vapour compression cycle giving clearly its flow diagram.
8. State the functions of the following parts of a simple vapour compression system
   (i) Compressor.   (ii) Condenser.
   (iii) Expansion valve and   (iv) Evaporator.
9. Show the vapour compression cycle on a Temperature-Entropy (T-s) diagram for the following cases
   (i) When the vapour is dry and saturated at the end of compression.
   (ii) When the vapour is superheated after compression.
   (iii) When the vapour is wet after compression.
10. What is the difference between 'Wet compression' and 'dry compression' ?
11. Write a short note on 'Pressure-Enthalpy (p-h) chart'.
12. Show the simple vapour compression cycle on a p-h chart.
13. Discuss the effect of the following on the performance of a vapour compression system
    (i) Effect of suction pressure.   (ii) Effect of delivery pressure.
    (iii) Effect of super-heating.   (iv) Effect of subcooling of liquid.
    (v) Effect of suction temperature and condenser temperature.
14. Show with the help of diagrams, the difference between theoretical and actual vapour compression cycles.
15. Define the terms 'Volumetric efficiency' and 'Clearance volumetric efficiency'.
16. Derive an expression for the 'Absolute volumetric efficiency'.
17. Explain in brief the term 'Total volumetric efficiency'.
18. Explain briefly 'simple vapour absorption system'.
19. Give the comparison between a vapour compression system and a vapour absorption system.
20. List the desirable properties of a good refrigerant.
21. Define 'Air-conditioning'.
22. Define an 'air-conditioning system'. Name its basic elements.
23. Explain with a neat sketch an 'air conditioning cycle'.
24. Enumerate the subdivisions of the equipment in the air-conditioning cycle.
25. How are air conditioning systems classified ?
26. Explain with a neat diagram the working of central system of air conditioning.
27. Write a short note on 'unit system of air conditioning'.
28. Define a 'Unitary system'. Where is it commonly preferred ? Explain a room air conditioner with a neat sketch.
29. State the advantages of central system over unitary system of air conditioning.

MODULE – 3

Chapter :
6. Power Transmission

6

Power Transmission

6.1. Introduction. 6.2. Belts and belt drives—Flat belts—V-belts—Round belts—Belt drive—Applications of belt drives—Velocity ratio of belt drive—Length of belt—Power transmitted by a belt—Ratio of tensions—Centrifugal tension—Condition for transmission of maximum (absolute) power—Initial tension—V-belt and rope drive. 6.3. Chains and chain drives—Roller chain drive—Silent chain drive—Advantages and disadvantages of chain drive. 6.4. Gear drive—Introduction—Advantages and disadvantages of toothed gearing—Definitions—Types of gears—Types of gear trains—Simple gear train—Compound gear train—Epicyclic (or planetary) gear train—Highlights—Objective Type Questions—Theoretical Questions—Unsolved Examples.

6.1. INTRODUCTION

The power from one shaft to another can be transmitted by the following means :
1. Belts and ropes
2. Chains
3. Gears
4. Clutches.

- *Belts and ropes* are used when the distance between the shaft centres is *large*.
- *Chains* and used when the distance between shaft centres is large and no slip is permitted.
- *Gears* are employed when the shaft distance is *adequately less*.

The use of *clutches* is restored to *when the shafts are co-axial*.

6.2. BELTS AND BELT DRIVES

A belt is a continuous band of flexible material passing over pulleys to transmit motion from one shaft to another.

Belts are available :

(*i*) with a narrow *rectangular cross-section*—**Flat belts** [Fig. 6.1 (*i*)].

(*ii*) with a *trapezoidal cross-section*—**V-belts** [Fig. 6.1 (*ii*)] and multiple **V-belts** [Fig. 6.1 (*iv*)].

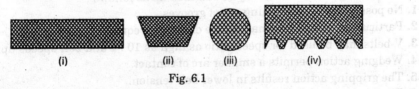

Fig. 6.1

(*iii*) *Round cross-section*—**Round belts** [Fig. 6.1 (*iii*)]

Chiefly used in machinery are *flat and V-belts*.

6.2.1. Flat Belts

- Flat belts are used for their simplicity and because they are subjected to minimum bending stress on the pulleys. The load capacity of flat belts is varied by varying their width, and only one is used in each drive. They are made of *leather, rubber, textile, balata and steel*.

- *Leather belts* have the best *pulling capacity*. Because of high cost of leather they are used very rarely.
- *Rubber belts* made of rubber on a cotton-duck base are used where the belt is exposed to the weather or steam, as they do not absorb moisture so readily as leather. They get destroyed if kept in contact with oil or grease.
- *Textile belts* are made of cotton and are used for rough and short service.
- *Balata belts* are acid and water proof and cannot withstand temperature higher than 100°C.
- *Steel belts* are claimed to transmit more horse power per cm width, and to remain unaffected by dampness or heat and be immune from stretching and slipping. The pulleys on which they are mounted do not have camber. Steel belts are sometimes used, the belt being subjected to considerable initial tension, to maintain the pressure on the pulley, on which the friction depends.

Note. The pulley of the flat belts is made *convex* at centre. This feature of the pulley is called **camber** or *crown* and due to it the lateral displacement belt is prevented.

6.2.2. V-Belts
(August 2000)

A V-belt is a belt of trapezoidal section running on pulleys with grooves cut to match the belt. The normal angle between the sides of the groove is 40 deg. Fig. 6.2 (*a, b*).

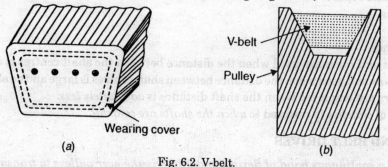

Fig. 6.2. V-belt.

- V-belts are usually made of fabric coated with rubber. They are silent and resilient. They are used when the distance between the shafts is too short for flat-belt drives. Owing to the *wedge action* between the belt and the sides of the groove in the pulley, the V-belt is *less likely to slip*, hence *more power can be transmitted for the same belt tension*.

The **advantages of V-belt,** may be summed up as follows :

1. No possibility of belt coming out of grooves.
2. Particularly suited for small centre distances requiring no idler.
3. V-belts may be used for speed ratio as high as 10 : 1 and belt speeds upto 2100 m/min.
4. Wedging action permits a smaller arc of contact.
5. The gripping action results in lower belt tension.
6. Power output can be increased by use of multiple belts.
7. In case of multiple-belt drive, if one belt fails, the machine does not come to a stop.
8. As V-belts are made endless the splicing problem is eliminated.
9. V-belts offer a *more positive drive because of reduced slippage.*
10. As these can be used over small pulleys, large reductions in speed are possible in a single drive.
11. V-belt drive may be inclined at any operating angle, slack side, top or bottom.

POWER TRANSMISSION

12. Drives are quiet at high speed.
13. The drive is capable of absorbing high shock.
14. Standardisation of V-belts results in better initial installation and replacement.

6.2.3. Round Belts

Round belts are employed to *transmit low power, mainly in instruments, table-type machine tools, machinery of the clothing industry and household appliances*. Round belts are used singly, as a rule. They may be made of *leather, canvas* and *rubber*. The diameter range is from 3 to 12 mm, usually from 4 to 8 mm. The minimum allowable ratio of the diameter of smaller pulley to the belt diameter is about 20, the recommended ratio is 30.

6.2.4. Belt Drive

A belt drive consists of the driving and driven pulleys and the belt which is mounted on the pulleys with a certain amount of tension and transmits peripheral force by friction.

- Belt drives may be :
(i) Open belt drive
(ii) Crossed belt drive.

Open belt drives [Fig. 6.3 (a)] are applied, as a rule, between parallel shafts which *rotate in the same direction*. Here the belt is subject to tension and bending.

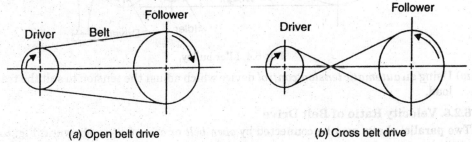

(a) Open belt drive   (b) Cross belt drive
Fig. 6.3. Belt drives.

In *crossed belt drives* [Fig. 6.3 (b)] the power is transmitted between small shafts *rotating in opposite direction*. Since *the angle of contact in this type of drive is more*, it can transmit *more power than open belt drive*. However there is *more wear and tear of the belt in this drive*.

6.2.5. Applications of Belt Drives

The main applications of belt drives are :

(i) To transmit power from low or medium capacity electric motors to operative machines.
(ii) To transmit power from small prime movers (internal combustion engines) to electric generators, agricultural and other machinery.

Advantages and disadvantages of belt drives :

Advantages :
(i) They can transmit motion over medium distances.
(ii) They give smooth operation (they cushion shocks and are silent).
(iii) They can operate at high speeds of rotation.
(iv) Their cost is relatively low.

Disadvantages :
(i) Their considerable overall size, usually several times larger than toothed gearing.
(ii) The inevitability of some elastic slipping (creep) of the belt.
(iii) The necessity for belt tensioning devices.

(*iv*) The necessity to keep oil from getting on the belt.

(*v*) The relatively short service life in high speed drives.

A primary condition in the operation of belt drives is proper belt tension which should be maintained in service.

Tensioning is accomplished by one of the following methods :

(*i*) Moving one of the pulleys to increase the distance between them.

(*ii*) Using an *idler pulley* which can be periodically adjusted, or one that automatically maintains constant tension by means of a weighted arm (Fig. 6.4) or by spring action.

(*iii*) Pretensioning the belt in mounting it on the pulleys and taking up slack after it stretches by shortening and replacing the belt (this is the least reliable method and is practically obsolete).

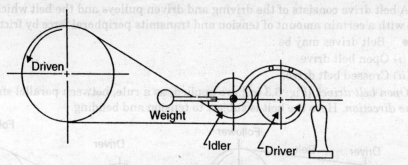

Fig. 6.4. Idler pulley.

(*iv*) Using an *automatic tension-control* device which adjust the tension to suit the transmitted load.

6.2.6. Velocity Ratio of Belt Drive

Two parallel shafts may be connected by *open belt* or *cross belt* as shown in Fig. 6.5 (*a*) and (*b*) respectively.

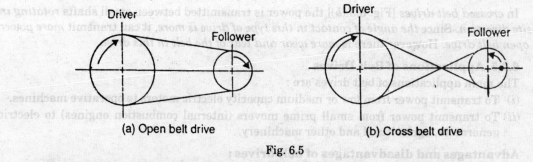

(a) Open belt drive   (b) Cross belt drive

Fig. 6.5

The velocity ratio between the two shafts depends upon the diameters of the respective pulleys. One of the pulleys may be called a *driver* and the other a *follower*.

In an *open belt drive* both the pulleys rotate in the *same direction*, whereas in a *crossed belt-drive* they rotate in *opposite direction*.

Let $d_1 (= 2r_1)$ = diameter of the driver pulley,

$d_2 (= 2r_2)$ = diameter of the driven pulley or follower,

ω_1 = *angular velocity of the driver*,

ω_2 = *angular velocity of the follower*,

POWER TRANSMISSION

N_1 = revolutions per minute (r.p.m.) of the driver, and
N_2 = revolutions per minute (r.p.m.) of the follower.

(*i*) *Assuming thickness of the belt to be negligible and no slip between the belt and the pulley,* the peripheral speed of the pulleys must be the same.

Velocity of the belt, $\quad v = \omega_1 r_1 = \omega_1 r_2$

$\therefore \qquad \dfrac{\omega_2}{\omega_1} = \dfrac{r_1}{r_2} = \dfrac{d_1}{d_2}$

$$\dfrac{2\pi N_2}{2\pi N_1} = \dfrac{d_1}{d_2} \quad \text{or} \quad \dfrac{N_2}{N_1} = \dfrac{d_1}{d_2} \qquad \ldots(6.1)$$

$\dfrac{N_2}{N_1}$ is called the *velocity ratio* of the rotating pulleys, in a belt drive, the speed of a pulley is *inversely proportional to its diameter.*

(*ii*) *Assuming no slip between the belt and the pulley but considering thickness of the belt :*

Let $\qquad t$ = thickness of the belt

Assuming no slip $\quad v = \omega_1\left(r_1 + \dfrac{t}{2}\right) = \omega_2\left(r_2 + \dfrac{t}{2}\right)$

or $\qquad \dfrac{\omega_2}{\omega_1} = \dfrac{r_1 + t/2}{r_2 + t/2} = \dfrac{d_1 + t}{d_2 + t} \quad \text{or} \quad \dfrac{N_2}{N_1} = \dfrac{d_1 + t}{d_2 + t} \qquad \ldots(6.2)$

(*iii*) *Considering slip between the belt and the pulley (and thickness negligible)*

Slip, in general, may be defined as the *relative motion between the pulley and belt*. The difference between the linear speeds of the pulley rim and belt is the measure of slip. One special reason for slip to occur is that there is a tendency for the belt to carry with it on the underside, between the pulley and the belt, a *thin layer of air which reduces the friction.* When friction is reduced the belt cannot take up more load. When load more than the permissible one is at once brought on the belt, the belt slips and may sometimes come off one of the pulleys. Assuming the belt remains on the pulley and slip occurs :

Let S_1 = percentage slip between driver and the belt, and
$\quad S_2$ = percentage slip between belt and follower.

Speed of the belt = $\omega_1 r_1 \left(\dfrac{100 - S_1}{100}\right)$

Peripheral speed of the follower = $\omega_2 r_2$

$\qquad = \left(\dfrac{100 - S_2}{100}\right) \times$ speed of the belt

i.e., $\qquad \omega_2 r_2 = \left(\dfrac{100 - S_2}{100}\right) \times \omega_1 r_1 \times \left(\dfrac{100 - S_1}{100}\right)$

or $\qquad \dfrac{\omega_2}{\omega_1} = \dfrac{N_2}{N_1}\left(\dfrac{100 - S_2}{100}\right) \times \left(\dfrac{100 - S_1}{100}\right) \times \dfrac{r_1}{r_2}$

or $\qquad \dfrac{N_2}{N_1} = \left(\dfrac{100 - S_2}{100}\right) \times \left(\dfrac{100 - S_1}{100}\right) \times \dfrac{d_1}{d_2} \qquad \ldots(6.3)$

If S = total percentage slip between the driver and follower,

$$\dfrac{N_2}{N_1} = \left(\dfrac{100 - S}{100}\right) \times \dfrac{d_1}{d_2} \qquad \ldots[6.3(a)]$$

(iv) Considering slip as well as thickness of the belt, we have

$$\frac{N_2}{N_1} = \left(\frac{100-S}{100}\right) \times \left(\frac{d_1+t}{d_2+t}\right) \qquad \ldots(6.4)$$

Example 6.1. *An engine shaft running at 240 r.p.m. is required to drive a machine shaft by means of a belt. The pulley on the engine shaft is of 1.5 metres diameter and that of the machine shaft is of 0.75 metre. If the belt thickness is 5 mm, find the speed of the belt when :*

(i) *There is no slip, and*   (ii) *There is slip of 2 per cent.*

Solution.

Speed of the shaft, $N_1 = 240$ r.p.m.
Diameter of the pulley on the engine shaft, $d_1 = 1.5$ m
Diameter of the pulley on the machine shaft, $d_2 = 0.75$ m
Thickness of the belt, $t = 5$ mm $= 0.005$ m

Speed of the machine shaft, N_2 :

(i) N_2 *with no slip :*

We know that, $\dfrac{N_2}{N_1} = \dfrac{d_1+t}{d_2+t}$ or $\dfrac{N_2}{240} = \dfrac{1.5+0.005}{0.75+0.005}$

or $N_2 = 240 \times \dfrac{1.505}{0.755} = \textbf{478.4 r.p.m. (Ans.)}$

(ii) N_2 *with a slip of 2 per cent :*

We know that, $\dfrac{N_2}{N_1} = \dfrac{d_1+t}{d_2+t}\left(\dfrac{100-S}{100}\right)$ or $\dfrac{N_2}{240} = \dfrac{1.5+0.005}{0.75+0.005}\left(\dfrac{100-2}{100}\right)$

or $N_2 = 240 \times \dfrac{1.505}{0.755} \times \dfrac{98}{100} = \textbf{468.8 r.p.m. (Ans.)}$

6.2.7. Length of Belt

Following two cases will be considered :

1. Open belt   2. Cross belt.

1. Open belt. Refer to Fig. 6.6.

Let 2α = angle subtended,

       d = distance between L and M, the centres of the two pulleys, and

       r_1, r_2 = respective radii of the pulleys.

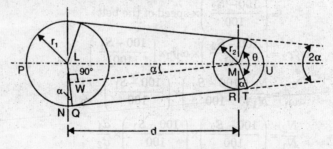

Fig. 6.6

Draw MW parallel to QT

$$\angle NLQ = \angle TMR = \angle LMW = \alpha$$

POWER TRANSMISSION

$$\angle PLQ = \frac{\pi}{2} + \alpha$$

$$\angle TMU = \frac{\pi}{2} - \alpha, \quad LW = r_1 - r_2$$

$$\sin \alpha = \frac{r_1 - r_2}{d}$$

Let l = length of the belt,

$$l = 2\,[\text{arc } PQ + QT + \text{arc } TU]$$

$$= 2\left[r_1\left\{\frac{\pi}{2} + \alpha\right\} + d\cos\alpha + r_2\left(\frac{\pi}{2} - \alpha\right) \right]$$

$$= 2\left[\frac{\pi}{2}(r_1 + r_2) + \alpha(r_1 - r_2) + d\cos\alpha \right]$$

$$= 2\left[\frac{\pi}{2}(r_1 - r_2) + \alpha(r_1 - r_2) + d\sqrt{1 - \sin^2\alpha} \right]$$

$$= 2\left[\frac{\pi}{2}(r_1 + r_2) + \alpha(r_1 - r_2) + \left\{d^2 - (r_1 - r_2)^2\right\}^{1/2} \right]$$

But $[d^2 - (r_1 - r_2)^2]^{1/2} = d\left\{1 - \left(\frac{r_1 - r_2}{d}\right)^2\right\}^{1/2} = d\left\{1 - \frac{(r_1 - r_2^2)}{2d^2} + \ldots\ldots\right\}$

α being very small, it can be replaced by $\sin \alpha = \dfrac{r_1 - r_2}{d}$

$$l = 2\left[\frac{\pi}{2}(r_1 + r_2) + \frac{(r_1 - r_2)^2}{d} + d\left\{1 - \frac{(r_1 - r_2)^2}{2d^2}\right\} \right] app.$$

$$= \pi(r_1 + r_2) + 2 \times \frac{(r_1 - r_2)^2}{d} + 2d - \frac{(r_1 - r_2)^2}{d} = \pi(r_1 + r_2) + \frac{(r_1 + r_2)^2}{d} + 2d$$

Hence length of the open belt is given by :

$$l = \pi(r_1 + r_2) + \frac{(r_1 - r_2)^2}{d} + 2d \qquad \ldots(6.5)$$

2. Cross belt. Refer to Fig. 6.7.

$$\angle LOQ = \alpha = \angle NLQ = \angle TMR$$

$$\angle PLQ = \frac{\pi}{2} + \alpha, \quad \angle TMU = \frac{\pi}{2} + \alpha$$

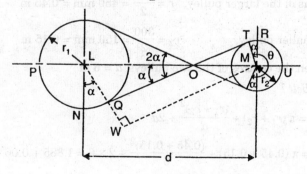

Fig. 6.7

Using the same notation as for open belt and drawing a line MW parallel to QT and producing LQ to meet MW at W, the length of the belt,

$$l = 2\,[\text{arc } PQ + QT + \text{arc } TU]$$

$$= 2\left[r_1\left(\frac{\pi}{2}+\alpha\right) + d\cos\alpha + r_2\left(\frac{\pi}{2}+\alpha\right)\right]$$

$$= 2\left[(r_1+r_2)\left(\frac{\pi}{2}+\alpha\right) + d\cos\alpha\right]$$

$$= 2\left[(r_1+r_2)\left(\frac{\pi}{2}+\alpha\right) + \sqrt{d^2-(r_1+r_2)^2}\right]$$

$$= 2\left[\frac{\pi}{2}(r_1+r_2) + \alpha(r_1+r_2) + \sqrt{d^2-(r_1+r_2)^2}\right]$$

Since $\quad \alpha = \sin\alpha = \dfrac{r_1+r_2}{d} \quad$ (because α is very small)... from $\triangle LMW$

But $\sqrt{d^2-(r_1+r_2)^2} = \{d^2-(r_1+r_2)^2\}^{1/2}$

$$= d\left\{1-\frac{(r_1+r_2)^2}{d^2}\right\}^{1/2} = d\left\{1-\frac{(r_1+r_2)^2}{2d^2}+\ldots\right\} = d - \frac{(r_1+r_2)^2}{2d}$$

Substituting the above value, we get

$$l = 2\left[\frac{\pi}{2}(r_1+r_2) + \frac{(r_1+r_2)^2}{d} + d - \frac{(r_1+r_2)^2}{2d}\right]$$

$$= \left[\pi(r_1+r_2) + \frac{(r+r_2)^2}{d} + 2d\right]$$

Hence length of the cross belt is given by

$$l = \pi(r_1+r_2) + \frac{(r_1+r_2)^2}{d} + 2d \qquad \ldots(6.6)$$

It is clear, from the above expression that the length of cross belt is constant if the sum of the radii of the corresponding pulleys is constant, this conclusion is *very useful* in determining the size of speed cones or stepped pulleys.

Example 6.2. *Two parallel shafts 6 m apart are provided with 900 mm and 300 mm diameter pulleys and are connected by means of a cross belt. The direction of rotation of the follower pulley is to be reversed by changing over to an open belt drive. How much length of the belt has to be reduced ?*

Solution. Radius of the larger pulley, $r = \dfrac{900}{2} = 450$ mm $= 0.45$ m

Radius of the smaller pulley, $\quad r_2 = \dfrac{300}{2} = 150$ mm $= 0.15$ m

Distance between the centres of the two pulleys, $d = 6$ m

Length of cross belt :

$$l_{cross} = \pi(r_1+r_2) + \frac{(r_1+r_2)^2}{d} + 2d$$

$$= \pi(0.45+0.15) + \frac{(0.45+0.15)^2}{6} + 2\times 6 = 1.885 + 0.06 + 12 = 13.945 \text{ m}$$

POWER TRANSMISSION

Length of open belt:

$$l_{open} = \pi(r_1 + r_2) + \frac{(r_1 - r_2)^2}{d} + 2d$$

$$= \pi(0.45 + 0.15) + \frac{(0.45 - 0.15)^2}{6} + 2 \times 6 = 1.885 + 0.015 + 12 = 13.9 \text{ m}$$

Hence the length of the belt to be reduced (to change the direction of rotation of the follower pulleys)

$$= l_{cross} - l_{open} = 13.945 - 13.9 = 0.045 \text{ m or } \textbf{45 mm (Ans.)}$$

6.2.8. Power Transmitted By a Belt

Fig. 6.8 shows a driving pulley (*i.e.*, driver) L and the driven pulley (*i.e.*, follower) M. The driving pulley pulls the belt from one side and delivers the same to the other. It is thus obvious, that the tension in the former side (*i.e.*, tight side) will be more than that in the latter side (*i.e.*, slack side) as shown in Fig. 6.8.

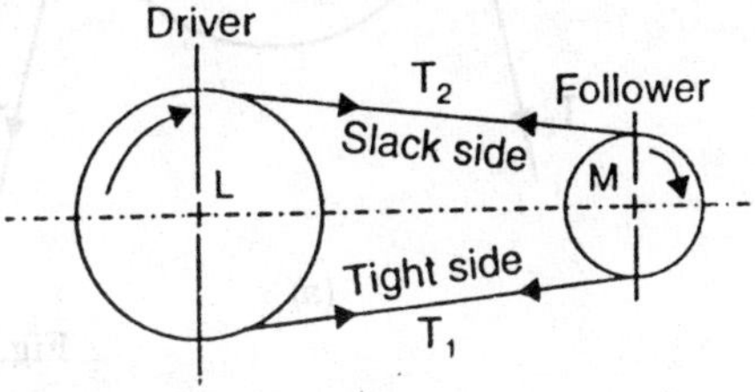

Let T_1 = tension in the tight side, N,
   T_2 = tension in the slack side, N, and
   v = velocity of the belt, m/s.

Now, effective turning force = $T_1 - T_2$

Work done = force × distance = $(T_1 - T_2) v$ Nm/s

and power = $(T_1 - T_2)v$ J/s = $(T_1 - T_2) v$ watts ...(6.7)

Fig. 6.8

Example 6.3. *The tensions in the two sides of the belt are 1500 N and 1200 N respectively. If the speed of the belt is 80 m/s, find the power transmitted by the belt.*

Solution. Tension on the tight side, T_1 = 1500 N
Tension on the slack side, T_2 = 1200 N
Speed of the belt, v = 80 m/s
Power transmitted by the belt, $P = (T_1 - T_2) v$ watts
   = (1500 - 1200) × 80 = 24000 Nm/s
   = 24000 W = **24 kW (Ans.)**

6.2.9. Ratio of Tensions

Let us consider a belt resting over the flat rim of a pulley which is not allowed to rotate (Fig. 6.9). Let a tension T_2 be applied at one end of the belt and the tension to the other end is steadily increased till its value is T_1 and the belt is about to slip over the rim. Such a state, when the belt tends to slip over is said to be in *limiting equilibrium*. T_1 is greater than T_2 on account of frictional resistance between the belt and the rim. It is required to establish a relation between T_1, T_2 ; μ (the co-efficient of friction between the belt and the rim and θ (the angle of contact or angle of lap).

Out of the belt ML, subtending an angle θ at the centre consider an elementary length QS subtending an angle $\delta\theta$ at the centre. Let T be the tension at the extremity Q and $(T + \delta T)$ at the other extremity S. The various forces acting on the elementary length of the belt are :

1. Tension T acting tangentially at Q
2. Tension $(T + \delta T)$ acting tangentially at S
3. The normal reaction R exerted by the rim
4. The frictional resistance μR acting against the tendency to slip and perpendicular to R.

Resolving all forces *horizontally* and equating the same,

$$T \cos \frac{\delta\theta}{2} + \mu R = (T + \delta T) \cos \frac{\delta\theta}{2}$$

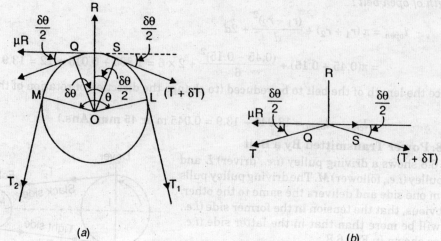

Fig. 6.9. Ratio of tensions.

or
$$T \cos \frac{\delta\theta}{2} + \mu R = T \cos \frac{\delta\theta}{2} + \delta T \cos \frac{\delta\theta}{2}$$

or
$$\mu R = \delta T \cos \frac{\delta\theta}{2}$$

Since $\frac{\delta\theta}{2}$ is very small, $\cos \frac{\delta\theta}{2} = 1$

∴
$$\mu R = \delta T \qquad \ldots(i)$$

Now, resolving all forces *vertically* and equating the same,

$$R = T \sin \frac{\delta\theta}{2} + (T + \delta T) \sin \frac{\delta\theta}{2}$$

$$= T \sin \frac{\delta\theta}{2} + T \sin \frac{\delta\theta}{2} + \delta T \sin \frac{\delta\theta}{2} = 2T \sin \frac{\delta\theta}{2}$$

$$\left(\text{since } \delta T \sin \frac{\delta\theta}{2} \text{ is negligible being very small} \right)$$

$$= 2T \cdot \frac{\delta\theta}{2} \qquad \left[\because \sin \frac{\delta\theta}{2} = \frac{\delta\theta}{2}, \delta\theta \text{ being very small} \right]$$

$$R = T.\delta\theta \qquad \ldots(ii)$$

Dividing (i) by (ii), we get $\mu = \dfrac{\delta T}{T \cdot \delta\theta}$

∴ $\mu \delta\theta = \dfrac{\delta T}{T}$

Integrating between the limits 0 to θ, and from T_2 to T_1

$$\int_0^\theta \mu \, d\theta = \int_{T_2}^{T_1} \frac{dT}{T} \quad \text{or} \quad \mu\theta = \log_e \frac{T_1}{T_2}$$

or
$$\frac{T_1}{T_2} = e^{\mu\theta} \qquad \ldots(6.8)$$

POWER TRANSMISSION

Note:
1. In the above expression θ is the angle of contact at the *smaller pulley*.

2. In an open belt drive, the angle of contact, $\theta = 180° - 2\alpha$ where $\alpha = \sin^{-1}\left(\dfrac{r_1 - r_2}{d}\right)$

3. In a cross-belt drive, the angle of contact, $\theta = 180° + 2\alpha$ where $\alpha = \sin^{-1}\left(\dfrac{r_1 - r_2}{d}\right)$

Example 6.4. *A pulley is driven by a flat belt running at a speed of 500 m / min. The co-efficient of friction between the pulley and the belt is 0.3, and the angle of lap is 160°. If the maximum tension in the belt is 700 N, find the power transmitted by the belt.*

Solution. Speed of the belt, $\quad v = 500$ m/min.
Co-efficient of friction, $\quad \mu = 0.3$
Angle of contact, $\quad \theta = 160° = 160 \times \pi/180 = 2.79$ radians.
Maximum tension in the belt, $\quad T_1 = 700$ N.

Power transmitted, P :

We know that the ratio of tension is given by,

$$\dfrac{T_1}{T_2} = e^{\mu\theta} \quad \text{or} \quad \dfrac{700}{T_2} = e^{0.3 \times 2.79} = 2.309$$

$$\therefore \quad T_2 = \dfrac{700}{2.309} = 303.2 \text{ N}$$

Hence, *power transmitted by the belt,*

$$P = (T_1 - T_2)\, v$$
$$= (700 - 303.2) \times \dfrac{500}{60} \text{ Nm/s} = 3306 \text{ W} = \textbf{3.306 kW (Ans.)}$$

Example 6.5. *Two pulleys of diameters 300 mm and 750 mm mounted on two parallel shafts 1.5 m apart, are connected by leather belt 150 mm wide. If the maximum safe tension of the belt is 14 N / mm width, determine the maximum power transmitted in case of (i) Open belt drive (ii) Cross-belt drive.*

Assume the speed of the belt as 540 m / min and the co-efficient of friction between the belt and the pulley rim, 0.25.

Solution. Radius of the longer pulley, $\quad r_1 = \dfrac{750}{2} = 375$ mm = 0.375 m.

Radius of the smaller pulley, $\quad r_2 = \dfrac{300}{2} = 150$ mm = 0.15 m.

Distance between the centres of the shafts, $d = 1.5$ m
Maximum safe tension $\quad = 14$ N/mm
Width of the leather belt $= 150$ mm
$\therefore$ Maximum tension, $T_1 = 14 \times 150 = 2100$ N
Speed of the belt, $\quad v = 540$ m/min.
Co-efficient of friction, $\mu = 0.25$

Maximum power transmitted, P :

(*i*) *Open belt drive* :

We know that $\qquad \dfrac{T_1}{T_2} = e^{\mu\theta}$ ...(*i*)

Let us first calculate the value of θ, the angle of lap or contact.

$$\sin\alpha = \frac{r_1 - r_2}{d} \quad\ldots\ldots \text{ in case of open belt drive}$$

or
$$\alpha = \sin^{-1}\left(\frac{0.375 - 0.15}{1.5}\right) = 8.6°$$

Now, $\theta = 180° - 2\alpha = 180° - 2 \times 8.6 = 162.8°$

Substituting various values in (i), we get

$$\frac{2100}{T_2} = e^{0.25 \times (162.8 \times \pi/180)} = 2.035$$

$$\therefore \quad T_2 = \frac{2100}{2.035} = 1031.9 \text{ N}$$

Hence maximum power transmitted, $P = (T_1 - T_2)v$

$$= (2100 - 1031.9) \times \frac{540}{60} \text{ Nm/s} = 9613 \text{ J/s}$$

$$= 9613 \text{ W} = \textbf{9.613 kW. (Ans.)}$$

(ii) Cross-belt drive :

In this case,
$$\sin\alpha = \frac{r_1 + r_2}{d} = \left(\frac{0.375 + 0.15}{1.5}\right) = 0.35 \quad \text{or} \quad \alpha = \sin^{-1}(0.35) = 20.5°$$

Now, $\theta = 180° + 2\alpha = 180° + 2 \times 20.5° = 221°$

Using the relation : $\dfrac{T_1}{T_2} = e^{\mu\theta}$

$$\frac{2100}{T_2} = e^{0.25 \times (221° \times \pi/180)} = 2.62$$

$$\therefore \quad T_2 = \frac{2100}{2.62} = 801.5 \text{ N}$$

Hence maximum power transmitted, $P = (T_1 - T_2)v$

$$= (2100 - 801.5) \times \frac{540}{60} = 11686 \text{ Nm/s} = 11686 \text{ W} = \textbf{11.686 kW. (Ans.)}$$

Note. It may be noted that the power transmitted by the crossed belt is *more* than the open belt.

6.2.10. Centrifugal Tension

Whenever a belt (or a rope) rotates in a circular path around the pulleys, because of its having mass, and speed, it is subjected to centrifugal force, which tends to lift the belt (or rope) from the rim, thereby reducing the normal reaction and hence the frictional resistance. This reduces the driving power capacity. The centrifugal force *produces additional tension* in the belt which is known as the **centrifugal tension**. At low speeds the centrifugal tension is negligible, but at speeds more than 750 m/min. it becomes very important factor.

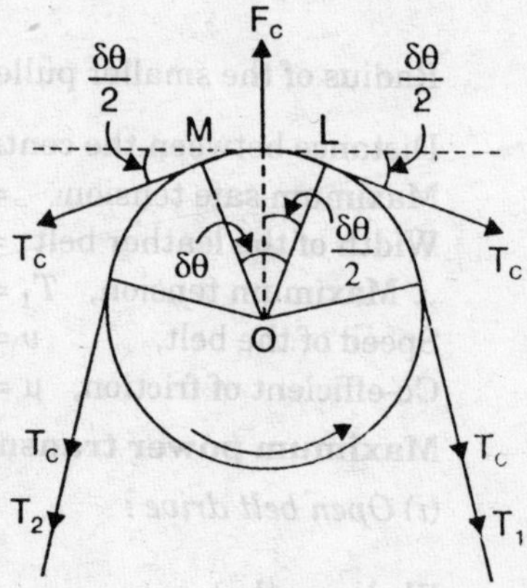

Consider a small portion LM of the belt shown in Fig. 6.10.

Let $\quad m$ = mass of the belt per unit length,

v = linear velocity of the belt,

r = radius of the pulley over which the belt runs,

T_c = centrifugal tension acting tangentially at L and M, and

$\delta\theta$ = angle subtended by the belt LM at the centre of the pulley.

Fig. 6.10. Centrifugal tension.

POWER TRANSMISSION

∴ Length of the belt $LM = r \cdot \delta\theta$

and mass $\qquad M = m \cdot r \cdot \delta\theta$

We know that centrifugal force of the belt LM,

$$F_c = \frac{M \cdot v^2}{r} = \frac{(m \cdot r \cdot \delta\theta) v^2}{r} = m \cdot \delta\theta \cdot v^2$$

Resolving the forces (*i.e.*, F_c and T_c) *vertically* and equating the same,

$$2T_c \sin\left(\frac{\delta\theta}{2}\right) = m \cdot \delta\theta \cdot v^2$$

Since $\delta\theta$ is very small, therefore substituting $\sin\left(\frac{\delta\theta}{2}\right) = \frac{\delta\theta}{2}$ in the above equation, we get

$$2T_c \left(\frac{\delta\theta}{2}\right) = m \cdot \delta\theta \cdot v^2$$

∴ $\qquad T_c = mv^2$ ...(6.9)

From the above expression of T_c, we find that the centrifugal tension in the elementary strip is independent of the angle of lap, and the original tension. Therefore, the same tension holds good for the two ends of the entire belt. There are already T_1 and T_2 tensions at the ends.

∴ Total tension on the light side = $(T_1 + T_c)$

Total tension on the slack side = $T_2 + T_c$

The effect of centrifugal tension is to reduce the driving power of the belt.

Example 6.6. *A pulley running at 200 r.p.m. drives a belt 250 mm × 6 mm. The diameter of the pulley is 900 mm and the density of the belt material is 1100 kg/m³. The permissible tension of the belt is 2 MN/m² and the friction tension ratio is 2, determine:*

(*i*) *The centrifugal tension* (*ii*) *Power transmitted in kW.*

Solution. Width of the belt, $\quad b = 250$ mm $= 0.25$ m

Thickness of the belt, $\quad t = 6$ mm $= 0.006$ m

Radius of the pulley, $\quad r = \dfrac{900}{2} = 450$ mm $= 0.45$ m

Density of the belt material, $\quad \rho = 1100$ kg/m³

Permissible tension of the belt $= 2$ MN/m²

Ratio of tensions, $\quad \dfrac{T_1}{T_2} = 2$

(*i*) **Centrifugal tension, T_c:**

Maximum safe tension of the belt,

$$T = 2 \times 10^6 \times b \times t = 2 \times 10^6 \times 0.25 \times 0.006 = 3000 \text{ N}$$

The centrifugal tension, $\quad T_c = mv^2$

Now, $\quad m$ = (mass of the belt per metre length)

$\qquad$ = area × length × density of the belt material

$\qquad$ = $(b \times t) \times (1) \times 1100 = 0.25 \times 0.006 \times 1 \times 1100 = 1.65$ kg/m

and $\quad v$ = velocity of the belt in m/s

$$= 2\pi \left(r + \frac{t}{2}\right) \times \frac{N}{60} = 2\pi \left(0.45 + \frac{0.006}{2}\right) \times \frac{200}{60} = 9.48 \text{ m/s}$$

Substituting these values, we get

$$\mathbf{T_c} = 1.65 \times (9.48)^2 = \mathbf{148.3 \text{ N}} \quad \textbf{(Ans.)}$$

(ii) Power transmitted in kW:

Total tension on the tight side $= T_1 + T_c$ and this should not be more than T

$\therefore \qquad T = T_1 + T_c$

$3000 = T_1 + 148.3$ or $T_1 = 2851.7$ N

Also $\qquad \dfrac{T_1}{T_2} = 2 \qquad$ (given)

or $\qquad T_2 = \dfrac{T_1}{2} = \dfrac{2851.7}{2} = 1425.8$ N

$\therefore$ Power transmitted $\quad P = (T_1 - T_2)v$

$\qquad = (2851.7 - 1425.8) \times 9.48$ Nm/s (or J/s or W)

$\qquad = 13517$ W $= \mathbf{13.5}$ **kW** **(Ans.)**

Example 6.7. *A flat belt is required to transmit 35 kW from a pulley of 1.5 m effective diameter running at 300 r.p.m. The angle of contact is spread over $\dfrac{11}{24}$ of the circumference and the co-efficient of friction between belt and pulley surface is 0.3. Taking centrifugal tension into account, determine the width of the belt. Take belt thickness as 9.5 mm, density 1100 kg/m³ and permissible stress as 2.5 MN/m².*

Solution. Power required to be transmitted, $P = 35$ kN

Effective diameter of the pulley, $\quad d = 1.5$ m

Speed of the pulley, $\quad N = 300$ r.p.m.

Angle of contact, $\quad \theta = \dfrac{11}{24} \times 2\pi = 2.88$ rad.

Co-efficient of friction, $\quad \mu = 0.3$

Thickness of the belt, $\quad t = 9.5$ mm $= 0.0095$ m

Density of the belt material, $\quad \rho = 1100$ kg/m³

Permissible stress, $\quad \sigma = 2.5$ MN/m²

Width of the belt, b :

Let $\qquad T_1$ = tension on the tight side of the belt, and

$\qquad T_2$ = tension on the slack side of the belt.

We know that velocity of the belt,

$$v = \dfrac{\pi d N}{60} \left(\text{or } \dfrac{2\pi r N}{60} \right) = \dfrac{\pi \times 1.5 \times 300}{60} = 23.56 \text{ m/s}$$

Also power transmitted (P),

$35 = (T_1 - T_2)v = (T_1 - T_2) \times 23.56$

$\therefore \qquad T_1 - T_2 = \dfrac{35}{23.56} = 1.486$ kN $= 1486$ N $\qquad$ ...(i)

We know that, $\qquad \dfrac{T_1}{T_2} = e^{\mu\theta}$

$\therefore \qquad \dfrac{T_1}{T_2} = e^{0.3 \times 2.88} = 2.37$ or $T_1 = 2.37 T_2$

Substituting the value of T_1 in (i), we get

$\qquad 2.37 T_2 - T_2 = 1486$

$\therefore \qquad T_2 = \dfrac{1486}{1.37} = 1084.7$ N

and $\qquad T_1 = 2.37 \times 1084.7 = 2570.7$ N

We know that maximum tension in the belt,
$$T = \sigma \cdot b \cdot t = 2.5 \times 10^6 \times b \times 0.0095 = 0.02375b \times 10^6 \text{ N},$$
where b is in metres and mass of the belt per metre length,
$$m = \text{area} \times \text{length} \times \text{density}$$
$$= (b \times 0.0095) \times 1 \times 1100 = 10.45\, b \text{ kg/m}$$

$\therefore$ Centrifugal tension, $T_c = mv^2$
$$= 10.45b \times (23.56)^2 = 5800.5\, b \text{ N}$$

and tension on the tight side of the belt (T_1)
$$2570.7 = T - T_c = 0.02375\, b \times 10^6 - 5800.5\, b$$
or
$$2570.7 = 23750\, b - 5800.5\, b = 17949.5\, b$$
$\therefore$
$$b = \frac{2570.7}{17949.5} = 0.143 \text{ m} \quad \text{or} \quad \textbf{143 mm (say 150 mm) (Ans.)}$$

6.2.11. Condition for Transmission of Maximum (Absolute) Power

It has already been pointed out that the maximum tension $T = T_1 + T_c$ and the effect of the centrifugal tension T_c is to *decrease the driving power of the belt*. The centrifugal tension increases with the increase of velocity. The power transmitted also increases with the increase of velocity but the value of T_1 and consequently the value of $(T_1 - T_2)$ diminishes. Hence the power transmitted increases with the increase of speed upto a certain value of velocity and after which it decreases. Allowing for the centrifugal tension, there is a certain velocity at which maximum power may be transmitted by a belt.

Let T = maximum permissible tension in a given belt,
T_1 = tension on the tight side of the belt,
T_2 = tension on the slack side of the belt,
T_c = centrifugal tension,

Power transmitted, $P = (T_1 - T_2)v = T_1\left(1 - \dfrac{T_2}{T_1}\right)v$

$$= T_1\left(1 - \frac{1}{e^{\mu\theta}}\right)v \qquad \left[\because \frac{T_1}{T_2} = e^{\mu\theta} \text{ or } \frac{T_2}{T_1} = \frac{1}{e^{\mu\theta}}\right]$$

But $T = T_1 + T_c$ or $T_1 = T - T_c$

$\therefore$
$$P = (T - T_c)\left(1 - \frac{1}{e^{\mu\theta}}\right)v$$
$$= (T - mv^2)\left(1 - \frac{1}{e^{\mu\theta}}\right)v = (Tv - mv^3)\left(1 - \frac{1}{e^{\mu\theta}}\right)$$
$$= (Tv - mv^3) \times k \qquad \left[\text{where } k \text{ is a constant} = \left(1 - \frac{1}{e^{\mu\theta}}\right)\right]$$

Differentiating P w.r.t. v and equating to zero,
$$\frac{dP}{dv} = (T - 3\, mv^2) = 0$$

or
$$T = 3\, mv^2 = 3T_c \qquad \ldots(6.10) \quad (\because T_c = mv^2)$$

from which
$$v = \sqrt{\frac{T}{3m}} \qquad \ldots(6.11)$$

Hence, the *power transmitted shall be maximum* when :
- The centrifugal tension is $\frac{1}{3}$ rd of the belt strength, or
- The belt is run at the velocity $\sqrt{\dfrac{T}{3m}}$.

Example 6.8. *A leather belt 200 m × 10 mm has a maximum permissible tension as 2 MN/m². If the ratio of tension is 1.8, determine at what velocity should it be run so as to transmit maximum power. Also determine the maximum value of the power.*

Take the density of belt material = 1100 kg/m³.

Solution. Width of the belt, $b = 200$ mm $= 0.2$ m
Thickness of the belt, $t = 10$ mm $= 0.01$ m
Maximum permissible tension $= 2$ MN/m²
Ratio of tension, $\dfrac{T_1}{T_2} = 1.8$
Density of belt material, $\rho = 1100$ kg/m³

Velocity of the belt, v :
Maximum permissible tension of the belt,
$$T = 2 \times 10^6 \times b \times t$$
$$= 2 \times 10^6 \times 0.2 \times 0.01 = 4000 \text{ N}$$

We know that for maximum transmission of power, $T = 3T_c$

$\therefore \quad T_c = \dfrac{T}{3} = \dfrac{4000}{3}$ N

But $\quad T_c = mv^2$

where, $\quad m = b \times t \times 1 \times \rho$
$$= 0.2 \times 0.01 \times 1 \times 1100$$
$$= 2.2 \text{ kg/m} \qquad \text{(considering 1 m length)}$$

$\therefore \quad \dfrac{4000}{3} = 2.2 \times v^2 \quad$ or $\quad \mathbf{v = 24.6 \text{ m/s}}$ **(Ans.)**

Now, $\quad T = T_1 + T_c$
$$T_1 = T - T_c$$
$$= T - \dfrac{T}{3} = \dfrac{2T}{3} \qquad \left(\because T_c = \dfrac{T}{3}\right)$$

or $\quad T_1 = \dfrac{2 \times 4000}{3} = \dfrac{8000}{3}$ N

Hence, maximum power transmitted,
$$P = (T_1 - T_2)v$$
$$= T_1\left(1 - \dfrac{T_2}{T_1}\right)v = T_1\left(1 - \dfrac{1}{1.8}\right) \times 24.6 \qquad \left[\because \dfrac{T_1}{T_2} = 1.8 \text{ and } \dfrac{T_2}{T_1} = \dfrac{1}{1.8}\right]$$
$$= \dfrac{8000}{3}\left(1 - \dfrac{1}{1.8}\right) \times 24.6$$
$$= 29155 \text{ Nm/s} \quad \text{or} \quad \text{J/s} \quad \text{or} \quad \text{W} = \mathbf{29.15 \text{ kW}} \text{ \textbf{(Ans.)}}$$

6.2.12. Initial Tension

To transmit the required power the belt must be given an initial tension T_0. When the power is supplied to one of the pulleys and transmitted to the other, the tensions in the two free lengths of

POWER TRANSMISSION

belt will be changed. The tight side of the belt stretches until the pull is *increased* from T_0 and T_1 and the slack side shortens until the pull is *decreased* from T_0 and T_2. Since the length of the belt remains unchanged, the mean tension must also remain unchanged and

$$T_1 - T_0 = T_0 - T_2 \text{ or } 2T_0 = T_1 + T_2$$

or
$$T_0 = \frac{T_1 + T_2}{2} \qquad \ldots(6.12)$$

If centrifugal tension is considered, $T_0 = \dfrac{T_1 + T_2 + 2T_c}{2}$ ...(6.13)

Example 6.9. *Initial tension in a belt connecting two pulleys is 1000 N. The minimum angle of embrace of the belt is 150° and the co-efficient of friction between the belt and the pulley rim is 0.25. If the speed of the belt is 500 metres/min, find:*

(i) *The value of the tension on the tight and the slack sides,*
(ii) *The power transmitted.*

Solution. Initial tension, $T_0 = 1000$ N
Angle of embrace, $\theta = 150°$
Co-efficient of friction, $\mu = 0.25$
Speed of the belt, $v = 500$ m/min.

(i) **Tension, $T_1 = ?$, T_2:**

We know that $T_0 = \dfrac{T_1 + T_2}{2}$ $\quad\left[\begin{array}{l}\text{where } T_1 = \text{tension on the tight side,}\\ \text{and } T_2 = \text{tension on the slack side}\end{array}\right]$

or $\quad 1000 = \dfrac{T_1 + T_2}{2}$ or $T_1 + T_2 = 2000$ ...(i)

Also, $\quad \dfrac{T_1}{T_2} = e^{\mu\theta} = e^{0.25 \times (150 \times \pi/180)} = 1.92$

∴ $\quad T_1 = 1.92 T_2$ ...(ii)

Substituting the above value of T_1 in (i), we get
$1.92 T_2 + T_2 = 2000$

or $\quad 2.92 T_2 = 2000$ or $T_2 = \mathbf{684.9\ N}$ **(Ans.)**
and $\quad T_1 = 1.92 \times 684.9 = \mathbf{1315\ N}$ **(Ans.)**

(ii) **The power transmitted, P = ?**

We know that $P = (T_1 - T_2) v$

$= (1315 - 684.9) \times \dfrac{500}{60}$ Nm/s or J/s or W

$= 5250$ W $= \mathbf{5.25\ kW}$ **(Ans.)**

Example 6.10. *Fig. 6.11 shows a band brake in which the drum is rotating anticlockwise. If the maximum value of the force P that can be applied = 400 N, the co-efficient of friction = 0.25 and the diameter of the drum = 0.6 m,*

(i) *Calculate the maximum bracking torque that can be developed.*

(ii) *If the drum is rotating in the clockwise direction what would be the maximum braking torque?*

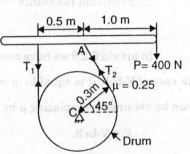

Fig. 6.11

Solution. (i) **Drum is rotating anticlockwise:**
Moments about A, $\quad M_A = 400 \times 1 - T_1 \times 0.5 = 0$
∴ $\quad T_1 = 800$ N

Angle of lap, $\theta = 180° + 45° = 225°$
Here $\mu = 0.25$ and $T_1 > T_2$

Using the relation, $\dfrac{T_1}{T_2} = e^{\mu\theta}$ or $\dfrac{800}{T_2} = e^{0.25 \times (225 \times \pi/180)} = 2.669$

$\therefore \quad T_2 = \dfrac{800}{2.669} = 299.7$ N

Maximum braking torque that can be developed,
$$M_c = T_1 \times 0.3 - T_2 \times 0.3$$
$$= 800 \times 0.3 - 299.7 \times 0.3 = \mathbf{150\ Nm} \quad \textbf{(Ans.)}$$

(*ii*) **Drum is rotating clock-wise :**
$$T_1 = 800\ \text{N}, \quad \theta = 225° \text{ and } \mu = 0.25$$

Here $\mathbf{T_2 > T_1}$

Now, $\dfrac{T_2}{T_1} = e^{\mu\theta} = 2.669$ (already calculated above)

$\therefore \quad T_2 = 2.669 T_1$

or $T_2 = 2.669 \times 800 = 2135.2$ N

$\therefore$ Maximum braking torque, $M_C = T_2 \times 0.3 - T_1 \times 0.3$
$$= 2135.2 \times 0.3 - 800 \times 0.3 = \mathbf{400.5\ Nm} \quad \textbf{(Ans.)}$$

6.2.13. V-belt and Rope Drive

6.2.13.1. V-belt drive

In Fig. 6.12 is shown a V-belt with a grooved pulley.

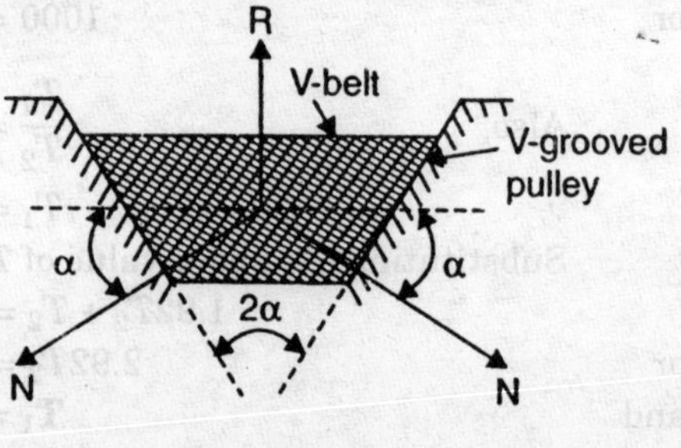

Let N = normal reaction between belt and sides of the V-grooved pulley,
2α = angle of groove,
μ = co-efficient of friction between belt and pulley, and
R = total reaction in the plane of groove.

Resolving forces *vertically*, we get
$$R = N \sin\alpha + N \sin\alpha$$
$$= 2N \sin\alpha$$

$\therefore \quad N = \dfrac{R}{2 \sin\alpha} = \dfrac{R}{2} \operatorname{cosec}\alpha$

$\therefore$ Frictional resistance $= \mu N + \mu N = 2\mu N$
$$= 2\mu \times \dfrac{R}{2} \operatorname{cosec}\alpha = \mu R \operatorname{cosec}\alpha. \quad \left(\because N = \dfrac{R}{2} \operatorname{cosec}\alpha\right)$$

Fig. 6.12

In article 6.2.9 we have seen that frictional force (or resistance) for a flat-belt is $\mu \times R$ whereas in case of V-belt it is equal to $(\mu \operatorname{cosec}\alpha) \times R$. Hence the ratio of tension $\left(i.e., \dfrac{T_1}{T_2}\right)$ in case of V-belt can be obtained by replacing μ by '$\mu \operatorname{cosec}\alpha$' in eqn. (6.8).

$\therefore$ For V-belt, $\dfrac{T_1}{T_2} = e^{(\mu \operatorname{cosec}\alpha) \times \theta}$

or $\dfrac{T_1}{T_2} = e^{\mu\theta \operatorname{cosec}\alpha}$ ...(6.14)

6.2.13.2. Rope drive

Refer Fig. 6.13. A rope-drive is generally used when the distance between the driving shaft and driven shafts is *large*. The ropes used are generally of circular section. Frictional grip in rope-drive is more than in V-belt drive.

The ratio of tension in this case will also be same as in the case of V-belt.

i.e., $\quad \dfrac{T_1}{T_2} = e^{\mu\theta \operatorname{cosec} \alpha}$ ...(6.15)

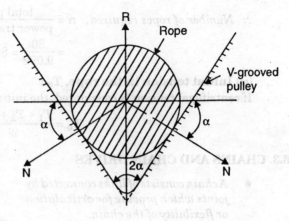

Fig. 6.13

Example 6.11. *A rope drive transmits 80 kW through a 1.5 m diameter 45° grooved pulley rotating at 200 r.p.m. Co-efficient of friction between the ropes and the pulley grooves is 0.3 and angle of lap is 160°. The mass of each rope is 0.6 kg/m and can safely take a pull of 800 N. Taking centrifugal tension into account, determine:*

  (i) *Number of ropes required for the drive.*
  (ii) *Initial tension in the rope.*

Solution.

Power to be transmitted by the rope, $P = 80$ kW
Diameter of the pulley, $d_1 = 1.5$ m
Speed of the pulley, $N_1 = 200$ r.p.m.
Angle of groove, $2\alpha = 45°$, $\therefore \alpha = 22.5°$
Angle of lap, $\theta = 160°$
Co-efficient of friction, $\mu = 0.3$
Mass of each rope, $m = 0.6$ kg/m
Safe pull, $T = 800$ N

(*i*) **Number of ropes required, n :**

Velocity of the rope is given by, $\quad v = \dfrac{\pi d_1 N_1}{60} = \dfrac{\pi \times 1.5 \times 200}{60} = 15.7$ m/s

Power transmitted by one rope :

Centrifugal tension in one rope is given by,

$$T_c = mv^2 = 0.6 \times (15.7)^2 = 147.9 \text{ N}$$

But $\qquad T = T_1 + T_c \quad \text{or} \quad T_1 = T - T_c$
$\qquad\qquad = 800 - 147.9 = 652.1$ N

Ratio of tensions in rope is given by,

$$\dfrac{T_1}{T_2} = e^{\mu\theta \operatorname{cosec} \alpha} = e^{0.3 \times (160 \times \pi/180) \times \operatorname{cosec}(22.5°)} = 8.93$$

$\therefore \qquad T_2 = \dfrac{T_1}{8.93} = \dfrac{652.1}{8.93} = 73$ N

$\therefore$ Power transmitted by one rope
$\qquad\qquad = (T_1 - T_2)v = (652.1 - 73) \times 15.7 = 9092$ W $= 9.092$ kW

But total power transmitted $= 80$ kW ...(Given)

$\therefore$ **Number of ropes required,** $n = \dfrac{\text{total power transmitted}}{\text{power transmitted by one rope}}$

$= \dfrac{80}{9.092} = 8.8$ say **9. (Ans.)**

(*ii*) **Initial tension in the rope, T_0 :**
If centrifugal tension is considered, the initial tension is given by :

$$T_0 = \dfrac{T_1 + T_2 + 2T_c}{2} = \dfrac{652.1 + 73 + 2 \times 147.9}{2} = 510.45 \text{ N} \quad \textbf{(Ans.)}$$

6.3. CHAINS AND CHAIN DRIVES

- A *chain consists of links connected by joints which provide for articulation or flexibility of the chain.*
- A chain drive consists of two sprockets and chain (Fig. 6.14). Chain drives, or transmissions, with several driven sprockets are also employed. Besides the enumerated components, chain drives may also include tensioning devices, lubricating devices and guards.

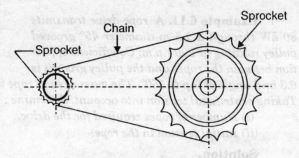

Fig. 6.14. Chain drive.

Chain drives are used for :
(1) *Medium centre to centre distances* which, in the case of a gear drive, would require idle gears, or intermediate stages not necessary to obtain the required speed ratio.
(2) Drives with strict requirements as to overall size or ones *requiring positive transmission without slippage* (preventing the use of V-belts drives).

There are two principal types of chain drives :
(*i*) Roller chain drive, and
(*ii*) Inverted tooth or silent chain drive.

6.3.1. Roller Chain Drive

Refer Fig. 6.15. It is quite popular in chain drives, a common form being the bicycle chain. In the roller construction shown here, the hollow rollers are held inside two flat link plates which are

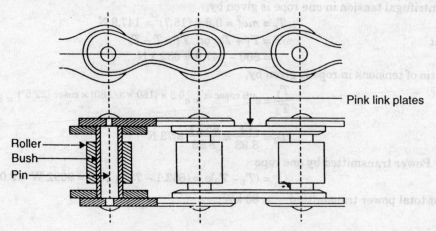

Fig. 6.15. Roller chain.

POWER TRANSMISSION

joined together by sleeves or bushing passing inside the rollers. Consecutive sets of such assemblies are connected together with another pair of plates called pin link plates, which in turn are held by central pins passing through the sleeves. Rivets (or split pins) are used to join the link plates. Sometimes the roller link plates are joined together by rollers themselves and no sleeves are used.

6.3.2. Silent Chain Drive

In the *silent chain drive* the chain (Fig. 6.16) consists of a number of flat links, tooth shaped at ends and joined together by long cross-pins. The sprockets are usually wider than those of the roller chain and have a central groove which holds retained plates provided in the central links for keeping the chain on to the sprocket securely. Silent chain justifies its name on account of its *quieter operation* and *lack of vibration,* even when running at high speed. Due to small size of links, this chain can be used on sprockets of *smaller diameter,* but the maximum ratio of number of teeth in two sprockets of a drive is restricted to *six*.

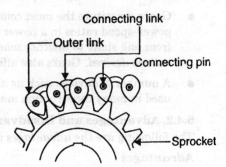

Fig. 6.16. Silent chain drive.

6.3.3. Advantages and Disadvantages of Chain Drive

A chain drive entails the following *advantages* and *disadvantages* :

Advantages :
 1. Efficiency of transmission is very high (upto 98%).
 2. More durable.
 3. Can carry heavy loads.
 4. Can be used where exact timing in movement is desired.
 5. No slip takes place.
 6. No initial tension required for operation.
 7. A single chain can be used for transmitting motion to several shafts.
 8. Smaller overall size than a belt drive.
 9. Small forces acting on the shafts since no initial tension is required.
 10. Easy replacement of the chain.
 11. These permit high speed ratio upto 8 in one step.
 12. These can be operated under adverse temperature and atmospheric conditions.

Disadvantages :
 1. Chain drive requires housing.
 2. It needs careful maintenance and accurate mounting.
 3. The velocity of the chain, especially with small numbers of teeth on the sprockets, is not constant leading to non-uniform rotation of sprockets.
 4. The chain gets stretched and requires a tensioning device.
 5. Cost is high comparatively.

Uses of chain drives :
Chain drives are extensively employed in :
 (*i*) Motor cycles   (*ii*) Bicycles   (*iii*) Automobiles
 (*iv*) Conveyers   (*v*) Agricultural machinery   (*vi*) Oil-well drilling rigs
 (*vii*) Machine tools etc.

6.4. GEAR DRIVE

6.4.1. Introduction

- A **gear** *is a wheel provided with teeth which mesh with the teeth on another wheel, or on to a rack, so as to give a positive transmission of motion from one component to another.*

- Gears constitute the most commonly used device for power transmission or for changing power-speed ratios in a power system. They are used for transmitting motion and power from one shaft to another *when they are not too far apart and when a constant velocity ratio is desired*. Gears also afford a convenient way of changing the direction of motion.

- A number of devices such as *differentials, transmission gear boxes, planetary drives* etc., used in many construction machines employ gears as basic component.

6.4.2. Advantages and Disadvantages of Toothed Gearing

The following are the *advantages* and *disadvantages* of toothed gearing/gear drive :

Advantages :
1. High efficiency
2. Long service life.
3. High reliability.
4. More compact.
5. Can operate at high speeds.
6. Can be used where precise timing is required.
7. Large power can be transmitted.
8. Constant speed ratio owing to absence of slipping.
9. Possibility of being applied for a wide range of torques, speeds and speed ratios.
10. The force required to hold the gears in position is much less than in an equivalent friction drive. This results in lower bearing pressure, less wear on the bearing surface and efficiency.

Disadvantages :
1. Special equipment and tools are required to manufacture the gears.
2. When one wheel gets damaged the whole set up is affected.
3. Noisy in operation at considerable speeds.

6.4.3. Definitions

Refer to Fig. 6.17.

1. **Pitch circle.** It is an imaginary circle which would transmit the same motion as the actual gear, by *pure rolling action*.
   The diameter of the pitch circle is known as *pitch circle diameter*.
2. **Addendum circle.** A circle concentric with the pitch circle and *bounding the outer ends* to the teeth is called an *addendum circle*.
   The diameter of the addendum circle is known as addendum circle diameter.
3. **Addendum.** It is the radial distance between the pitch circle and addendum circle.
4. **Dedendum (or root) circle.** It is a circle concentric with the pitch circle and bounding the bottom of the tooth.
5. **Dedendum.** It is the radial distance between the pitch circle and the dedendum circle.
6. **Clearance.** The difference between the dedendum (of one gear) and addendum (of the mating gear) is called as *clearance*.

POWER TRANSMISSION

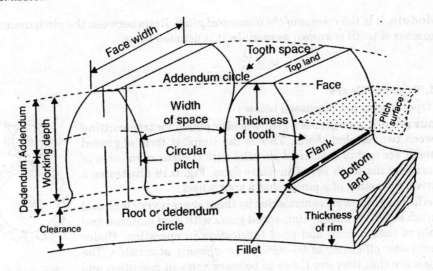

Fig. 6.17. Terms of gears.

7. **Working depth.** It is the sum of the addenda of the two mating gears.
8. **Circular thickness (or Thickness of tooth).** The length of arc between the sides of a gear tooth, measured on the pitch circle is known as *circular thickness (or thickness of tooth)*.
9. **Tooth space.** It is the width of the recess between two adjacent teeth measured along pitch circle.
10. **Backlash.** It is the difference between the tooth space and the tooth thickness.
11. **Face.** It is the action or working surface of the addendum.
12. **Flank.** The working face of the dedendum is called the *flank*.
13. **Top land.** It is the surface of the top of the tooth.
14. **Bottom land.** It is the surface of the bottom of the tooth space.
15. **Whole depth.** It is the total depth of the tooth space, equal to addendum plus dedendum ; also it is equal to the working depth plus clearance.
16. **Tooth fillet.** It is the radius which connects the root circle to the tooth profile.
17. **Circular pitch.** The distance measured along the pitch circle from a point on one tooth to the corresponding point on an adjacent tooth is called *circular pitch*. It is represented by p_c

$$p_c = \frac{\pi D}{T} \qquad \ldots(6.16)$$

where D = pitch diameter, T = number of teeth.

18. **Pitch diameter.** It is the diameter of a pitch circle. It is usually represented by d_p or d_g for pinion and gear respectively.
19. **Diametral pitch.** Number of teeth on a wheel per unit of its pitch diameter is called the *diametral pitch*. It is denoted by p_d

$$\therefore \quad p_d = \frac{T}{D} \qquad \ldots(6.17)$$

From eqns. (6.16) and (6.18), we have

$$p_c \cdot p_d = \pi \qquad \ldots(6.18)$$

20. **Module.** It is the *reverse of the diametral pitch*. Ratio between the pitch diameter and the number of teeth is known as *module*, it is denoted by m.

$$\therefore \quad m = \frac{D}{T} \qquad \qquad \ldots(6.19)$$

6.4.4. Types of Gears

The types of gear are discussed below :

1. **Spur gear.** A spur gear is a gear wheel or pinion for transmitting motion between two *parallel shafts*. This is the simplest form of geared drive. The teeth are cast or machined *parallel with the axis of rotation of the gear*. Normally the teeth are of involute form. Fig. 6.18 illustrates a spur gear drive, consisting of a pinion and a spur wheel.

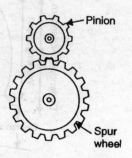

Fig. 6.18. Spur gear.

The efficiency of power transmission by these gears is very high and may be as much as 99% in case of high-speed gears with good material and workmanship of construction and good lubrication in operation. Under average conditions, efficiency of 96–98% are commonly attainable. The *disadvantages* are that they are liable to be *more noisy* in operation and may *wear out* and develop *backlash* more readily than the other types.

2. **Helical gear.** Refer Fig. 6.19, helical gear is one in which teeth instead, of being parallel with shaft as in ordinary spur gears, are *inclined*. This ensures *smooth action* and *more accurate maintenance of velocity ratio*.

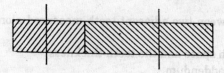

Fig. 6.19. Helical gear.

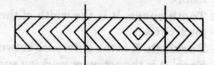

Fig. 6.20. Double Helical gear.

A *disadvantage* is that the inclination of the teeth sets up a *lateral thrust*. A method of neutralising this lateral or axial thrust is to use *double-helical gears* (also known as *Herring bone gears*) shown in Fig. 6.20.

3. **Bevel gear.** Refer Fig. 6.21. A bevel gear transmits motion between two shafts which *intersect*. If the shafts are at right angles and wheels equal in size, they are called *mitre gears* ; if the shafts are not at right angles, they are sometimes called angle bevel gears. Spiral toothed bevel gears are preferred to straight-toothed bevels in certain applications, because they will run more smoothly and make less noise at high speeds.

4. **Worm gear.** Refer Fig. 6.22. Worm gears connect two *non-parallel, non-intersecting shafts* which are usually at *right angles*. One of the gears is called the *'worm'*. It is essentially part of a screw, meshing with the teeth on a gear wheel, called the *"worm wheel"*. The gear ratio is the ratio of number of teeth on the wheel to the number of threads on the worm.

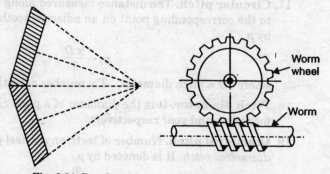

Fig. 6.21. Bevel gear.   Fig. 6.22. Worm gear.

One of the great advantages of worm gearing in that high gear ratios (*i.e.*, ratio of rotational speed of worm to that of worm wheel) are easily obtained. Worm gearing is *smooth and quiet*.

5. **Rack and pinion.** Refer Fig. 6.23. A rack is a spur gear of infinite diameter, thus it assumes the shape of a straight gear. The rack is generally used with a pinion to *convert rotary motion into rectilinear motion*.

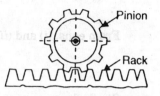

Fig. 6.23. Rack and pinion.

6.4.5. Types of Gear Trains

The combination of gear wheels by means of which motion is transmitted from one shaft to another shaft is called a **gear train**.

The gear trains are of the following types:
1. Simple gear train
2. Compound gear train
3. Epicyclic gear train.

6.4.6. Simple Gear Train

A *simple gear train* is one in which each shaft carries one wheel only, (Fig. 6.24). Simple gear trains are employed where a *small velocity ratio* is required. The gear train in which the driving and the driven shafts are co-axial or coincident is known as the *reverted gear train* (Fig. 6.25).

Refer to Fig. 6.24. 1 is the driving wheel and 4 the driven wheel.

Let N_1 = speed of driver in r.p.m.,

N_2 = speed of the idle gear 2 in r.p.m.,

N_3 = speed of the idle gear 3 in r.p.m.,

N_4 = speed of the driven (or follower) in r.p.m.

and T_1, T_2, T_3 and T_4 be the number of teeth on the gears 1, 2, 3 and 4 respectively.

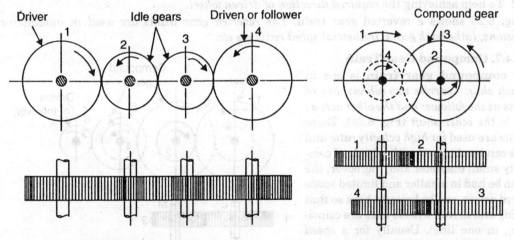

Fig. 6.24. Simple gear train. Fig. 6.25. Reverted gear train.

Let D_1 and D_2 be the pitch diameters of wheels 1 and 2.

Since gears 1 and 2 are meshing together, therefore,

$$\pi D_1 N_1 = \pi D_2 N_2$$

$$\therefore \quad \frac{N_1}{N_2} = \frac{D_2}{D_1} \quad \ldots(i)$$

and diametral pitch of gear 1 = diametral pitch of gear 2

$$\therefore \quad \frac{T_1}{D_1} = \frac{T_2}{D_2} \quad \text{or} \quad \frac{T_2}{T_1} = \frac{D_2}{D_1} \qquad \ldots(ii)$$

From eqns. (i) and (ii), we have

$$\frac{N_1}{N_2} = \frac{T_2}{T_1} \qquad \ldots(1)$$

Similarly

$$\frac{N_2}{N_3} = \frac{T_3}{N_2} \qquad \ldots(2)$$

$$\frac{N_2}{N_4} = \frac{T_4}{T_3} \qquad \ldots(3)$$

Multiplying eqns. (1), (2) and (3), we get

$$\frac{N_1}{N_2} \times \frac{N_2}{N_3} \times \frac{N_3}{N_4} = \frac{T_2}{T_1} \times \frac{T_3}{T_2} \times \frac{T_4}{T_3} \quad \text{or} \quad \frac{N_1}{N_4} = \frac{T_4}{T_1} \qquad \ldots(6.20)$$

$\therefore$ Speed (or velocity) ratio $= \dfrac{\text{speed of the driver}}{\text{speed of the driven}} = \dfrac{\text{no. of teeth on driven}}{\text{no. of teeth on driver}}$

Train value. It is the *reciprocal of velocity ratio*

$$= \frac{\text{speed of driven}}{\text{speed of driver}} = \frac{\text{no. of teeth on driver}}{\text{no. of teeth on driven}}$$

Similarly, it can be proved that the above equation holds good even if there are any number of intermediate gears. These intermediate gears are called *idle gears*, as they *do not affect the speed ratio or train value of the system*. In simple train of gears (as seen above) the speed ratio and train value is *independent of the size and number of intermediate / idle gears*.

The idle gears are provided for the following *purposes* :

1. To *bridge the distance* between the driving and driven wheels of moderate sizes instead of providing two wheels (driving and driven) of extra-ordinary big sizes.
2. To help achieving the *required direction of driven wheel*.

Fig. 6.25 shows a reverted gear train. *The reverted gear trains are used in automotive transmissions, lathe back gears, industrial speed reducers etc.*

6.4.7. Compound Gear Train

A **compound gear train** is one in which *each shaft ; carries two wheels, one of which acts as the follower and the other acts as a driver to the other shaft* (Fig. 6.26). These gear trains are used for *high velocity ratio* and the same can be obtained with wheels of comparatively small diameter and, moreover, the driver can be had in smaller and limited space and if need arises, can be brought back so that the driving and driven wheels axes are coincident (*i.e.*, in one line). Usually for a *speed reduction in excess of 7 to 1* a compound train or worm gearing is employed (instead of a simple train).

Refer to Fig. 6.26.

The gear 1 is driving gear mounted on shaft L, gears 2 and 3 are compound gears which are mounted on shaft M. The gears 4

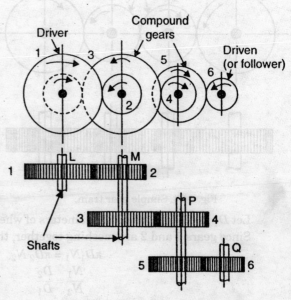

Fig. 6.26. Compound gear train.

and 5 are also compound gears which are mounted on shaft P and gear 6 is the driven gear mounted on shaft Q.

Let N_1 = speed of driving gear 1 in r.p.m.,
T_1 = no. of teeth on driving gear 1,
N_2, N_3, N_4, N_5, N_6 = speed of respective gears in r.p.m., and
T_2, T_3, T_4, T_5, T_6 = no. of teeth on respective gears.

Since gear 1 meshes with gear 2, therefore its speed (or velocity) ratio is

$$\frac{N_1}{N_2} = \frac{T_2}{T_1} \qquad \ldots(i)$$

Similarly, for gears 3 and 4, speed ratio is

$$\frac{N_3}{N_4} = \frac{T_4}{T_3} \qquad \ldots(ii)$$

and for gears 5 and 6, speed ratio is $\quad \dfrac{N_5}{N_6} = \dfrac{T_6}{T_5} \qquad \ldots(iii)$

The speed ratio of the compound gear train is obtained by multiplying eqns. (i), (ii) and (iii)

$$\therefore \quad \frac{N_1}{N_2} \times \frac{N_3}{N_4} \times \frac{N_5}{N_6} = \frac{T_2}{T_1} \times \frac{T_4}{T_3} \times \frac{T_6}{T_5}$$

But $\quad N_2 = N_3 \quad$ ($\because$ gears 2 and 3 are mounted on shaft M)

$\quad N_5 = N_4 \quad$ ($\because$ gears 5 and 6 are mounted on shaft P)

$$\therefore \quad \frac{N_1}{N_6} = \frac{T_2}{T_1} \times \frac{T_4}{T_3} \times \frac{T_6}{T_5}$$

i.e., Speed (or velocity) ratio $= \dfrac{\text{speed of the first driver}}{\text{speed of the last driven or follower}}$

$= \dfrac{\text{product or the number of teeth on drivens}}{\text{product of the numbers teeth on the drivers}}$

Train value $\left(= \dfrac{N_6}{N_1}\right) = \dfrac{\text{speed of the last driven or follower}}{\text{speed of the first driver}}$

$= \dfrac{\text{product of the number of teeth on the drivers}}{\text{product of the number of teeth on the drivens}}$.

Example 6.12. *A toothed gear has 72 teeth and circular pitch of 26 mm, find the following :*
(i) *Pitch diameter* (ii) *Diametral pitch*
(iii) *Module of the gear.*

Solution. Number of teeth, $\quad T = 72$
Circular pitch, $\quad p_c = 26$ mm

(i) **Pitch diameter, D :**

$$p_c = \frac{\pi D}{T}$$

$\therefore \quad 26 = \dfrac{\pi \times D}{72} \quad$ or $\quad D = \dfrac{26 \times 72}{\pi} =$ **595.87 mm (Ans.)**

(ii) **Diametral pitch, p_d :**

$$p_c \cdot p_d = \pi$$

$\therefore \quad p_d = \dfrac{\pi}{p_c} = \dfrac{\pi}{26} =$ **0.12 teeth/mm (Ans.)**

(*iii*) **Module, m :**

$$m = \frac{D}{T} = \frac{595.87}{72} = 8.27 \text{ mm/tooth} \quad \text{(Ans.)}$$

Example 6.13. *Fig. 6.27 shows the gearing of a machine tool. The gear A is connected to the motor shaft which rotates at 1000 r.p.m. Find the speed of the gear F mounted on the output shaft L.*

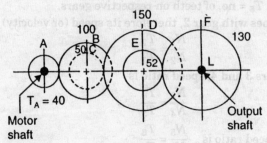

Fig. 6.27

Solution. Number of teeth of gear *A*, $T_A = 40$
Number of teeth of gear *B*, $T_B = 100$
Number of teeth of gear *C*, $T_C = 50$
Number of teeth of gear *D*, $T_D = 150$
Number of teeth of gear *E*, $T_E = 52$
Number of teeth of gear *F*, $T_F = 130$
Speed of the motor shaft, $N_A = 1000$ r.p.m.

Speed of the output shaft, N_P :

We know that, $\dfrac{N_F}{N_A} = \dfrac{T_A \times T_C \times T_E}{T_B \times T_D \times T_F}$ or $\dfrac{N_F}{1000} = \dfrac{40 \times 50 \times 52}{100 \times 150 \times 130}$

∴ $N_F = 53.33$ **r.p.m.** (Ans.)

6.4.8. Epicyclic (or Planetary) Gear Train

So far we have discussed those gear trains in which axes of the wheels remain *fixed* relative to one another. But there is another system of gear train in which there is *relative motion* between two or more of the axes of the wheels (constituting the train) ; such an arrangement of wheels is known as *"epicyclic gear train"*. The wheels are usually carried on an arm or link pivoted about a fixed centre and itself capable of rotating. For example, in Fig. 6.28 gear rolls around the outside of the stationary gear 2 as the arm *A* revolves. Epicyclic trains are sometimes called as planetary gear trains because of the fact that gear 1 goes round and round the gear 2 just like a planet moving round the sun. The motion of the planets around the sun is called planetary motion, so the motion of gear 1 around gear 2 is called planetary motion.

Epicyclic gear trains are also *simple* as well as *compound* exactly in the same manner as explained earlier.

The following points are worth noting :

1. The *epicyclic gear trains are useful for transmitting very high velocity ratios, with gears of moderate size in a comparatively lesser space.*
2. These trains are of great practical importance and find use in almost all kinds of workshop and electrical machines ; *e.g.* back gear of lathe, differential gears and gear boxes for motor vehicles, cyclometers etc.

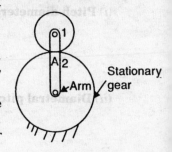

Fig. 6.28

POWER TRANSMISSION

HIGHLIGHTS

1. The power from one shaft to another can be transmitted by the following means :
   (i) Belts and ropes   (ii) Chains
   (iii) Gears   (iv) Clutches.
2. A **belt** is a continuous bond of flexible material passing over pulleys to transmit motion from one shaft to another.
3. A **V-belt** is a belt of trapezoidal section running on pulleys with grooves cut to match the belt.
4. **Belt-drives** may be (i) open belt-drive, (ii) crossed belt-drive.
5. Velocity ratio in case of a belt drive is given by : $\dfrac{N_2}{N_1} = \left(\dfrac{d_1 + t}{d_2 + t}\right)\left(\dfrac{100 - S}{100}\right)$
6. Length of the belt :

   (i) Open belt : $l = \pi(r_1 + r_2) + \dfrac{(r_1 - r_2)^2}{d} + 2d$

   (ii) Cross-belt : $l = \pi(r_1 + r_2) + \dfrac{(r_1 + r_2)^2}{d} + 2d$

7. Power transmitted by the belt, $P = (T_1 - T_2)v$
8. Ratio of tension :

   (i) Flat belt : $\dfrac{T_1}{T_2} = e^{\mu\theta}$

   (ii) V belt or rope : $\dfrac{T_1}{T_2} = e^{\mu\theta \, \text{cosec}\, \alpha}$ (where 2α is the angle of groove.)

9. Angle of contact in belt drive (θ) :

   Open belt : $\theta = 180 - 2\alpha$, where $\alpha = \sin^{-1}\left(\dfrac{r_1 - r_2}{d}\right)$

   Cross-belt : $\theta = 180 + 2\alpha$, where $\alpha = \sin\left(\dfrac{r_1 + r_2}{d}\right)$.

10. Centrifugal tension, $T_c = mv^2$.
11. Initial tension in the belt $T_0 = \dfrac{T_1 + T_2 + 2T_c}{d}$ ($T_c = 0$ if centrifugal tension is not considered)
12. A **chain** consists of links connected by joints which provide for articulation or flexibility of the chain.
13. There are two principal types of chain drives :
    (i) Roller chain drive   (ii) Inverted tooth or silent chain drive.
14. A **gear** is a wheel provided with teeth which mesh with the teeth on another wheel, or on to a rack, so as to give a positive transmission of motion from one component to another.
15. *Circular pitch* (p_c). The distance measured along the pitch circle from a point on one tooth to the corresponding point on an adjacent tooth is called *circular pitch*.

    $$p_c = \dfrac{\pi D}{T}$$

    where D = pitch diameter, T = number of teeth.

    Diametral pitch (p_d). Number of teeth on a wheel per unit of its pitch diameter is called the *diametral pitch*.

    $$p_d = \dfrac{T}{D}$$

    Also $p_c \cdot p_d = \pi$

16. *Types of gears*
    (i) Spur gears (ii) Helical gears (iii) Bevel gears
    (iv) Worm gears (v) Rack and pinion.
17. *Types of gear trains.*
    The gear trains are of the following types :
    (i) Simple gear train (ii) Compound gear train (iii) Epicyclic gear train
    Velocity ratio = $\dfrac{\text{speed of the driver}}{\text{speed of the driven}}$; Train value = $\dfrac{\text{speed of the driven}}{\text{speed of the driver}}$
    (*i.e.*, reciprocal of velocity ratio).

OBJECTIVE TYPE QUESTIONS

Choose the Correct Answer :

1. Belts and ropes are used when the distance between the shaft centres is
    (a) very small (b) small
    (c) large (d) none of the above.
2. ...... are used when the distance between the shaft centres is large and no slip is required.
    (a) Belts and ropes (b) Gears
    (c) Chains (d) Clutches.
3. A ...... is a continuous band of flexible material passing over pulleys to transmit motion from one shaft to another.
    (a) belt (b) rope
    (c) chain (d) none of the above.
4. Which of the following are used when the shafts are co-axial ?
    (a) Belts (b) Gears
    (c) Chains (d) Clutches.
5. Which of the following are used when the shaft distance is adequately less ?
    (a) Belts (b) Chains
    (c) Gears (d) Clutches.
6. ...... belts have the best pulling capacity.
    (a) Leather (b) Rubber
    (c) Textile (d) Balata.
7. ...... belts are acid and water proof.
    (a) Leather (b) Rubber
    (c) Balata (d) Textile.
8. ...... belts are employed to transmit low power, mainly in instruments, table-type machine tools, machinery of clothing industry and household appliances.
    (a) Flat (b) V
    (c) Round (d) Steel.
9. Which of the following is an advantage of V-belt ?
    (a) No possibility of belt coming out of grooves
    (b) The gripping action results in lower belt tension
    (c) Power output can be increased by use of multiple belts
    (d) Drives are quiet at high speed.
    (e) All of the above.
10. In cross belt drives the power is transmitted between pulleys rotating in
    (a) same direction (b) opposite direction
    (c) either of the above (d) none of the above.

11. Power transmitted by the belt is given by
    (a) $(T_1 - T_2) v$ watts
    (b) $(T_1 + T_2) v$ watts
    (c) $\left(\dfrac{T_1 + T_2}{2}\right) v$ watts
    (d) none of the above
    [where T_1 = tension on right side (N) T_2 = tension on slack side (N) and v = velocity of belt (m/s)]

12. The ratio of tensions in case of a belt drive is given by
    (a) $\dfrac{T_1}{T_2} = e^{\theta}$
    (b) $\dfrac{T_1}{T_2} = e^{\mu\theta}$
    (c) $\dfrac{T_1}{T_2} = e^{2\theta}$
    (d) none of the above.

13. A circle concentric with pitch circle and bounding the outer ends of the teeth is called
    (a) dedendum circle
    (b) addendum circle
    (c) either of the above
    (d) none of the above.

14. The reverse of the diametral pitch is called
    (a) module
    (b) addendum
    (c) dedendum
    (d) clearance.

15. Number of teeth on a wheel per unit of its pitch diameter is called
    (a) module
    (b) diametral pitch
    (c) circular pitch
    (d) none of the above.

16. A ...... is a gear wheel or pinion for transmitting motion between two parallel shafts.
    (a) spur gear
    (b) bevel gear
    (c) worm gear
    (d) none of the above.

17. ...... gears connect two non-parallel, non-intersecting shafts which are usually at right angles.
    (a) Spur
    (b) Bevel
    (c) Worm
    (d) none of the above.

18. ......gear train is one in which each shaft carries one wheel only
    (a) Simple
    (b) Compound
    (c) Epicyclic
    (d) none of the above.

19. The gear train in which the driving and driven shafts are co-axial or coincident is called ...... gear train.
    (a) compound
    (b) reverted
    (c) epicyclic
    (d) none of the above.

20. In a simple gear train, if the number of idle gears is odd, then the motion of driven gear will
    (a) be opposite as that of driving gear
    (b) be same as that of driving gear
    (c) depend upon the number of teeth on the driving gear
    (d) none of the above.

21. The train value of a gear train is
    (a) always less than unity
    (b) always greater than unity
    (c) equal to velocity ratio of a gear train
    (d) reciprocal of velocity ratio of gear train.

22. In a clock mechanism ...... gear train is used to connect minute hand to hour hand.
    (a) simple
    (b) compound
    (c) reverted
    (d) epicyclic.

Answers

1. (c) 2. (c) 3. (a) 4. (d) 5. (c) 6. (a) 7. (c)
8. (c) 9. (e) 10. (b) 11. (a) 12. (b) 13. (b) 14. (a)
15. (b) 16. (a) 17. (c) 18. (a) 19. (b) 20. (b) 21. (d)
22. (c).

THEORETICAL QUESTIONS

1. Enumerate the means by which power can be transmitted from one shaft to another.
2. Define a belt.
3. Write a short-note on flat belts.
4. How does a flat belt differ from that of a V-belt.
5. Name the materials from which belts are made.
6. What is a camber or crown, with reference to a pulley ?
7. What are the advantages of V-belts ?
8. Where are round belts usually used ?
9. List the advantages and disadvantages of belt drives.
10. Drive an expression for the length of an open belt.
11. Drive an expression for the ratio of tensions in case of a belt drive.
12. What is centrifugal tension ?
13. What is the condition for transmission of maximum (absolute) power ?
14. Describe neat stretches the following types of chain drives :
    (i) Roller chain drive (ii) Inverted tooth or silent chain drive.
15. What are the advantages and advantages of chain drives ?
16. List the uses of chain drives.
17. What is a gear ?
18. What are the advantages and disadvantages of toothed gearing ?
19. Define the following terms of gears :
    (i) Addendum (ii) Diametral pitch
    (iii) Circular pitch (iv) Back lash
    (v) Module (vi) Working depth.
20. Describe briefly various types of gears.
21. What do you understand by gear train ? Discuss the various types of gear trains.
22. Explain briefly an epicyclic gear train. What are the special advantages of epicyclic gear trains ?

UNSOLVED EXAMPLES

1. A shaft running at 200 r.p.m is to drive a parallel shaft at 300 r.p.m. They pulley on the driving shaft is 600 mm diameter. Calculate the diameter of the pulley on the driven shaft (i) neglecting belt thickness, (ii) taking belt thickness into account which is 5 mm, (iii) assuming in case (ii) a total slip of 4 per cent.
   [**Ans.** (i) 400 mm, (ii) 398.3 mm, (iii) 382.2 mm]
2. Two parallel shafts 6 m apart are to be connected by a belt running over pulleys of diameters 600 mm and 400 mm respectively. Determine the length of the belt required : (i) if the belt is open (ii) if the belt is crossed. [**Ans.** (i) 13.57 m (ii) 13.61 m]
3. Two pulleys on parallel shafts are connected by a cross-belt. The diameters of the pulleys are 450 mm and 200 mm. The shafts are 1.95 m apart. Find :
   (i) The length of the belt required, and
   (ii) The angle of contact between the belt and each pulley.
   (iii) Power transmitted when the larger pulley rotates at 100 r.p.m., if the maximum permissible tension in the belt is 1000 N and the co-efficient of friction between the belt and pulley is 0.25.
   [**Ans.** (i) 4.975 m (ii) 199.2° (iii) 1.37 kW]
4. The maximum allowable tension in a flat belt is 1500 N. The angle co-efficient of friction between the belt and material of the pulley is 0.27. Neglecting the effect of centrifugal tension, calculate the net driving tension and power transmitted if the belt speed is 2 m/s. [**Ans.** 826. 7 N, 1.65 kW]

POWER TRANSMISSION 229

5. The following data pertains to a belt drive :
   Width of the belt = 100 mm
   Thickness of the belt = 8 mm
   Speed of the belt = 31.67 m/s
   Angle of contact/lap = 165°
   Co-efficient of friction between the belt and the pulley = 0.3
   The mass of the belt = 0.9 kg/m
   Maximum permissible stress in the belt = 2 MN/m^2
   Find : (i) Maximum power transmitted (ii) Initial tension in the belt   [**Ans.** (i) 14.82 kW (ii) 1.32 kN]

6. The following data relates to an open belt drive :
   Distance between the centre lines of the two parallel shafts = 4.8 m
   Diameter of the larger pulley = 1.5 m
   Diameter of the smaller pulley = 1 m
   Initial tension in the belt = 3.0 kN
   Mass of the belt material = 1.5 kg/m
   Co-efficient of friction between the belt and the pulley = 0.3
   Speed of the smaller pulley = 400 r.p.m.
   Calculate the power transmitted.   [**Ans.** 53.6 kW]

7. The following data relates to a V-belt drive :
   Angle of lap = 170°
   Angle of the groove = 45°
   Co-efficient of friction
   between the belt and the pulley = 0.17
   Maximum permissible tension in the belt = 2.2 kN
   Mass of the belt = 0.9 kg/m
   Find : (i) Velocity of the belt for maximum power. (ii) Power transmitted at this velocity.
   [**Ans.** (i) 28.54 m/s, (ii) 30.67 kW]

8. A rope-drive transmits 230 kW from a pulley of 1 m diameter running at 450 r.p.m. The safe pull in each rope is 800 N and mass of the rope is 0.46 kg per metre length. The angle of lap is 160° whereas the angle of groove is 45°. The co-efficient of friction between the belt and the pulley is 0.3. Find the number of ropes required.   [**Ans.** 21]

9. Find the pitch diameter, diametral pitch and module of a toothed gear having 36 teeth and circular pitch of 13 mm.   [**Ans.** 149 mm. 0.242 teeth/mm, 4.14 mm/tooth]

10. The common module of two mating spur gears, having 70 and 30 teeth respectively, is 5. Find the centre to centre distance between gears axes.   [**Ans.** 250 mm]

11. Two parallel shafts are to be connected by spur gearing. The approximate distance between the shafts is 600 mm. If one shaft runs at 120 r.p.m. and the other at 360 r.p.m., find :
    (i) The number of teeth on each wheel, if the module is 8 mm ;
    (ii) The exact distance apart of the shafts.   [**Ans.** (i) 114,38, (ii) 608 mm]

12. In a compound train of wheels spur wheel 1, with 40 teeth and rotating at 16 r.p.m. drives a pinion 6 with 15 teeth at 300 r.p.m. across two intermediate spindles. The first intermediate spindle has wheels 2 and 3 and the second carries wheels 4 and 5 ; wheel 2 gearing with wheel 1 while 3 gearing with wheel 4 and 5 gearing with wheel 6. If wheels 2, 4 and 5 have respectively 20, 15 and 30 teeth, calculate the number of teeth on wheel 3.   [**Ans.** 75]

MODULE – 4

Chapter :

7. Power Plants, Non-conventional Energy Sources, Hydraulic and Steam Turbines

MODULE – 4

7
Power Plants, Non-Conventional Energy Sources, Hydraulic and Steam Turbines

7.1. **Power plants** : Sources of energy—Types of power plants—Steam power plant—Hydro-electric power plant—Nuclear power plant—Simple gas turbine plant—Diesel engine power plant. 7.2. **Non-conventional energy sources** : Introduction—Description of non-conventional sources of energy—7.3. **Hydraulic turbines** : Introduction to hydraulic turbines—Classification of hydraulic turbines—Impulse turbines—Pelton wheel—Reaction turbines—Tubular or bulb turbines—Runaway speed—Draft tube—Specific speed—Cavitation—Selection of hydraulic turbines. 7.4. **Steam turbines** : Introduction to steam turbines—Classification of steam turbines—Advantages of steam turbine over the steam engines—Description of common types of steam turbines—Methods of reducing wheel or rotor speed (compounding methods)—Differences between impulse and reaction turbines—Highlights—Objective Type Questions—Theoretical Questions.

7.1. POWER PLANTS

7.1.1. Sources of Energy
The various sources of energy are :

1. Fuels — Solids—Coal, coke, anthracite etc.
   — Liquids—Petroleum and its derivatives.
   — Gases—Natural gas, blast furnace gas etc.

2. Energy stored in water   3. Nuclear energy
4. Wind power   5. Solar energy
6. Tidal power   7. Geothermal energy
8. Thermoelectric power.

7.1.2. Types of Power Plants
The various types of power plants in common use are :
1. Steam thermal power plant ;
2. Hydro-electric power plant ;
3. Nuclear power plant ;
4. Gas turbine power plant ;
5. Diesel engine power plant.

7.1.3. Steam Power Plant
A steam power plant converts the chemical energy of the fossil fuels (coal, oil, gas) into mechanical / electrical energy.

7.1.3.1. Layout
Refer to Fig. 7.1. The layout of a modern steam power plant comprises of the following four circuits :
    1. Coal and ash circuit   2. Air and gas circuit

3. Feed water and steam flow circuit   4. Cooling water circuit.

1. Coal and ash Circuit. Coal arrives at the storage yard and after necessary landing, passes on to the furnaces through the *fuel feeding device*. Ash resulting from combustion of coal collects at the back of the boiler and is removed to the ash storage yard through *ash handling equipment*.

2. Air and gas Circuit. Air is taken in from atmosphere through the action of a forced or induced draught fan and passes on to the furnace through the *air preheater*, where it has been heated by the heat of flue gases which pass to the chimney via the preheater. The flue gases after passing around boiler tubes and superheater tubes in the furnace pass, through a *dust* catching device or precipitator, then through the economiser, and finally through the air preheater before being exhausted to the atmosphere.

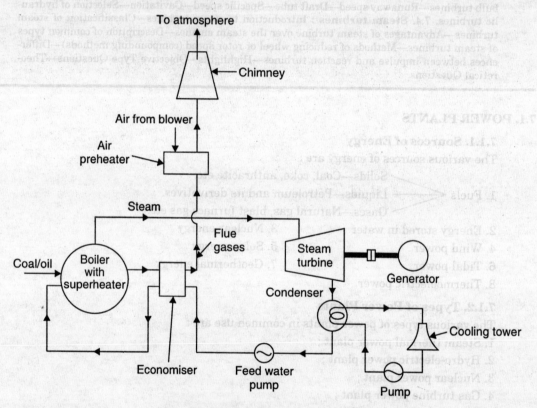

Fig. 7.1. Layout of a steam power plant.

3. Feed water and steam flow circuit. In the water and steam circuit condensate leaving the condenser is first heated in a closed feed water heater through extracted steam from the lowest pressure extraction point of the turbine. It then passes through the *deaerator* and a few more water heaters before going into the boiler through *economiser*.

A part of steam and water is lost while passing through different components and this is compensated by supplying additional feed water. This feed water should be purified before hand, to avoid the scaling of the tubes of the boiler.

4. Cooling water circuit. The cooling water supply to the condenser helps in maintaining a low pressure in it. The water may be taken from a natural source such as river, lake or sea or the same water may be cooled and circulated once again. In the latter case the cooling arrangement is made through spray pond or cooling lower.

7.1.3.2. Components of a modern steam power plant

A modern steam power plant comprises of the following components :

1. Boiler
    (i) Superheater
    (ii) Reheater
    (iii) Economiser
    (iv) Air-heater
2. Steam turbine
3. Generator
4. Condenser
5. Cooling towers
6. Circulating water pump
7. Boiler feed pump
8. Wagon tippler
9. Crusher house
10. Coal mill
11. Induced draught fans
12. Ash precipitators
13. Boiler chimney
14. Forced draught fans
15. Water treatment plant
16. Control room
17. Switch yard.

Functions of some important parts of a steam power plant :

1. Boiler. Water is converted into wet steam.

2. Superheater. It converts wet steam into superheated steam.

3. Turbine. Steam at high pressure expands in the turbine and drives the generator.

4. Condenser. It condenses steam used by the steam turbine. The condensed steam (known as *condensate*) is used as a feed water.

5. Cooling tower. It cools the condenser circulating water. Condenser cooling water absorbs heat from steam. This heat is discharged to atmosphere in cooling water.

6. Condenser circulating water pump. It circulates water through the condenser and the cooling tower.

7. Feed water pump. It pumps water in the water tubes of boiler against boiler steam pressure.

8. Economiser. In economiser heat in flue gases is partially used to heat incoming feed water.

9. Air preheater. In air preheater heat in flue gases (the products of combustion) is partially used to heat incoming air.

7.1.4. Hydro-electric Power Plant

In **hydro-electric plants** *energy of water is utilised to move the turbines which in turn run the electric generators.*

Classification :

Hydro-electric power stations may be classified as follows :

A. According to availability of head :

1. High head power plants
2. Medium head power plants
3. Low head power plants.

B. According to the nature of load :

1. Base load plants
2. Peak load plants.

C. According to the quantity of water available :

1. Run-of-river plant without pondage
2. Run-of-river plant with pondage
3. Storage type plants
4. Pump storage plants
5. Mini and micro-hydel plants.

A. According to availability of head :

The following figures give a rough idea of the heads under which the various types of plants work :

   (i) High head power plants   ...... 100 m and above
  (ii) Medium head power plants   ...... 30 to 500 m
 (iii) Low head power plants   ...... 25 to 80 m.

Note. It may be noted that figures given above overlap each other. Therefore it is difficult to classify the plants directly on the basis of head alone. The basis, therefore, technically adopted is the *specific speed* of the turbine used for a particular plant.

1. High head power plants :

Water is usually stored up in lakes on high mountains during the rainy season or during the season when the snow melts. The rate of flow should be such that water can last throughout the year.

Fig. 7.2 shows high head power plant layout. Surplus water discharged by the spillway cannot endanger the stability of the main dam by erosion because they are separated. The tunnel through the mountain has a surge chamber excavated near the exit. Flow is controlled by head gates at the tunnel intake, butterfly valves at the top of the penstocks, and gate valves at the turbines. This type of site might also be suitable for an underground station.

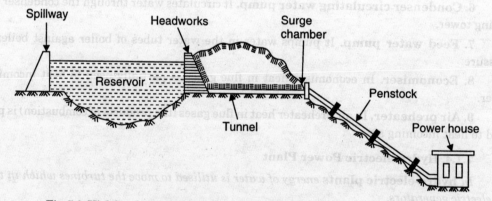

Fig. 7.2. High head power plant layout. The main dam, spillway, and power house stand at widely separated locations. Water flows from the reservoir through a tunnel and penstock to the turbines.

The Pelton *wheel* is the common prime-mover used in high head power plants.

2. Medium head power plants :

Refer to Fig. 7.3. This type of plant commonly uses *Francis turbines*. The forebay provided at the beginning of the penstock serves as water reservoir. In such plants, the water is generally carried in open canals from main reservoir to the forebay and then to the powerhouse through the penstock. The forebay itself works as a surge tank in this plant.

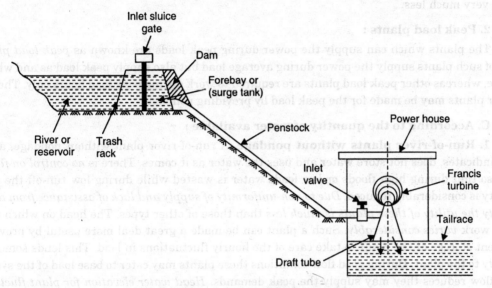

Fig. 7.3. Medium head power plant layout.

3. Low head power plants :

Refer to Fig. 7.4. These plants usually consist of a dam across a river. A sideway stream diverges from the river at the dam. Over this stream the power house is constructed. Later this channel joins the river further downstream. This type of plant uses vertical shaft Francis turbine or Kaplan turbine.

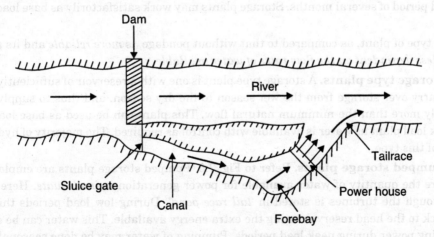

Fig. 7.4. Low head power plant layout.

B. According to the nature of load :

1. Base load plants :

The plants which cater for the base load of the system are called *base load plants*. These plants are required to *supply a constant power when connected to the grid*. Thus they *run without* stop and are often remote-controlled with which least staff is *required for such plants*. Run-of-river plants without pondage may sometimes work as base load plant, but the firm capacity in such cases will be very much less.

2. Peak load plants :

The plants which can supply the power during peak loads are known as *peak load plants*. Some of such plants supply the power during average load but also supply peak load as and when it is there, whereas other peak load plants are required to work during peak load hours only. The run-of-river plants may be made for the peak load by providing pondage.

C. According to the quantity of water available :

1. Run-of-river plants without pondage. A run-of-river plant without pondage, as the name indicates, does not store water and uses the water as it comes. There is *no control on flow of water* so that during high floods or low loads water is wasted while during low run-off the plant capacity is considerably reduced. *Due to non-uniformity of supply and lack of assistance from a firm capacity the utility of these plants is much less* than those of other types. The head on which these plants work *varies considerably*. Such a plant can be made a great deal more useful by providing sufficient storage at the plant to take care of the hourly fluctuations in load. This lends some *firm capacity* to the plant. During good flow conditions these plants may cater to base load of the system, when flow reduces they may supply the peak demands. *Head water elevation for plant fluctuates with the flow conditions*. These plants without storage may sometimes be made to supply the base load, but the firm capacity depends on the minimum flow of river. The run-of-river plant may be made for load service with pondage, though storage is usually seasonal.

2. Run-of-river plant with pondage. *Pondage* usually refers to the collection of water behind a dam at the plant and increases the stream capacity for a short period, say a week. *Storage* mean collection of water in upstream reservoirs and this increases the capacity of the stream over an extended period of several months. Storage plants may work satisfactorily as base load and peak load plants.

This type of plant, as compared to that without pondage, is *more reliable* and its generating capacity *is less dependent on the flow rates of water available*.

3. Storage type plants. A storage type plant is one with a reservoir of sufficiently large size to permit carry-over storage from the wet season to the dry season, and thus to supply firm flow substantially more than the minimum natural flow. This plant can be used as base load plant as well as peak load plant as water is available with control as required. The majority of hydro-electric plants are of this type.

4. Pumped storage plants. Refer to Fig. 7.5. Pumped storage plants are employed at the places where the quantity of water available for power generation is *inadequate*. Here the water passing through the turbines is stored in '*tail race pond*'. During low load periods this water is pumped back to the head reservoir using the extra energy available. This water can be again used for generating power during peak load periods. Pumping of water may be done seasonally or daily depending upon the conditions of the site and the nature of the load on the plant.

Such plants are *usually interconnected* with steam or diesel engine plants so that off peak capacity of interconnecting stations is used in pumping water and the same is used during peak load periods. Of course, the energy available from the quantity of water pumped by the plant is *less* than the energy input during pumping operation. Again while using pumped water the *power available is reduced* on account of losses occurring in prime-movers.

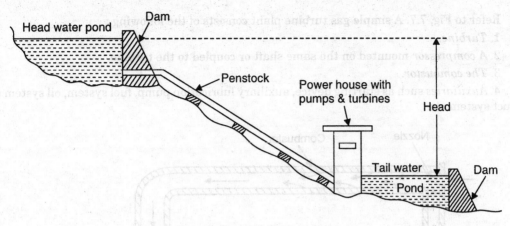

Fig. 7.5. Pumped storage plant.

7.1.5. Nuclear Power Plant

Fig. 7.6 shows schematically a nuclear power plant.

The main components of a nuclear power plant are :

1. Nuclear reactor
2. Heat exchanger (steam generator)
3. Steam turbine
4. Condenser
5. Electric generator.

In a nuclear power plant the reactor performs the same function as that of the furnace of steam power plant (*i.e.*, produces heat). The heat liberated in the reactor as a result of the nuclear fission of the fuel is taken up by the coolant circulating through the reactor core. Hot coolant leaves the reactor at the top and then flows through the tubes of steam generator and passes on its heat to the feed water. The steam so produced expands in the steam turbine, producing work, and thereafter is condensed in the condenser. The steam turbine in turn runs an electric generator thereby producing electrical energy. In order to maintain the flow of coolant, condensate and feed water pumps are provided as shown in Fig. 7.6.

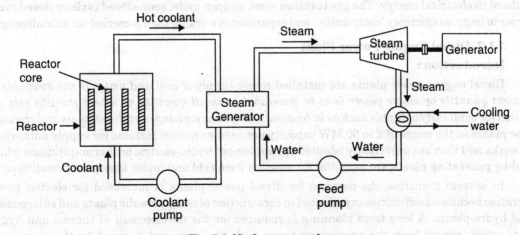

Fig. 7.6. Nuclear power plant.

7.1.6. Simple Gas Turbine Plant

A **gas turbine plant** may be defined as *"a plant in which the principal prime-mover is of the turbine type and the working medium is permanent gas"*.

Refer to Fig. 7.7. A simple gas turbine plant consists of the following :
1. *Turbine*.
2. *A compressor* mounted on the same shaft or coupled to the turbine.
3. *The combustor*.
4. *Auxiliaries* such as starting device, auxiliary lubrication pump, fuel system, oil system and the duct system etc.

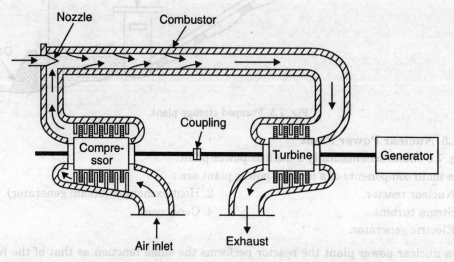

Fig. 7.7. Arrangement of a simple gas turbine plant.

A modified plant may have in addition to above an *intercooler, regenerator, a reheater* etc.

The working fluid is compressed in a compressor which is generally rotary, multistage type. Heat energy is added to the compressed fluid in the combustion chamber. This high energy fluid, at high temperature and pressure, then expands in the turbine unit thereby generating slower. Part of the power generated is consumed in driving the generating compressor and accessories and the rest is utilised in electrical energy. The gas turbines work on open cycle, semi-closed cycle or closed cycle. In order to improve efficiency, compression and expansion of working fluid is carried out in multistages.

7.1.7. Diesel Engine Power Plant
Introduction :

Diesel engine power plants are installed *where supply of coal and water is not available in sufficient quantity* or *where power is to be generated in small quantity* or *where standby sets are required for continuity of supply such as in hospitals, telephone exchanges, radio stations and cinemas*. These plants in the range of 2 to 50 MW capacity are used as *central stations* for supply authorities and works and they are universally adapted to supplement hydro-electric or thermal stations where stand-by generating plants are essential for starting from cold and under emergency conditions.

In several countries, the demand for diesel power plants is increased for electric power generation because of difficulties experienced in construction of new hydraulic plants and enlargement of old hydro-plants. A long term planning is required for the development of thermo and hydro-plants which cannot keep the pace many times with the increased demand by the people and industries.

The diesel units used for electric generation are *more reliable* and *long-lived piece of equipment* compared with other types of plants.

POWER PLANTS, NON-CONVENTIONAL ENERGY SOURCES, HYDRAULIC AND STEAM TURBINES

Essential components of a diesel power plant :

Refer to Fig. 7.8. The essential components of a diesel power plant are listed and discussed below :

1. Engine
2. Air intake system
3. Exhaust system
4. Fuel system
5. Cooling system
6. Lubrication system
7. Engine starting system
8. Governing system.

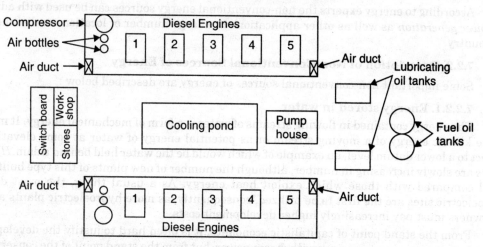

Fig. 7.8. Schematic arrangement of a diesel power plant.

7.2. NON-CONVENTIONAL ENERGY SOURCES

7.2.1. Introduction

The various *sources of energy* are :

1. Conventional Sources

Fuels :

(*i*) **Solids**—Coal, coke, anthracite etc.

(*ii*) **Liquids**—Petroleum and its derivatives

(*iii*) **Gases**—Natural gas, blast furnace gas etc.

2. Non-Conventional Sources

A plenty of energy is needed to sustain industrial growth and agricultural production. The existing sources of energy such as coal, oil, uranium etc. may not be adequate to meet the ever increasing energy demands. These conventional sources of energy are also depleting and may be exhausted at the end of the century or beginning of the next century. Consequently sincere and untiring efforts shall have to be made by the scientists and engineers in exploring the possibilities of harnessing energy from several non-conventional energy sources. The various *non-conventional energy sources* are as follows :

(*i*) Solar energy
(*ii*) Wind energy
(*iii*) Energy from biomass and biogas
(*iv*) Ocean thermal energy conversion
(*v*) Tidal energy
(*vi*) Geothermal energy
(*vii*) Hydrogen energy
(*viii*) Fuel cells

(*ix*) Magneto-hydrodynamic generator (*x*) Thermionic converter
(*xi*) Thermo-electric power.

Advantages of non-conventional energy sources :
The leading advantages of non-conventional energy sources are :
1. They do not pollute the atmosphere.
2. They are available in large quantities.
3. They are well suited for decentralised use.

According to energy experts the non-conventional energy sources can be used with advantage for *power generation* as well as other applications in a large number of locations and situations in our country.

7.2.2. Description of Non-Conventional Sources of Energy

Some important non-conventional sources of energy are described below :

7.2.2.1. Energy stored in water

The energy contained in flowing streams of water is a form of mechanical energy. It may exist as the kinetic energy of a moving stream or as potential energy of water at some elevation with respect to a lower datum level, an example of which would be the water held behind a dam. *Hydraulic plants* are slowly increasing in number, although the number of new plants of this type built is quite small compared with those which exploit heat energy. As a usual thing, the most desirable hydroelectric sites are the first to be utilized, consequently, as more hydroelectric plants are built, the owners must pay increasingly higher development costs.

From the stand point of capitalistic economics, it is often hard to justify the development of hydroelectric power in comparison with steam power, but from the stand point of the conservation of a fixed natural resource, namely, its mineral fuels, it is obvious that every effort should be made to harness the water power of the country, since if unharnessed it goes to waste, whereas fuel, if unmined, remains intact and undiminished in value in the ground.

Water power is quite cheap where water is available in abundance. Although *capital cost of hydroelectric power plants is higher as compared to other types of power plants yet their operating costs are quite low.*

7.2.2.2. Wind power

The man has been served by the power from winds for many centuries but the total amount of energy generated in this manner is small. The expense of installation and variability of operation have tended to limit the use of the windmill to intermittent services where *its variable output has no serious disadvantage.* The principal services of this nature are the *pumping of water into storage tanks and the charging of storage batteries.*

Windmill power equipment may be classified as follows :
1. *The multi-bladed turbine wheel.* This is the foremost type in use and its efficiency is about 10 per cent of the kinetic energy of the wind passing through it.
2. *The high-speed propeller type.*
3. The rotor.

The propeller and rotor types are *suitable for the generation of electrical energy,* as both of them possess the ability to start in very low winds. The *propeller type is more likely to be used in small units* such as the driving of small battery charging generators, whereas the *rotor,* which is rarely seen, is more practical *for large installations,* even of several hundred kilowatts' capacity.

In india, the wind velocity along coastline has a range 10-16 kmph and a survey of wind power has revealed that wind power is capable of exploitation for pumping water from deep wells for generating small amounts of electric energy.

Modern windmills are capable of working on velocities as low as 3-7 kmph while *maximum efficiency is attained at 10-12 kmph.*

A normal working life of 20 to 25 years is estimated for windmills.

The great advantage of this source of energy is that *no operator is needed and no maintenance and repairs are necessary for long intervals.*

Characteristics of wind power/energy. Some characteristics of wind energy are given below :

1. No fuel provision and transport are required in wind energy systems.
2. It is a renewable source of energy.
3. Wind power systems are non-polluting.
4. Wind power systems, upto a few kW, are less costly, but on a large scale, costs can be competitive with conventional electricity. Lower costs can be achieved by mass production.

Problems Associated with wind energy :

1. Wind energy systems are noisy in operation.
2. Large areas are needed to install *wind farms* for electrical power generators.
3. Wind energy available is dilute and fluctuating in nature. Because of dilute form, conversion machines have to be necessarily large.
4. Wind energy needs storage means because of its irregularity.

7.2.2.3. Solar energy

General aspects :

Solar Constant. Solar constant is the energy from the sun, per unit time, received on a unit area of surface perpendicular to the radiation, in space, at the earth's mean distance from the sun. According to Thekaekara and Drummond (1971) the value of the solar constant is 1353 W/m² (1.940 cal/cm² min, or 4871 kJ/m² hr.).

Beam Radiation. The solar radiation received from the sun without change of direction is called beam radiation.

Diffuse Radiation. It is the solar radiation received from the sun after its direction has been changed by reflection and scattering by the atmosphere.

Air Mass. It is the path length of radiation through the atmosphere, considering the vertical path at sea level as unity.

Zenith Angle. It is the angle between the beam from the sun and the vertical.

Solar Altitude. It is the angle between the beam from the sun and horizontal *i.e.*, (90-zenith angle).

Solar or Short-wave Radiation. It is the radiation originating from the sun, at a source temperature of about 6000°K and in the wavelength range of 0.3 to 3.0 μm.

Long-wave Radiation. Radiation originating from sources at temperatures near ordinary ambient temperatures and thus substantially all at wavelength greater than 3 μm.

Declination. It is the angular position of the sun at solar noon with respect to the plane of the equator (north positive).

- The surface of the earth receives from the sun about 10^{14} kW of solar energy which is approximately five order of magnitude greater than currently being consumed from all resources. It is evident that sun will last for 10^{11} years. Even though the sun light is filtered by the atmosphere one square metre of the land exposed to direct sun light receives the energy equivalent of about 1 H.P. or 1 kW. However, this vast amount of solar energy reaching earth is not easily convertible and certainly is not "free".

- There are two obvious obstacles to harnessing solar energy. *Firstly* it is not constantly available on earth. Thus some form of storage is needed to sustain a solar power system through the night and during periods when local weather conditions obscure the sun. *Second* the solar energy is diffused. Although the total amount of energy is enormous, the collection and conservation of solar energy into useful forms must be carried out over a large area which entails a large capital investment for the conversion apparatus.

- Solar energy, therefore, most likely will be developed not because it is cheaper than alternative energy sources but because these alternative sources sooner or later (*i*) will be exhausted, (*ii*) will become increasingly move expensive, (*iii*) will continue to political and economical control by the nations possessing them and (*iv*) will produce undesirable yet incompletely understood environmental consequences, especially on large scale that will be required to meet projected demands even with controlled growth.

- Solar energy has some good advantages in comparison to the other sources of power. Solar radiation does not contaminate environment or endanger ecological balance. It avoids major problems like exploration, extraction and transportation.

Collectors in various ranges and applications :

Following list gives the thermal applications of solar energy and possible temperature ranges :

1. *Long temperature*
   ($t = 100°C$)
   (*i*) Water heating
   (*ii*) Space heating   ... Flat plate
   (*iii*) Space cooling
   (*iv*) Drying

2. *Medium temperature*
   (100 to 200°C)
   (*i*) Vapour engines and turbines
   (*ii*) Process heating   ... Cylindrical Parabola
   (*iii*) Refrigeration
   (*iv*) Cooking

3. *High temperature*
   (> 200°C)
   (*i*) Steam engines and turbines
   (*ii*) Stirling engine   ... Parabolloid Mirror arrays
   (*iii*) Thermo-electric generator

The above classification of low, medium and high temperature ranges is somewhat arbitrary. Heating water for domestic applications, space heating and cooling and drying of agricultural products (and industrial products) is generally at temperature below 100°C, achieved using flat plate collectors with one or two glass plate covers. Refrigeration for preservation of food products, heating for certain industrial processes, and operation of engines and turbines using low boiling organic vapours is possible at some what higher temperature of 100 to 200°C and may be achieved using focusing collectors with cylindrical-parabola reflectors requiring only one directional diurnal tracking. Conventional steam engines and turbines, stirling hot air engines, and thermoelectric generators require the solar collectors to operate at high temperatures. Solar collectors operating at temperature above 200°C generally consist of parabolloid reflector as an array of mirrors reflecting to a central target, and requiring two directional diurnal tracking.

The concentrators or focusing type collectors can give high temperatures than flat plate collectors, but they entail the following *shortcomings/limitations*.

1. Non-availability and high cost of materials required. These materials must be easily shapeable, yet have a long life; they must be light weight and capable of retaining their brightness in tropical weather. Anodised aluminium and stainless steel are two such materials but they are expensive and not readily available in sufficient quantities.

2. They require direct light and are not operative when the sun is even partly covered with clouds.

3. They need tracking systems and reflecting surfaces undergo deterioration with the passage of time.

4. These devices are also subject to similar vibration and movement problems as radar antenna dishes.

Comparison between flat plate and focusing collectors :

1. The absorber area of a concentrator system is smaller than that of a flat-plate system of the same solar energy collection area and the *insolation intensity is therefore greater*.

2. Because the area from which heat is lost to the surroundings per unit of the solar energy collecting area is less than that for a flat plate collector and because the insolation on the absorber is more concentrated, the working fluid can attain higher temperatures in a concentrating system than in a flat-plate collector of the same solar energy collecting surface.

3. Owing to the small area of absorber per unit of solar energy collecting area, selective surface treatment and/or vacuum insulation to reduce heat losses and *improve collector efficiency are economically feasible*.

4. Since higher temperatures can be achieved, the focusing collector *can be used for power generation*.

5. *Little or no anti-freeze is required* to protect the absorber in a concentrator system whereas the entire solar energy collection surface requires anti-freeze protection in a flat-plate collector.

6. Out of the beam and diffuse solar radiation components, only beam component is collected in case of focusing collectors because diffuse component cannot be reflected and is thus lost.

7. *Costly orienting systems* have to be used to track the sun.

8. Non-uniform flux on the absorber whereas flux in flat-plate collectors is uniform.

Flat plate collectors :

Description

Fig. 7.9 shows a Flat plate collector which consists of four essential components :

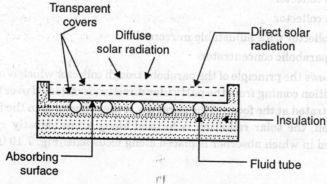

Fig. 7.9. Flat plate solar collector.

1. *An absorber plate.* It intercepts and absorbs solar radiation. This plate is usually metallic (Copper, aluminium or steel), although plastics have been used in some low temperature applications.

In most cases it is coated with a material to enhance the absorption of solar radiation. The coating may also be tailored to minimise the amount of infrared radiation emitted.

A heat transport fluid (usually air or water) is used to extract the energy collected and passes over, under or through passages which forms an integral part of the plate.

2. *Transparent covers.* These are one or more sheets of solar radiation transmitting materials and are placed above the absorber plate. They allow solar energy to reach the absorber plate while reducing convection, conduction and re-radiation heat losses.

3. *Insulation beneath the absorber plate.* It minimises and protects the absorbing surface from heat losses.

4. *Box like structure.* It contains the above components and keeps them in position.

- Various types of flat plate collectors have been designed and studied. These include tube in plate, corrugated type, spiral wound type etc. Other criteria is single exposure, double exposure or exposure and reflector type. The collector utilizes sheets of any of the highly conducting material *viz.* copper, aluminium, or galvanized iron. The sheets are painted dead black for increasing the absorbtivity. The sheets are provided with one or more glass or plastic covers with air gap in between to reduce the heat transfer losses. The sides which are not exposed to solar radiation are well insulated. The whole assembly is fixed in air tight wooden box which is mounted on simple device to give the desired angle of inclination. The dimensions of collectors should be such as to make their handling easy. The collector will absorb the sun energy (direct as well as diffused) and transfer it to the fluid (air, water or oil) flowing within the collector. Basically, a flat plate collector is *effective* most of time, *reliable* for good many years and also *inexpensive*.

- Use of flat mirrors in the flat plate collectors improves the output, permitting higher temperatures of operation. Side mirrors are used either at north and south edges or at east and west edges of the collector or a combination of both. The mirrors may be of reversible or non-reversible type.

Focusing (or concentrating) collectors :

The main types of focusing or concentrating collectors are as follows :

1. Parabolic trough collector
2. Mirror strip collector
3. Fresnel less collector
4. Flat plate collector with adjustable mirrors
5. Compound parabolic concentrator.

Fig. 7.10 (*a*) shows the principle of the parabolic trough collector which is often used in focusing collectors. Solar radiation coming from the particular direction is collected over the area of reflecting surface and is concentrated at the focus of the parabola, if the reflector is in the form of a trough with parabolic cross-section, the solar radiation is focused along a line. Mostly *cylindrical parabolic concentrators* are used in which absorber is placed along focus axis [Fig. 7.10 (*b*)].

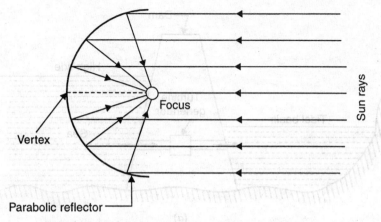

Fig. 7.10. (*a*) Cross-section of parabolic trough collector.

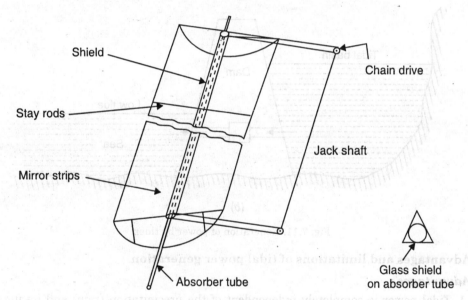

Fig. 7.10. (*b*) Cylindrical parabolic system.

7.2.2.4. Tidal power

The rise and fall of tides offers a means for storing water at the rise and discharging the same at fall. Of course the head of water available under such cases is very low but with increased catchment area considerable amounts of power can be generated at a negligible cost.

The use of tides for electric power generation is practical in a few favourably situated sites where the geography of an inlet of bay favours the construction of a large scale hydroelectric plant. To harness the tides, a dam would be built across the mouth of the bay in which large gates and low head hydraulic turbines would be installed. At the time of high tide the gates are opened and after storing water in the tidal basin the gates are closed. After the tide has receded, there is a working hydraulic head between the basin water and open sea/ocean and the water is allowed to flow back to the sea through water turbines installed in the dam. With this type of arrangement, the generation of electric power is *not continuous*. However by using reversible water turbine the turbine can be run continuously as shown in Fig. 7.11.

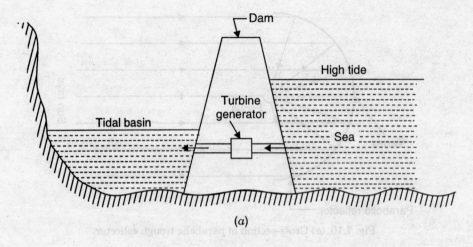

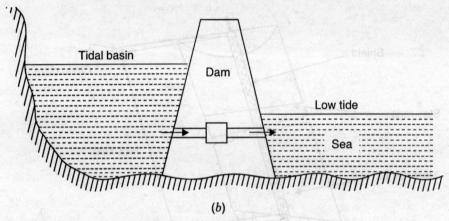

Fig. 7.11. Generation of power by tides.

Advantages and limitations of tidal power generation

Advantages :

1. Tidal power is completely independent of the precipitation (rain) and its uncertainty, besides being inexhaustible.
2. Large area of valuable land is not required.
3. When a tidal power plant works in combination with thermal or hydro-electric system peak power demand can be effectively met with.
4. Tidal power generation is free from pollution.

Limitations :

1. Due to variation in tidal range the output is not uniform.
2. Since the turbines have to work on a wide range of head variation (due to variable tidal range) the plant efficiency is affected.
3. There is a fear of machinery being corroded due to corrosive sea water.
4. It is difficult to carry out construction in sea.

POWER PLANTS, NON-CONVENTIONAL ENERGY SOURCES, HYDRAULIC AND STEAM TURBINES 249

5. As compared to other sources of energy, the tidal power plant is costly.
6. Sedimentation and silteration of basins are the problems associated with tidal power plants.
7. The power transmission cost is high because the tidal power plants are located away from load centres.
— The first commercial tidal power station in the World was constructed in France in 1965 across the mouth of La Rance Estuary. It has a high capacity of 240 MW. The average tidal range at La Rance is 8.4 m and the dam built across the estuary encloses an area of 22 km^2.

7.2.2.5. Geothermal energy

Geothermal energy is the heat from the high pressure steam coming from within the earth. It is a renewable source of energy derived from the rainwater in the earth heated to over 180°C by subterranean hot rocks. Geothermal sources show up hot water springs with temperature of steam upto 150°C ; these are widely distributed over the globe.

Fig. 7.12 shows a schematic diagram depicting how hot springs are produced through hot magma (molten mass), the fractured crystalline rocks, the permeable rocks and percolating ground water.

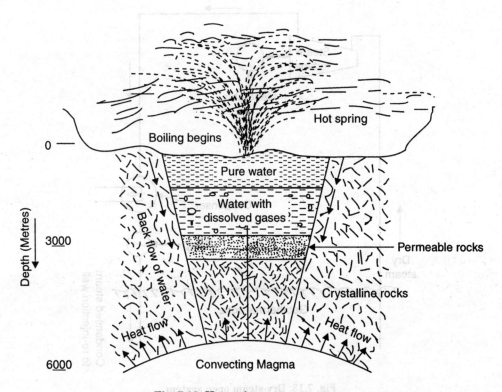

Fig. 7.12. Hot spring system structure.

Geothermal sources :

The following five general categories of geothermal sources have been identified :

1. Hydrothermal convective systems
   - (i) Vapour-dominated or dry steam fields.
   - (ii) Liquid-dominated system or wet steam fields.
   - (iii) Hot-water fields.
2. Geopressure resources.
3. Petro-thermal or hot dry rocks (HDR).
4. Magma resources.
5. Volcanoes.

The hydro-thermal convective systems are best resources for geothermal energy exploitation at present. Hot dry rock is also being considered.

Fig. 7.13 shows a dry-steam open system used in Larderello (Italy) and Greyser (U.S.A.).

Fig. 7.14 shows a flash steam open type system used in Cerro Prietol Mexico, Otake (Japan).

Fig. 7.15 shows a hot water closed (Binary) system developed in U.S.S.R. and U.S.A.

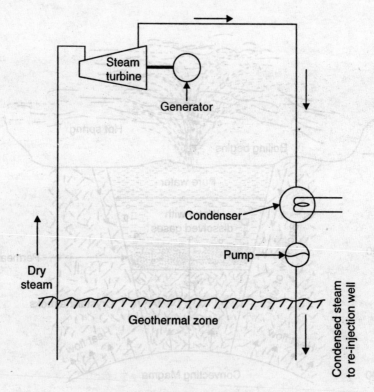

Fig. 7.13. Dry-steam open system.

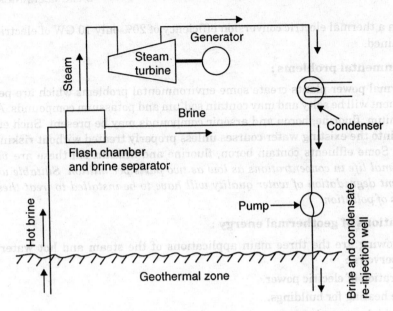

Fig. 7.14. Flash steam open type system.

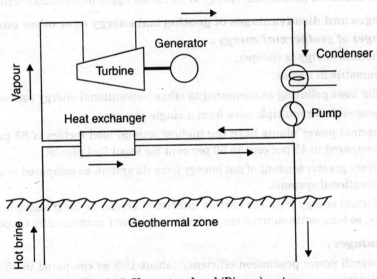

Fig. 7.15. Hot water closed (Binary) system.

Geothermal power estimates :

Although geothermal power estimates vary very widely yet rough estimate is given below :

Depth	Total stored energy (approximately)
3 km	8×10^{21} joules
10 km	4×10^{22} joules

— The energy stored in hot springs = 10 per cent of the above quantities.
— If the above energy is extracted from a 3 km belt, with 1% thermal energy recovery factor, at a uniform rate of over 50 years period, thermal power of 50 GW is obtained.

With a thermal electric conversion efficiency of 20% only 10 GW of electric power will be obtained.

Environmental problems :

Geothermal power plants create some environmental problems which are peculiar to them alone. The effluent will be salty and may contain sodium and potassium compounds. Additionally, in some cases lithium, fluorine, boron and arsenic compounds may be present. Such effluents cannot be discharged into the existing water courses unless properly treated without risking severe pollution problems. Some effluents contain boron, fluorine and arsenic. All these are *very harmful to plants and animal life in concentrations as low as two parts per million. Suitable waste treatment plants to prevent degradation of water quality will have to be installed to treat these new and increased sources of pollution.*

Applications of geothermal energy :

The following are the three main applications of the steam and hot water from the wet geothermal reservoirs :

1. Generation of electric power.
2. Space heating for buildings.
3. Industrial process heat.

The major benefit of geothermal energy is its *varied application and versatility*.

Advantages and disadvantages of geothermal energy over other energy forms

Advantages of geothermal energy :

1. Geothermal energy is cheaper.
2. It is versatile in its use.
3. It is the least polluting as compared to other conventional energy sources.
4. It is amenable for multiple uses from a single resource.
5. Geothermal power plants have the highest annual load factors of 85 per cent to 90 per cent compared to 45 per cent to 50 per cent for fossil fuel plants.
6. It delivers greater amount of net energy from its system as compared to other alternative or conventional systems.
7. Geothermal energy from the earth's interior is almost as inexhaustible as solar or wind energy, so long as its sources are actively sought and economically tapped.

Disadvantages :

1. Low overall power production efficiency (about 15% as compared to 35 to 40% for fossil fuel plants).
2. Drilling operation is noisy.
3. Large areas are needed for exploitation of geo-thermal energy.
4. The withdrawal of large amounts of steam or water from a hydro-thermal reservoir may result in surface subsidence or settlement.

7.2.2.6. Thermo-electric power

According to *Seebeck effect,* when the two ends of a loop of two dissimilar metals are held at a different temperature, an electromotive force is developed and the current flows in loop. This method by selection of suitable materials, can also be used for power generation. This method involves *low initial cost and negligible operating cost.*

POWER PLANTS, NON-CONVENTIONAL ENERGY SOURCES, HYDRAULIC AND STEAM TURBINES

7.3. HYDRAULIC TURBINES

7.3.1. Introduction to Hydraulic Turbines

A **hydraulic turbine** is a *prime mover* (a machine which uses the raw energy of a substance and converts into mechanical energy) *that uses the energy of flowing water and converts it into the mechanical energy (in the form of rotation of the runner)*. This mechanical energy is used in running an electric generator which is directly coupled to the shaft of the hydraulic turbine ; from this electric generator, we get electric power which can be transmitted over long distances by means of transmission lines and transmission towers.

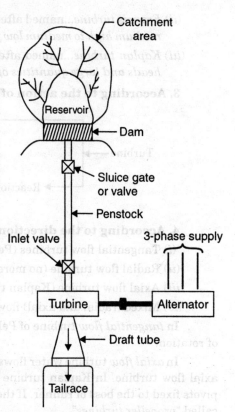

Fig. 7.16. Flow sheet of hydroelectric power plant.

- The hydraulic turbines are also known as 'water turbines' since the fluid medium used in them is water.
- First hydroelectric station was probably started in America in 1882 and thereafter development took place very rapidly. In India, the first major hydroelectric development of 4.5 MW capacity named as Sivasamudram Scheme in Mysore was commissioned in 1902.
- Hydro (water) power is a conventional renewable source of energy which is clean, free from pollution and generally has a good environmental effect. However the following factors are major obstacles in the utilisation of hydropower resources :

(i) Large investments,
(ii) Long gestation period, and
(iii) Increased cost of power transmission.

Fig. 7.16 shows the flow sheet of hydroelectric power plant.

7.3.2. Classification of Hydraulic Turbines

The hydraulic turbines are classified as follows :

1. According to the head and quantity of water available
2. According to the name of the originator
3. According to the action of water on moving blades
4. According to the direction of flow of water in the runner
5. According to the disposition of the turbine shaft
6. According to the specific speed N.

1. According to the head and quantity of water available :

(i) *Impulse turbine*...requires *high head* and *small quantity of flow*.
(ii) *Reaction turbine*...requires *low head* and *high rate of flow*.

Actually there are two types of reaction turbines, one for medium head and medium flow and the other for low head and large flow.

2. According to the name of the originator :

(i) *Pelton turbine*...named after Lester Allen Pelton of California (U.S.A.). It is an impulse type of turbine and is used for *high head* and *low discharge*.

(*ii*) *Francis turbine*...named after James Bichens Francis. It is a reaction type of turbine from *medium high to medium low heads* and *medium small to medium large quantities of water*.

(*iii*) *Kaplan turbine*...named after Dr. Victor Kaplan. It is a *reaction type of turbine for low heads and large quantities of flow*.

3. According to the action of water on moving blades :

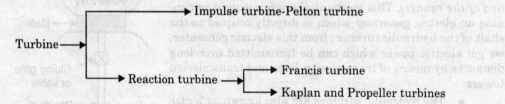

4. According to the direction flow of water in the runner :

 (*i*) Tangential flow turbines (Pelton turbine)

 (*ii*) Radial flow turbine (no more used)

 (*iii*) Axial flow turbine (Kaplan turbine)

 (*iv*) Mixed (radial and axial) flow turbine (Francis turbine).

In *tangential flow* turbine of Pelton type the water strikes the runner tangential to the path of rotation.

In *axial flow* turbine water flows parallel to the axis of the turbine shaft. Kaplan turbine is an axial flow turbine. In Kaplan turbine the runner blades are *adjustable and can be rotated* about pivots fixed to the boss of runner. If the runner blades of the axial flow turbines are *fixed*, these are called "*propeller turbines*".

In *mixed flow* turbines the water enters the blades radially and comes out axially, parallel to the turbine shaft. *Modern Francis turbines have mixed flow runners*.

5. According to the disposition of the turbine shaft :

Turbine shaft may be either vertical or horizontal. In the modern practice, Pelton turbines usually have horizontal shafts whereas the rest, especially the large units, have vertical shafts.

6. According to the specific speed :

The *specific speed* of a turbine is defined as the speed of a geometrically similar turbine that would develop 1 kW *under 1 m head*. All geometrically similar turbines (irrespective of the sizes) will have the same specific speeds when operating under the same head.

$$\text{Specific, } N_s = \frac{N\sqrt{P}}{H^{5/4}}$$

where N = The normal working speed,

 P = Power output of the turbine, and

 H = The net or effective head in metres.

Turbines with low specific speeds work under high head and low discharge conditions, while high specific speed turbines work under low head and high discharge conditions.

The following table gives the comparison between the impulse and reaction turbines with regard to their operation and application.

Table 7.1. Comparison between Impulse and Reaction Turbines

S. No.	Aspects	Impulse turbine	Reaction turbine
1.	Conversion of fluid energy	The available fluid energy is converted into K.E. by a nozzle.	The energy of the fluid is partly transformed into K.E. before it (fluid) enters the runner of the turbine.
2.	Changes in pressure and velocity	The pressure remains same (atmospheric/throughout the action of water on the runner.	After entering the runner with an excess pressure, water undergoes changes both in velocity and pressure while passing through the runner.
3.	Admittance of water over the wheel	Water may be allowed to enter a part or whole of the wheel circumference.	Water is admitted over the circumference of the wheel.
4.	Water-tight casing	Required	Not necessary.
5.	Extent to which the water fills the wheel / turbine	The wheel/turbine does not run full and air has a free access to the buckets.	Water completely fills all the passages between the blades and while flowing between inlet and outlet sections does work on the blades.
6.	Installation of unit	Always installed above the tail race. No draft tube is used.	Unit may be installed above or below the tail race-use of a draft tube is made.
7.	Relative velocity of water	Either remaining constant or reduces slightly due to friction.	Due to continuous drop in pressure during flow through the blade, the relative velocity increases.
8.	Flow regulation	—By means of a needle valve fitted into the nozzle. —Impossible without loss.	—By means of a guide-vane assembly. —Always accompanied by loss.

7.3.3. Impulse Turbines—Pelton Wheel

In an impulse turbine the *pressure energy of water is converted into kinetic energy* when passed through the nozzle and forms the high velocity jet of water. The formed water jet is used for driving the wheel.

Pelton wheel (named after the American engineer Lester Allen Pelton), among the various impulse turbines that have been designed and utilized is by far the important. The Pelton wheel or Pelton turbine is a *tangential flow impulse turbine*.

Important Pelton turbine Installation in India :

S. No.	Scheme/Project	Location (State)	Source of water
1.	Koyana hydroelectric project	Koyana (Maharashtra)	Koyana river
2.	Mahatma Gandhi hydroelectric works	Sharavathi (Karnataka)	Sharavathi river
3.	Mandi hydroelectric scheme	Joginder Nagar (Himachal Pradesh)	Uhl river
4.	Pallivasal power station	Pallivasal (Kerala)	Mudirapuzle river
5.	Pykara hydroelectric scheme	Pykara (Tamil Nadu)	Pykara river.

Construction and working of Pelton wheel/turbine :

A Pelton wheel/turbine consists of a **rotor**, at the periphery of which are mounted equally spaced *double hemispherical or double ellipsoidal buckets*. Water is transferred from a high head source through penstock which is fitted with a **nozzle,** through which the water flows out as a *high speed jet.* A **needle spear** moving inside the nozzle *controls the water flow through the nozzle* and at the same time, provides a smooth flow with negligible energy loss. All the available *potential energy is thus converted into kinetic energy* before the jet strikes the **buckets** of the **runner.** *The pressure all over the wheel is constant and equal to atmosphere, so that energy transfer occurs due to purely impulse action.*

The pelton turbine is provided with a **casing** the function of which is to prevent the splashing of water and to discharge water to the tail race.

When the nozzle is completely closed by moving the spear in the forward direction the amount of water striking the runner is reduced to zero but the runner due to inertia continues revolving for a long time. In order to bring the runner to rest in a short time, a nozzle (brake) is provided which directs the jet of **water** on the back of buckets ; this jet of water is called **braking jet.**

Speed of the turbine runner is kept constant by a **governing mechanism** *that automatically regulates the quantity of water flowing through the runner in accordance with any variation of load.*

Fig. 7.17 shows a schematic diagram of a Pelton wheel, while Fig. 7.18 shows two views of its bucket.

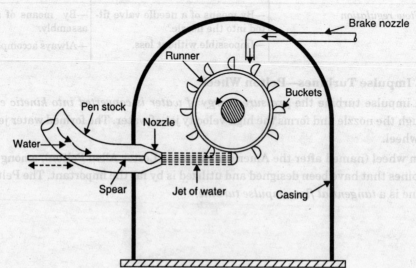

Fig. 7.17. Pelton wheel.

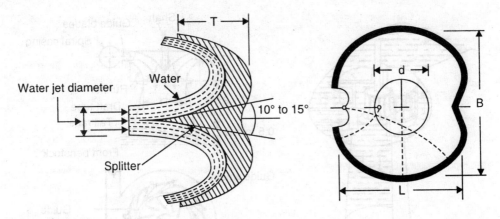

Fig. 7.18. The bucket dimensions.

The jet emerging from the nozzle hits the splitter symmetrically and is equally distributed into the two halves of hemispherical bucket as shown. The bucket centre line cannot be made exactly like a mathematical cusp, *party because of manufacturing difficulties and party because the jet striking the cusp invariably carries particles of sand and other abrasive material which tend to wear it down.* The inlet angle of the jet is therefore between 1° and 3°, but it is always assumed to be zero in all calculations. Then the relative velocity of the jet leaving the bucket would be opposite in direction to the relative velocity of the entering jet : This cannot be achieved in practice since the jet leaving the bucket would then strike *the back of the succeeding bucket to cause splashing and interference so that overall turbine efficiency would fall to low values.* Hence, in practice, the angular deflection of the jet in the bucket is limited to about 165° or 170°, and the bucket is therefore *slightly smaller* than a hemisphere in size.

7.3.4. Reaction Turbines

In reaction turbines, the *runner utilizes both potential and kinetic energies.* As the water flows through the stationary parts of the turbine, whole of its pressure energy is *not* transformed to kinetic energy and when the water flows through the moving parts, there is a *change both in pressure and in the direction and velocity of flow water.* As the gives up its energy to the runner, both its pressure and absolute velocity get reduced. The water which acts on the runner blades is under a pressure *above atmospheric and the runner passages are always completely filled with water.*

7.3.4.1. Francis turbine

Fig. 7.19 shows a schematic diagram of a Francis turbine. The *main parts of a Francis turbine are* :

1. *Penstock* ... It is a large size conduit which conveys water from the upstream of the dam/reservoir to the turbine runner.
2. *Spiral / scroll casing* ... It constitutes a closed passage whose cross-sectional area gradually decreases along the flow direction, area is maximum at inlet and nearly zero at exit.
3. *Guide vanes / wicket gates* ... These vanes direct the water onto the runner at an angle appropriate to the design. The motion to them is given by means of a hand wheel or automatically by a governor.
4. *Governing mechanism* ... It changes the position of the guide blades/vanes to affect a variation in water flow rate, when the load conditions on the turbine change.

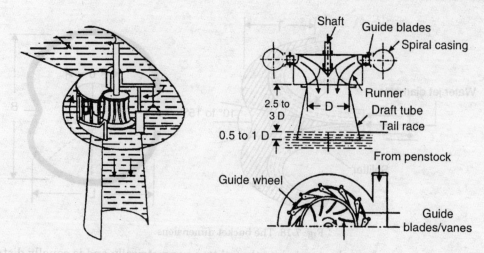

Fig. 7.19. Schematic diagram of a Francis turbine.

5. *Runner and runner blades* ... — The driving force on the runner is both due to impulse and reaction effects ;

— The number of runner blades usually varies between 16 to 24

6. *Draft tube* ... It is a gradually expanding tube which discharges water, passing through the runner, to the tail race.

The modern Francis turbine is an *inward mixed flow reaction turbine* (in the earlier stages of development, Francis turbine had a purely radial flow runner) *i.e., water under pressure, enters the runner from the guide vanes towards the centre in radial direction and discharges out of the runner axially.* The Francis turbine operates under *medium heads* and also requires *medium quantity* of water. It is employed in the medium head power plants. This type of turbine covers a wide range of heads. Water is brought down to the turbine through a *penstock* and directed to a number of stationary orifices fixed all around the circumference of the *runner*. These stationary orifices are commonly called as guide *vanes* or *wicket gates.*

The head acting on the turbine is partly transformed into kinetic energy and the rest remains as pressure head. There is a difference of pressure between the guide vanes and the runner which is called the *reaction pressure* and is responsible for the motion of the runner. This is why a Francis turbine is also known *as reaction turbine.*

In Francis turbine *the pressure at inlet is more than that at the outlet*. This means that the water in the turbine must flow in a closed conduit. Unlike the Pelton type, where the water strikes only a few of the runner buckets at a time, in the Francis turbine the *runner is always all of water. The moment of runner is affected by the change of both the potential and kinetic energies of water.* After doing the work the water is discharged to the tail race through a closed tube of gradually enlarging section. This is known as *draft tube*. It does not allow water to fall freely to tail race level as in the Pelton turbine. The free end of the draft tube is submerged deep to tail water making, thus, the entire water passage, right from the head race upto the tail race, *totally enclosed.*

Fig. 7.20 shows general layout of a hydroelectric power plant *using a reaction turbine.*

POWER PLANTS, NON-CONVENTIONAL ENERGY SOURCES, HYDRAULIC AND STEAM TURBINES

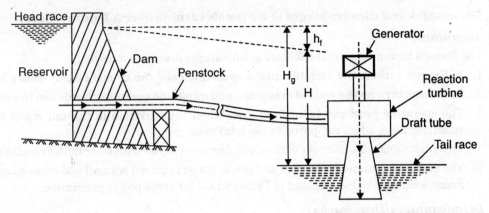

Fig. 7.20. General layout of a hydroelectric power plant using a reaction turbine.

Important Francis turbine installations in India :

S. No.	Scheme/Project	Location (State)	Source of water
1.	Bhakhra dam project	Bhakra (Punjab)	Sutlej river
2.	Cauvery hydroelectric scheme	Siva Samudram (Karnataka)	Cauvery river
3.	Chambal hydroelectric scheme	Gandhi sagar (Rajasthan)	Chembal river
4.	Hirakud dam project	Hirakud (Orissa)	Hirakud river
5.	Rihand dam project	Rihand (Uttar Pradesh)	Rihand river

Important differences between inward and outward flow reaction turbines :

The following are the important differences between inward and outward flow reaction turbines :

S. No.	Aspects	Inward flow reaction turbine	Outward flow reaction turbine
1.	*Entry of water*	Water enters at the outer periphery, flows inward and towards the centre of the turbine and discharges at the outer periphery.	Water enters at the inner periphery flows outward and discharges at the outer periphery.
2.	*Centrifugal head imparted*	Negative (negative centrifugal head reduces the relative velocity of water at the outlet)	Positive (Positive centrifugal head increases the relative velocity of water at the outlet).
3.	*Discharge*	Does not increase.	The discharge increases.
4.	*Speed control*	Easy and effective.	Very difficult.
5.	*Tendency of the wheel to race*	Nil. The turbine adjusts the speed by itself.	If the turbine speed increases the wheel tends to race ; the turbine cannot adjust the speed by itself.
6.	*Suitability*	Quite suitable for medium high heads ; best suitable for large outputs and units.	Quite suitable for low or medium heads.
7.	*Application*	For power projects.	Practically obsolete.

Advantages and disadvantages of a Francis turbine over a Pelton wheel :

Advantages :
The Francis turbine claims the following *advantages over Pelton wheel* :
1. In Francis turbine the variation in the operating head can be more easily controlled.
2. In Francis turbine the ratio of maximum and minimum operating heads can be even two.
3. The operating head can be utilized even when the variation in the tail water level is relatively large when compared to the total head.
4. The mechanical efficiency of Pelton wheel decreases faster with wear than Francis turbine.
5. The size of the runner, generator and power house required is small and economical if the Francis turbine is used instead of Pelton wheel for same power generation.

Disadvantages/Drawbacks :
As compared with Pelton wheel, the Francis turbine has the following *drawbacks/shortcomings* :
1. Water which is not clean can cause very rapid wear in high head Francis turbine.
2. The overhaul and inspection is much more difficult comparatively.
3. Cavitation is an ever-present danger.
4. The water hammer effect is more troublesome with Francis turbine.
5. If Francis turbine is run below 50 per cent head for a long period it will not only lose its efficiency but also the cavitation danger will become more serious.

7.3.4.2. Propeller and Kaplan turbines–Axial flow reaction turbines

It has been observed that with *increasing specific speed* flow tends to be *axial*. If water flows parallel to the axis of the rotation of the shaft, the turbine is known as *axial flow turbine* ; when the head at inlet of the turbine is the sum of pressure energy and kinetic energy and during the flow of water through runner a part of the pressure energy is converted into kinetic energy, the turbine is known as *reaction turbine*. The shaft of an axial flow reaction turbine is *vertical*. The lower end of the shaft is made larger which is known as '*hub*' or '*boss*'. The vanes are fixed on the hub and its acts as runner for axial flow reaction turbine. Two important axial flow reaction turbines are : (*i*) Propeller turbine, and (*ii*) Kaplan turbine.

In these turbines all parts such as *spiral casing, stay vanes, guide vanes, control vanes,* and *draft tube* are similar to mixed-flow turbines in design. But the *water enters the runner in an axial direction* and during the process of energy transfer, it travels across the blade passage in axial direction and *leaves axially. The pressure at the inlet of the blades is larger than the pressure at the exit of the blades. The energy transfer is due to the reaction effect, i.e., the change in the magnitude of relative velocity across the blades.*

In an axial flow turbine the number of blades are *fewer* and hence the *loading on the blade is larger. Smaller contact area causes less frictional loss compared to mixed flow turbines,* but the peripheral speed of the turbine is *larger*. Axial flow rotors *do not have a rim at the outer end* like the Francis rotors ; but the *blades are enclosed in a cylindrical casing.*

The tip clearance between the blades and the cylindrical casing is small ; hence the flow past blades can be considered *two dimensional*. The water coming out from the guide vanes undergoes a whirl which is assumed to satisfy the law of free vortex ($V_w = C/r$). Accordingly the *whirl is largest near the hub and smallest at the outer end of blade*. Hence the *blade is twisted along its axis.*

Propeller turbine :

The need to utilize *low heads where large volume of water is available* makes it essential to provide a large flow area and to run the machine at very low speeds. The propeller turbine is a reaction turbine used for heads between 4 m and 80 m. It is purely *axial-flow* device providing the largest possible flow area that will *utilize a large volume of water and still obtain flow velocities which are not too large.*

The propeller turbine (Fig. 7.21) consists of an axial-flow runner with four to six or at the most ten blades of air-foil shape. The runner is generally kept horizontal, *i.e.,* the shaft is vertical. The blades resemble the propeller of a ship. In the propeller turbine, as in Francis turbine, the runner blades are fixed and *non-adjustable*. The *spiral casing* and *guides blades* are similar to those in Francis turbine. The *guide mechanism* is similar to that in a Francis turbine.

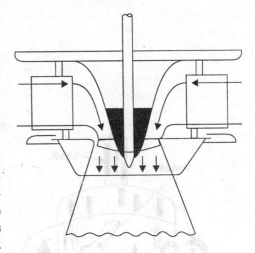

Fig. 7.21. Propeller turbine.

Kaplan turbine :

A propeller turbine is quite suitable when the *load on the turbine remains constant.* At *part load* its efficiency is very low ; since the blades *are fixed,* the water enters with shock (at part load) and eddies are formed which reduces the efficiency. This defect of the propeller turbine is removed in Kaplan turbine. In a Kaplan turbine the *runner blades are adjustable* and *can be rotated about pivots fixed to the boss of the runner.* The *blades are adjusted automatically by servomechanism so that at all loads the flow enters them without shock.* Thus *a high efficiency is maintained even at part load.* The servomotor cylinder is usually accommodated in the hub. Figs. 7.22 and 7.23 show the Kaplan turbine runner and Kaplan turbine (schematic diagram) respectively.

The Kaplan turbine has *purely axial flow.* Usually it has 4 to 6 blades having no outside rim. It is also known as a *variable-pitch propeller turbine* since the pitch of the turbine can be changed because of adjustable vanes. The Kaplan turbine behaves like a propeller turbine at full-load conditions.

The scroll casing, guide mechanism and draft tube are similar to that in the Francis turbine. The shape of runner blades is different from that of Francis turbine. The blades of Kaplan turbine are made of *stainless steel*.

Kaplan turbine, like every propeller turbine, is a *high speed turbine* and is used for smaller heads ; as the speed is high, the number of runner-vanes is small.

Kaplan turbines have taken the place of Francis turbines for certain medium head installations. Kaplan turbines with sloping guide vanes to reduce the overall dimensions are being used.

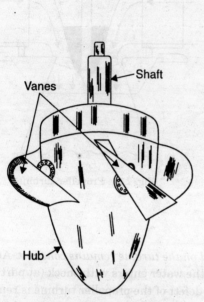

Fig. 7.22. Kaplan turbine runner.   Fig. 7.23.

Important Kaplan turbine installations in India :

S. No.	Scheme/Project	Location (State)	Source of water
1.	Bhakra-Nangal project	Gangwal and Kota (Punjab)	Nangal hydel
2.	Hirakud dam project	Hirakud (Orissa)	Mahanadi river
3.	Nizam Sagar project	Nizam Sagar (Andhra Pradesh)	Nanjira river
4.	Radhanagri hydroelectric scheme	Kolhapur (Maharashtra)	Bhagvati river
5.	Tungbhadra hydroelectric scheme	Tungbadhra (Karnataka)	Tungbadhra river

Differences between Francis turbine and Kaplan turbine :

S. No.	Aspects	Francis turbine	Kaplan turbine
1.	Type of turbine	Radially inward or mixed flow	Partially axial flow
2.	Disposition of shaft	Horizontal or vertical	Only vertical
3.	Adjustability of runner	Runner vanes are not adjustable	Runner vanes are adjustable
4.	Vanes	Large, 16 to 24 blades	Small 3 to 8 blades
5.	Number of vanes Resistance to be overcome	Large, (owing to large number of vanes and greater area of contact with water)	Less (owing to fewer number of vanes and less wetted area)
6.	Head	Medium (60 m to 250 m)	Low (upto 30 m)
7.	Flow rate	Medium	Large
8.	Specific speed	50–250	250–850
9.	Type of governor	Ordinary	Heavy duty

Deriaz turbine :

Fig. 7.24 shows a schematic view of a Deriaz (or diagonal) turbine which is a reaction turbine. It is named in the honour of its inventor P. Deriaz. This turbine is intermediate between the mixed-flow and the axial-flow turbines, because the flow of water as it passes through the runner is at an angle of 45° to axis and hence it is also known as *Diagonal turbine*. Deriaz turbine has the following *features* :

- It can be employed for the heads varying from 30 m to 150 m.
- The blades of the runner are pivoted to the hub and unlike in the Kaplan turbines, the axes of the blades are inclined to the axis of shaft. The direction of flow of water is as in the Francis runner.
- Guide vanes are provided ahead of the blades to regulate and direct the flow.
- The runner has no outer rim connecting all the blades as these blades are movable.
- The casing of the turbine (not shown in Fig. 7.24) is so shaped that there is only small clearance between the blade tips and the casing to reduce leakage loss.

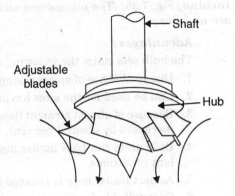

Fig. 7.24. Deriaz turbine.

The runner of the Deriaz turbine is so shaped that it can be used both as a *turbine as well as a pump* and it may be classified as a *reversible type turbine*. As such Deriaz turbines are amply suitable for *pumped storage hydropower plants*.

Advantages of Deriaz turbine :

The Deriaz turbine entails the following *advantages* :

1. Improved part load efficiency.
2. Can be conveniently used as a pump-turbine unit.
3. By adjusting the runner to shut position the starting torque under water can be reduced.
4. Unlike axial flow turbines at shut position the flow area is completely closed.
5. Due to oblique location of blades the loading on the outer trunnion journal bearing is reduced.
6. The arrangement for varying the blade angle can be housed with greater convenience as compared to the Kaplan turbine.

7.3.5. Tubular or Bulb Turbines

Invariably the electric generator coupled to the Kaplan turbine is enclosed and works *inside a straight passage* having the shape of a *bulb*. The *water tight bulb* is submerged directly into a stream of water, and the bends at inlet to casing, draft tube etc. which are *responsible for the loss of head are dispensed with*. The unit then needs *less installation space* with a consequent *reduction in excavation* and *other civil engineering works*. These turbines are referred to as **tubular or bulb turbines**. The tubular turbine, a modified axial flow turbine, was developed in Germany by Arno Fischer in 1973. The economical harnessing of fairly low heads on major rivers is now possible with high-output bulb turbines. The following **features** are worthnoting :

- A *tubular bulb turbine* is an axial flow turbine with either adjustable or non-adjustable runner vanes (and hence similar to Kaplan or propeller turbines).

- In such a turbine the scroll casing is *not* provided but the runner is placed in a tube extending from head water to the tail water (and hence it is tubular turbine).
- It is a low head turbine and is employed for heads varying from 3 m to 15 m.
- The disposition of shaft in a tubular turbine may be vertical, or inclined or horizontal.

The turbo-generator set using tubular turbine has an outer casing having the shape of a bulb. Such a set is now termed as **bulb set** and the turbine used for the set is called a **bulb turbine** (Fig. 7.25). The *advantages* and *disadvantages* of bulb sets *compared to Kaplan turbines* are as follows :

Advantages :

The bulb sets claim the following *advantages* over the Kaplan turbines :
1. Due to absence of spiral casing the plant width is small.
2. Can be used for the sites having very low head.
3. Because of almost straight flow and straight draft tube the maximum turbine efficiency is increased by about 3 per cent.
4. Bulb units can pass higher discharge (than conventional Kaplan turbine) under equivalent conditions.
5. At part loads there is reduced loss of efficiency.
6. Quit suitable for operation on widely varying heads.
7. Because of small dimension of the power house there is saving in excavation and civil engineering works.

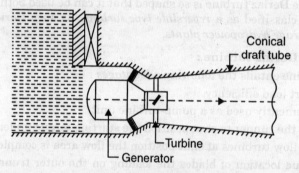

Fig. 7.25. Bulb turbine.

Disadvantages :
1. Leakage of water into generator chamber and condensation are source of trouble (leading to gradual deterioration of electrical insulation).
2. The erection techniques may be time consuming.

The use of bulb turbines offers the saving in the equipment of low head developments and great flexibility of operation and hence are highly suitable for **tidal power station.**

7.3.6. Runaway Speed

'Runaway speed' *is the maximum speed, governor being disengaged, at which a turbine would run when there is no external load but operating under design head and discharge.* All the rotating parts including the rotor of alternator should be designed for the centrifugal stresses caused by this maximum speed.

The practical values of runaway speeds for various turbines with respect to their *rated speed* N are as follows :

Pelton wheel	... 1.8 to 1.9 N
Francis turbine (mixed flow)	... 2.0 to 2.2 N
Kaplan turbine (axial flow)	... 2.5 to 3.0 N

7.3.7. Draft Tube

In the case of mixed and axial flow turbines only a part of available energy is converted into velocity energy at the inlet to the runner ; the rest is in the form of pressure energy. This residual pressure is converted into velocity in the runner, as consequence of which the outlet velocity increases. With increase in the value of specific speed N_s, the exit velocity energy $\frac{V_2^2}{2g}$ increases compared with H (the available energy).

In the *Pelton Wheel* all the available energy is converted into velocity energy before it strikes the wheel. As such it works under *atmospheric conditions* and the wheel *has to be placed above the maximum tail water level*. The loss of energy due to exit velocity varies from 1 to 4%.

In the case of *mixed and axial flow turbines* a large portion of the energy is associated with the water as it leaves the runner. This exit energy varies from 4 to 25% for mixed flow turbines and from 20 to 50% of the total head for axial flow turbines. As this energy cannot be used in the runner, therefore, it becomes necessary to find a way out to extract this energy. *An expanding pressure conduit hermetically fixed at runner outlet and having the other end below the minimum tail water level helps to convert the velocity head into pressure or potential head. This expanding device is called* **draft tube.** Draft tube is an *integral part* of mixed and axial flow turbines. Because of the draft tube *it is possible to have the pressure at runner outlet much below the atmospheric pressure*.

The draft tube serves the following two *purposes* :

1. *It allows the turbine to be set above tail-water level, without loss of head, to facilitate inspection and maintenance.*
2. *It regains, by diffuse action, the major portion of the kinetic energy delivered to it from the runner.*

At rated load, the velocity at the upstream end of the tube for modern units ranges from 7 to 9 m/s, representing from 2.7 to 4.8 m head. As the specific speed (it is the speed of a geometrically similar turbine running under a unit head and producing unit power) is increased and the head reduced, it becomes increasingly important to have an efficient draft tube. Good practice limits the velocity at the discharge end of the tube from 1.5 to 2.1 m/s, representing less than 0.3 m velocity head loss.

Types of draft tubes :

The following two types of draft tubes are commonly used :

1. The straight conical or concentric tube,
2. The elbow type.

Properly designed, the two types are about equally efficient, *over 85 per cent*.

1. Conical type. The conical type draft tube is *generally* used on *low-powered units for all specific speeds, frequently, on large-head units. The side angle of flare ranges from 4 to 6°*, the length from 3 to 4 times the diameter and the discharge area from four to five times the throat area. Fig. 7.26 show that straight *conical* draft tube.

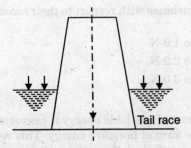

Fig. 7.26. Straight conical draft tube.

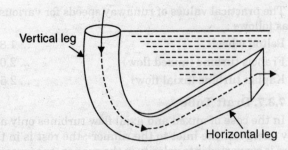

Fig. 7.27. Elbow type draft tube.

2. Elbow type. The elbow type of tube is used with *most* turbine installations. This type of draft tube is designed to turn the water from the *vertical to the horizontal direction with a minimum depth of excavation and at the same time having a high efficiency*. The transition from a circular section in the vertical leg to a rectangular section in the horizontal leg takes place in the bend. The horizontal portion of the draft tube is generally *inclined upwards to lead the water gradually to the level of the tail race and to prevent entry of air from the exit end*. The *exit end of the draft tube must be totally immersed in water*. Fig. 7.27 shows an elbow type draft tube. One or two vertical piers are placed in the horizontal portion of the tube, for structural and hydraulic reasons.

Moody spreading draft tube. Fig. 7.28 shows a *Moody's spreading draft tube*. It is provided with a *solid central core* of conical shape which *reduces whirling action of discharge water*. The efficiency of such a draft tube is about 85%. It is suited particularly for *helical flows* which occur when the water leaves the runner with a whirl component.

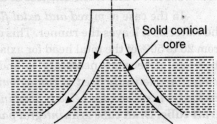

Fig. 7.28. Moody's spreading draft tube or 'Hydrocone.'

7.3.8. Specific Speed

The specific speed of a turbine is defined as the *speed of a turbine which is identical in shape, geometrical dimensions, blade angles, gate opening etc. which would develop unit power when working under a unit head.*

$$\text{Specific speed, } N_s = \frac{N\sqrt{P}}{H^{5/4}}$$

where N = speed of actual turbine (in r.p.m.),

P = shaft power (in kW), and

H = head under which turbine is working (in m).

Specific speed plays an important role in the *selection of the type of turbine*. By knowing the specific speed of turbine the *performance of the turbine can also be predicted*.

If a runner of *high specific speed* is used for a given head and power output, the *overall cost of installation* is *lower*. The selection of too high specific speed reaction runner would reduce the size of the runner to such an extent that the discharge velocity of water into the throat of draft tube would be excessive. This is objectionable because a *vacuum* may be created in the extreme case.

The runner of *too high specific speed* with high available head *increases the cost of turbine* on account of high mechanical strength required. The runner of *too low specific speed* with low available head increases the cost of generator due to the low turbine speed.

An increase in specific speed of turbine is accompanied by lower maximum efficiency and greater depth of excavation of the draft tube. In choosing a high specific speed turbine, an increase in cost of excavation of foundation and draft tube should be considered in addition to the efficiency. The weighted *efficiency over the operating range of turbine is more important in the selection of a turbine instead of maximum efficiency.*

7.3.9. Cavitation

The formation, growth, and collapse of vapour filled cavities or bubbles in a flowing liquid due to local fall in fluid pressure is called cavitation. When the pressure at any point in a flow field equals the vapour pressure of the liquid at that temperature vapour cavities (bubbles of vapour) begin to appear. It is presumed that a vapour cavity is formed around a dust nuclei which is in the liquid (The vapour pressure values of water at 15° and 20°C are 1.74 m and 2.38 m of water column absolute). The cavities thus formed, due to motion of liquid, are carried to high pressure regions where the vapour condenses and they *suddenly collapse*. The adjoining liquid rushes with a very great velocity (and hence with very great force) to occupy the empty spaces thus created, *causes series of violent, irregular, spherical shock waves*. When these irregular implosions occur on the metallic surface, they produce *noise and vibration*.

When the cavities collapse (the collapsing pressure is of the order of 100 times the atmospheric pressure) on the surface of a body, due to repeated 'hammering' action, the metal particle gives way ultimately due to fatigue and *indentations* are formed ; this erosion of material is called **pitting** (Fig. 7.29).

In reaction turbines the cavitation may occur at the *runner exit* or the *draft tube inlet* where the *pressure is negative*. The hydraulic machinery is affected by the cavitation in the following three ways :

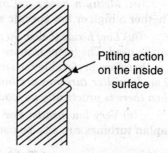

Fig. 7.29. Pitting action on the inside surface (shown on large scale).

1. Roughening of the surface takes place due to loss of material caused by pitting.
2. Vibration of parts is caused due to irregular collapse of cavities.
3. The *actual volume of liquid flowing through the machine is reduced* (since the volume of cavities is many times more than the volume of water from which they are formed) causing sudden drop in output and efficiency.

7.3.10. Selection of Hydraulic Turbines

The following points should be considered while selecting right type of hydraulic turbines for hydroelectric power plant :

1. **Specific speed.** High specific speed is essential where head is low and output is large, because otherwise the rotational speed will be low which means cost of turbo-generator and power house will be high. On the other hand there is practically no need of choosing a high value of specific speed for high installations, because even with low specific speed high rotational speed can be attained with medium capacity plants. Refer Table 7.2.

2. **Rotational speed.** Rotational speed depends on specific speed. Also the rotational speed of an electrical generator with which the turbine is to be directly coupled, depends on the frequency and number of pair of poles. The *value of specific speed adopted should be such that it will give the synchronous speed of the generator*.

3. **Efficiency.** The turbine selected should be such that it gives the *highest overall efficiency for various operating conditions*.

4. **Partload operation.** In general the efficiency at partloads and overloads is less than normal. For the sake of economy the turbine should always run with maximum possible efficiency to get more revenue.

When the turbine has to run at part or overload conditions *Deriaz turbine* is employed. Similarly, for low heads, Kaplan turbine will be useful for such purposes in place of propeller turbine.

5. Cavitation. The installation of water turbines of reaction type over the tailrace is effected by *cavitation*. The critical value of cavitation factor must be obtained to see that the turbine works in *safe zone*. Such a value of cavitation factor also affects the design of turbine, especially of Kaplan propeller and bulb types.

6. Disposition of turbine shaft. Experience has shown that the *vertical shaft* arrangement is better for large-sized reaction turbines, therefore, it is *almost universally adopted*. In case of *large size impulse turbines, horizontal shaft arrangement* is mostly employed.

7. Head. (*i*) **Very high heads** *(350 m and above).* For heads greater than 350 m, Pelton turbine is generally employed and there is practically no choice except in very special cases.

(*ii*) **High heads** *(150 m to 350 m).* In this range either Pelton or Francis turbine may be employed. *For higher specific speeds Francis turbine is more compact and economical than the Pelton turbine* which for the same working conditions would have to be much bigger and rather cumbersome.

(*iii*) **Medium heads** *(60 m to 150 m).* A Francis turbine is usually employed in this range. Whether a high or low specific speed unit would be used depends on the selection of the speed.

(*iv*) **Low heads** *(below 60 m).* Between 30 m and 60 m heads both Francis and Kaplan turbines may be used. The latter is more expensive but yields a higher efficiency at part loads and overloads. It is therefore preferable for *variable loads*. Kaplan turbine is generally employed for heads under 30 m. Propeller turbines are however, commonly used for heads upto 15 m. They are adopted only when there is practically no load variations.

(*v*) **Very low heads.** For very low heads *bulb turbines* are employed these days. Although Kaplan turbines can also be used for heads from 2 m to 15 m but they are *not economical*.

Table 7.2. Criteria for Selection of Turbines

S. No.	Types of turbine	Head H(m)	Specific speed (N_s)	Speed ratio (K_u)	Maximum hydraulic efficiency (%)	Remarks
1.	Pelton : 1 jet 2 jets 4 jets	upto 2000 upto 1500 upto 500	12 to 30 17 to 50 24 to 70	0.43–0.48	89	Employed for very high head.
2.	Francis : High head Medium head Low head	upto 300 50 to 150 30 to 60	80 to 150 150 to 250 250 to 400	0.6 to 0.9	93	Full load efficiency high ; part load efficiency lower than Pelton wheel.
3.	Propeller and Kaplan	4 to 60	300 to 1000	1.4 to 2	93	High part load efficiency; high discharge with low head.
4.	Bulb or tubular turbines	3 to 10	1000 to 1200	6 to 8	91	Employed for very low head-tidal power plants.

Overall efficiency (η_0) of all turbines $\simeq$ 85 percent.

7.4. STEAM TURBINES

7.4.1. Introduction to Steam Turbines

The steam turbine is a prime-mover in which the potential energy of the steam is transformed into kinetic energy, and latter in its turn is transformed into the mechanical energy of rotation of the turbine shaft. The turbine shaft, directly or with the help of a reduction gearing, is connected with the driven mechanism. Depending on the type of the driven mechanism a steam turbine may be *utilised in most diverse fields of industry, for power generation* and *for transport.* Transformation of the potential energy of steam into the mechanical energy of rotation of the shaft is brought about by different means.

7.4.2. Classification of Steam Turbines

There are several ways in which the steam turbines may be classified. The most important and common division being with respect to the *action of the steam*, as :

(a) Impulse.
(b) Reaction.
(c) Combination of impulse and reaction.

Other classification are :

1. According to the number of pressure stages :

(i) Single stage turbines with one or more velocity stages usually of small-power capacities ; these turbines are mostly used for driving centrifugal compressors, blowers and other similar machinery.

(ii) Multistage impulse and reaction turbines ; they are made in a wide range of power capacities varying from small to large.

2. According to the direction of steam flow :

(i) Axial turbines in which steam flows in a direction parallel to the axis of the turbine.

(ii) Radial turbines in which steam flows in a direction perpendicular to the axis of the turbine ; one or more low-pressure stages in such turbines are made axial.

3. According to the number of cylinders :

(i) Single cylinder turbines.
(ii) Double cylinder turbines.
(iii) Three cylinder turbines.
(iv) Four cylinder turbines.

Multi-cylinder turbines which have their rotors mounted on one and the same shaft and coupled to a single generator are known as *single shaft turbines* ; turbines with separate rotor shafts for each cylinder placed parallel to each other are known as **multiaxial turbines.**

4. According to the method of governing :

(i) *Turbines with throttle governing* in which fresh steam enters through one or more (depending on the power developed) simultaneously operated throttle valves.

(ii) *Turbines with nozzle governing* in which fresh steam enters through two or more consecutively opening regulators.

(iii) *Turbines with by pass governing* in which steam turbines besides being fed to the first stage is also directly fed to one, two or even three intermediate stages of the turbine.

5. According to heat drop process :

(i) *Condensing turbines with generators* ; in these turbines steam at a pressure less than atmospheric is directed to a condenser ; besides, steam is also extracted from intermediate stages for feed water heating, the number of such extractions usually being from 2-3 to as much 8-9. The latent heat of exhaust steam during the process of condensation is completely lost in these turbines.

(ii) *Condensing turbines with one or two intermediate stage extractions* at specific pressures for industrial and heating purposes.

(iii) *Back pressure turbines*, the exhaust steam from which is utilised for industrial or heating purposes ; to this type of turbines can also be added (in a relative sense) turbines with deteriorated vacuum, the exhaust steam of which may be used for heating and process purposes.

(iv) *Topping turbines* ; these turbines are also of the back pressure type with the difference that the exhaust steam from these turbines is further utilised in medium and low pressure condensing turbines. These turbines, in general, operate at high initial conditions of steam pressure and temperature, and are mostly used during extension of power station capacities, with a view to obtain better efficiencies.

(v) *Back pressure turbines with steam extraction from intermediate stages at specific pressure* ; turbines of this type are meant for supplying the consumer with steam of various pressures and temperature conditions.

(vi) *Low pressure turbines* in which the exhaust steam from reciprocating steam engines, power hammers, presses, etc., is utilised for power generation purposes.

(vii) *Mixed pressure turbines* with two or three pressure stages, with supply of exhaust steam to its intermediate stages.

6. According to steam conditions at inlet to turbine :

(i) *Low pressure turbines*, using steam at a pressure of 1.2 to 2 ata.

(ii) *Medium pressure turbines*, using steam at pressures of upto 40 ata.

(iii) *High pressure turbines*, utilising pressures above 40 ata.

(iv) *Turbines of very high pressures*, utilising steam at pressures of 170 ata and higher and temperatures of 550°C and higher.

(v) *Turbines of supercritical pressures*, using steam at pressures of 225 ata and above.

7. According to their usage in industry :

(i) *Stationary turbines with constant speed of rotation* primarily used for driving alternators.

(ii) *Stationary steam turbines* with variable speed meant for driving turbo-blowers, air circulators, pumps, etc.

(iii) *Non-stationary turbines* with variable speed ; turbines of this type are usually employed in steamers, ships and railway locomotives.

7.4.3. Advantages of Steam Turbine over the Steam Engines

The following are the principal advantages of steam turbine over steam engines :

1. The thermal efficiency of a steam turbine is much higher than that of a steam engine.

2. The power generation in a steam turbine is at a uniform rate, therefore necessity to use a flywheel (as in the case of steam engine) is not felt.
3. Much higher speeds and greater range of speed is possible than in case of a steam engine.
4. In large thermal stations where we need higher outputs, the steam turbines prove very suitable as these can be made in big sizes.
5. With the absence of reciprocating parts, (as in steam engine) the balancing problem is minimised.
6. No internal lubrication is required as there are no rubbing parts in the steam turbine.
7. In a steam turbine there is no loss due to initial condensation of steam.
8. It can utilise high vacuum very advantageously.
9. Considerable overloads can be carried at the expense of slight reduction in overall efficiency.

7.4.4. Description of Common Types of Steam Turbines

The common types of steam turbines are :
1. Simple impulse turbine.
2. Reaction turbine.

The main difference between these turbines lies *in the way in which the steam is expanded while it moves through them. In the former type steam expands in the nozzles and its pressure does not alter as it moves over the blades while in the latter type the steam expands continuously as it passes over the blades and thus there is gradual fall in the pressure during expansion.*

1. Simple impulse turbines

Fig. 7.30 shows a simple impulse turbine diagrammatically. The top portion of the figure exhibits a longitudinal section through the upper half of the turbine, the middle portion shows one set of nozzles which is followed by a ring of moving blades, while lower part of the diagram indicates approximately changes in pressure and velocity during the flow of steam through the turbine. This turbine is called '*simple*' impulse turbine since the expansion of the steam takes place in one set of the nozzles.

As the steam flows through the nozzle its pressure falls from steam chest pressure to condenser pressure (or atmospheric pressure if the turbine is non-condensing). Due to this relatively higher ratio of expansion of steam in the nozzles the steam leaves the nozzle with a very high velocity. Refer Fig. 7.30, it is evident that the velocity of the steam leaving the moving blades is a large portion of the maximum velocity of the steam when leaving the nozzle. The loss of energy due to this higher exit velocity is commonly the **"carry over loss"** or **"leaving loss"**.

The principal example of this turbine is the well known **"De laval turbine"** and in this turbine the 'exit velocity' or 'leaving velocity' or 'lost velocity' may amount to 3.3 per cent of the nozzle outlet velocity. Also since all the kinetic energy is to be absorbed by one ring of the moving blades only, the velocity of wheel is too high (varying from 25000 to 30000 r.p.m.). This wheel or rotor speed however, can be reduced by different methods (discussed in the following article).

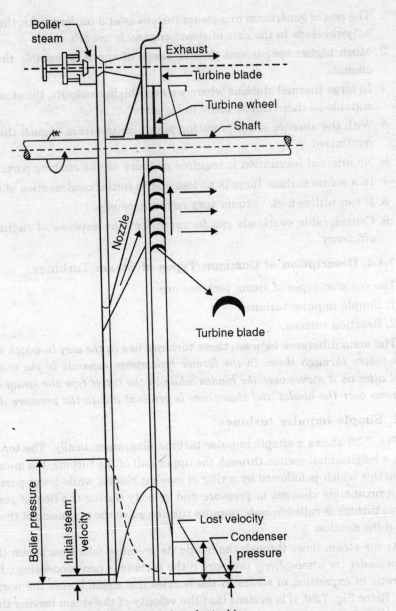

Fig. 7.30. Simple impulse turbine.

2. Reaction turbine

In this type of turbine, *there is a gradual pressure drop and takes place continuously over the fixed and moving blades*. The function of the fixed blades is (the same as the nozzle) that they alter the direction of the steam as well as allow it expand to a larger velocity. As the steam passes over the moving blades its kinetic energy (obtained due to fall in pressure) is absorbed by them. Fig. 7.31 shows a *multi-stage* reaction turbine. The changes in pressure and velocity are also shown there in.

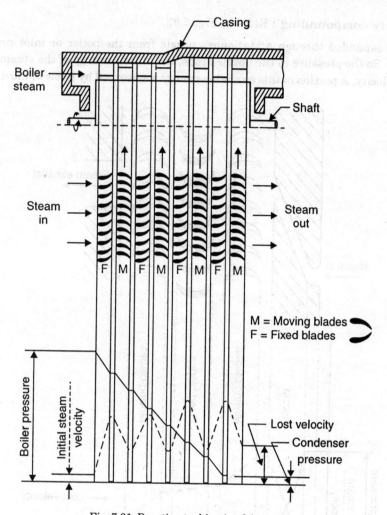

Fig. 7.31. Reaction turbine (multi-stage).

As the volume of steam increases at lower pressures therefore, the diameter of the turbine must increase after each group of blade rings. It may be noted that in this turbine since the pressure drop per stage is small, therefore the number of stages required is much higher than an impulse turbine of the same capacity.

7.4.5. Methods of Reducing Wheel or Rotor Speed (Compounding methods)

As already discussed under the heading 'simple impulse turbine' that if the steam is expanded from the boiler pressure to condenser pressure in one stage the speed of the rotor becomes *tremendously high which crops up practical complicacies*. There are several methods of reducing this speed to lower value ; all these methods utilise a multiple system of rotor in series, keyed on a common shaft and the steam pressure or jet velocity is absorbed in stages as the steam flows over the blades. This is known as '**compounding**'. The different methods of compounding are :

1. Velocity compounding.
2. Pressure compounding.
3. Pressure velocity compounding.
4. Reaction turbine.

1. Velocity compounding : Refer to Fig. 7.32.

Steam is expanded through a stationary nozzle from the boiler or inlet pressure to condenser pressure. So the pressure in the nozzle drops, the kinetic energy of the steam increases due to increase in velocity. A portion of this available energy is absorbed by a row of moving blades. The

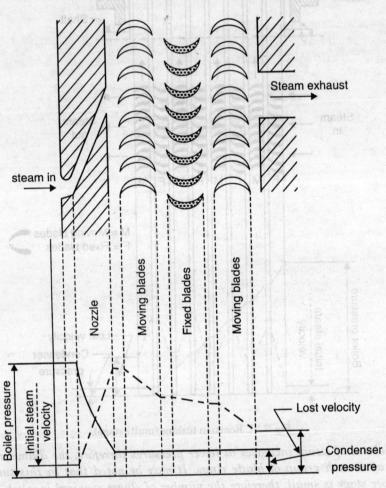

Fig. 7.32. Velocity compounding.

steam (whose velocity has decreased while moving over the moving blades) then flows through the second row of blades which are fixed. The function of these fixed blades is to re-direct the steam flow without altering its velocity to the following next row moving blades where again work is done on them and steam leaves the turbine with a low velocity. Fig. 7.32 shows a cut away section of such a stage and changes in pressure and velocity as the steam passes through the nozzle, fixed and moving blades.

Though this method has the advantage that the initial cost is low due to lesser number of stages yet its efficiency is low.

2. Pressure compounding :

Fig. 7.33 shows rings of fixed nozzles incorporated between the rings of moving blades. The steam at boiler pressure enters the first set of nozzles and expands partially. The kinetic energy of the steam thus obtained is absorbed by the moving blades (stage 1). The steam then expands partially in the second set of nozzles where its pressure again falls and the velocity increases ; the kinetic energy so obtained is absorbed by the second ring of moving blades (stage 2). This is repeated in stage 3 and steam finally leaves the turbine at low velocity and pressure. The number of stages (or pressure reductions) depends on the number of rows of nozzles through which the steam must pass.

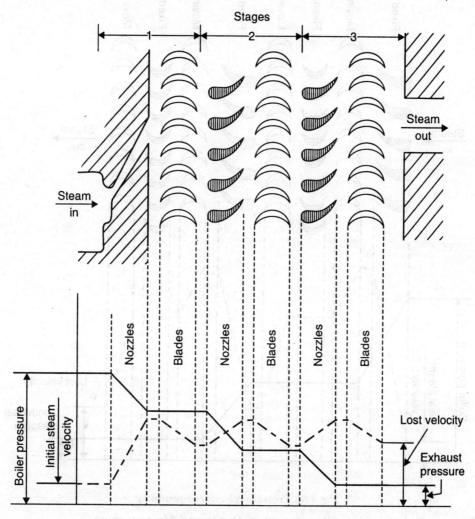

Fig. 7.33. Pressure compounding.

This method of compounding is used in *Rateau and Zoelly turbine. This is most efficient turbine since the speed ratio remains constant but it is expensive owing to a large number of stages.*

3. Pressure velocity compounding :

This method of compounding is the combination of two previously discussed method. The total drop in steam pressure is divided into stages and the velocity obtained in each stage is also compounded. The rings of nozzles are fixed at the beginning of each stage and pressure remains constant during each stage. The changes in pressure and velocity are shown in Fig. 7.34.

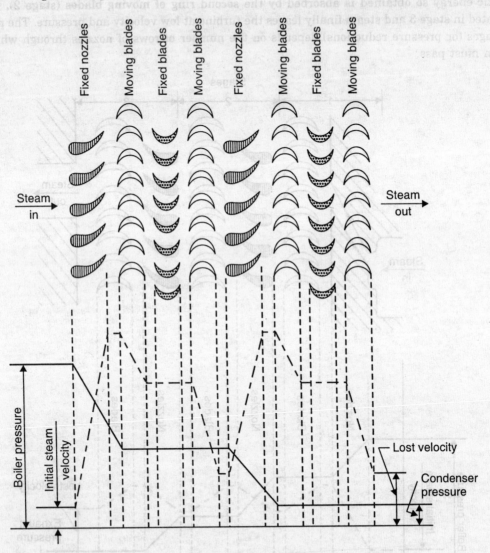

Fig. 7.34. Pressure velocity compounding.

This method of compounding is used in *Curits and Moore turbine*.

4. Reaction turbine
It has been discussed in Article 7.4.4.

7.4.6. Differences between Impulse and Reaction Turbines

S. No.	Particulars	Impulse turbine	Reaction turbine
1.	Pressure drop	Only in nozzles and not in moving blades.	In fixed blades (nozzles) as well as in moving blades.
2.	Area of blade channels	Constant.	Varying (converging type).
3.	Blades	Profile type.	Aerofoil type.
4.	Admission of steam	Not all round or complete.	All round or complete.
5.	Nozzles / fixed blades	Diaphragm contains the nozzle.	Fixed blades similar to moving blades attached to the casing serve as nozzles and guide the steam.
6.	Power	Not much power can be developed.	Much power can be developed.
7.	Space	Requires less space for same power.	Requires more space for same power.
8.	Efficiency	Low.	High.
9.	Suitability	Suitable for small power requirements.	Suitable for medium and higher power requirements.
10.	Blade manufacture	Not difficult.	Difficult.

HIGHLIGHTS

1. The various types of power plants are :
   steam power plant ; Hydro-electric power plant ; Nuclear power plants ; Gas turbine power plants ; Diesel engine power plants.
2. The various sources of energy are :
   (i) *Conventional sources.* Fuels
   (ii) *Non-conventional sources.* Solar energy, wind energy, energy from biomass and biogas, ocean thermal energy conversion, tidal energy, geothermal energy, fuel cells, magneto-hydrodynamic generator, thermionic converters, thermo-electric power.
3. A *hydraulic turbine* is a prime mover that uses the energy of flowing water and converts it into the mechanical energy (in the form of rotation of runner).
   - In an *impulse turbine* the pressure energy of water is converted into kinetic energy when passed through the nozzle and forms the high velocity jet of water.
   - In *reaction turbines*, the runner utilizes both potential and kinetic energies.
   - The modern reaction turbine is an inward flow reaction turbine.
   - A *propeller turbine* has fixed blades and is suitable when the *load on the turbine remains constant*. The blades of a Kaplan turbine are adjustable and can be rotated about pivots fixed to the boss of the runner, subsequently high efficiency is maintained even at part loads.
   - The *Deriaz turbine* is intermediate between the mixed-flow and axial-flow turbines, because the flow of water as it passes through the runner is at an angle of 45° to axis and hence it is also known as *Diagonal turbine*.
   - A *tubular or bulb turbine* is an axial flow turbine with either adjustable or non-adjustable runner vanes.
   - '*Runaway speed*' is the maximum speed, governor being disengaged, at which a turbine would run when there is no external load but operating under design head and discharge.
   - A *draft tube* is an expanding pressure conduit hermetically fixed at runner outlet and having the other end below the minimum tail water level which helps to convert the velocity head into pressure or potential head.

- The *specific speed of a turbine* is defined as the speed of a turbine which is identical in shape, geometrical dimensions, blades angles, gate opening etc. which would develop unit power when working under a unit head.

4. A *steam turbine* is a prime-mover in which the potential energy of steam is transformed into kinetic energy and latter in its turn is transformed into the mechanical energy of rotation of the turbine shaft.
   - Steam turbines are classified as follows :
   (*i*) Impulse turbines ; (*ii*) Reaction turbines ; (*iii*) Combination of impulse and reaction.

OBJECTIVE TYPE QUESTIONS

Fill in the blanks or say "Yes" or "No" :

1. A ...... power plant converts the chemical energy of fossil fuels into mechanical/electrical energy.
2. Non-conventional energy sources do not pollute the atmosphere.
3. A chemical fuel is a substance which releases heat energy on ....
4. Peat is the first stage in the formation of coal from wood.
5. Anthracite is a soft coal and has a dull black lustre.
6. Wood charcoal is obtained by destructive distillation of ....
7. In ...... analysis, individual elements are not determined ; only the percentage of moisture, volatile matter, fixed carbon and ash are determined.
8. The chief source of liquid fuels (chemical) is ....
9. The main constituents of natural gas are methane and ethane.
10. Coke-oven gas is obtained during the production of coke, by heating the bituminous coal.
11. Producer gas is produced by blowing steam into white hot coke or coal.
12. Complete fission of one kg of uranium contains the energy equivalent of 4500 tonnes of coal.
13. The energy liberated by the complete oxidation of a unit mass or volume of a fuel is called ...... of the fuel.
14. The operating cost of hydroelectric power plants is quite high.
15. The normal working life of windmills range from 20 to 25 years.
16. A hydraulic turbine is a prime-mover that uses the energy of flowing water and converts it into ...... energy.
17. Pelton wheel is used for high head and high discharge.
18. Kaplan turbine is a reaction type of turbine for low heads and large quantities of flow.
19. If the runner blades of the axial flow turbines are fixed, these are called ...... turbines.
20. Turbines with low specific speed work under high head and low discharge conditions, while high specific speed turbines work under low head and high discharge conditions.
21. In an ...... turbine the pressure energy of water is converted into kinetic energy when passed through the nozzle and forms the high velocity jet of water.
22. In ...... turbines, the runner utilizes both potential and kinetic energies.
23. ...... is a large size conduit which conveys water from the upstream of the dam/reservoir to the turbine runner.
24. ...... is a gradually expanding tube which discharges water, passing through the runner, to the tail race.
25. The mechanical efficiency of Pelton wheel decreases faster with wear than Francis turbine.
26. In Francis turbine the variation in the operating head can be more easily controlled.
27. A propeller turbine is quite suitable when the load on the turbine varies considerably.
28. The blades of the propeller turbine are fixed.
29. In case of a Kaplan turbine a high efficiency is maintained.
30. The ...... turbine has purely axial flow.
31. Deriaz turbine is also known as ...... turbine.
32. ...... speed is the maximum speed, governor being disengaged, at which a turbine would run when there is no external load but operating under design head and discharge.
33. Because of the ...... it is possible to have the pressure at runner outlet much below the atmospheric pressure.

POWER PLANTS, NON-CONVENTIONAL ENERGY SOURCES, HYDRAULIC AND STEAM TURBINES 279

34. The ...... speed of a turbine is defined as the speed of a turbine which is identical in shape, geometrical dimensions, blade angles, gate opening, etc. which would develop unit power when working under a unit head.
35. The formation, growth, and collapse of vapour filled cavities or bubbles in a flowing liquid due to local fall in fluid pressure is called ....
36. The steam turbine is a prime-mover in which the ...... energy of the steam is transformed into ...... energy.
37. In axial flow turbines steam flows in a direction parallel to the axis of the turbine.
38. In radial turbines steam flows ...... to the axis of the turbine.
39. Low pressure turbines use steam at a pressure of 1.2 to 2 ata.
40. Medium pressure turbines use steam at a pressure upto ...... ata.
41. High pressure turbines utilise pressure above 40 ata.
42. The thermal efficiency of a steam turbine is much higher than that of a steam engine.
43. In steam turbines internal lubrication is not required.
44. The steam turbines can utilise high vacuum very advantageously.
45. In ...... turbine, there is a gradual pressure drop and takes place continuously over the fixed and moving blades.
46. In case of reaction turbine blade manufacture is difficult.
47. In ...... turbine area of blade channels is constant.
48. In ...... turbine the admission of steam is all round or complete.
49. As compared to a reaction turbine, the efficiency of an impulse turbine is ....
50. Impulse turbine is suitable for small power requirements.

Answers

1. steam   2. Yes   3. combustion   4. Yes   5. No
6. wood   7. proximate   8. petroleum   9. Yes   10. Yes
11. No   12. Yes   13. calorific value 14. No   15. Yes
16. mechanical 17. No   18. Yes   19. propeller   20. Yes
21. impulse   22. reaction   23. Penstock   24. Draft tube 25. Yes
26. Yes   27. No   28. Yes   29. Yes   30. Kaplan
31. diagonal   32. Runaway   33. draft tube   34. specific   35. cavitation
36. potential, kinetic 37. Yes 38. perpendicular   39. Yes   40. No
41. Yes   42. Yes   43. No   44. Yes   45. reaction
46. Yes   47. impulse   48. reaction   49. low   50. Yes.

THEORETICAL QUESTIONS

1. Describe briefly, with a neat diagram, the working of a "Steam power plant".
2. How are hydro-electric power plants classified ?
3. Enumerate various sources of energy.
4. Enlist the various non-conventional energy sources.
5. What are the advantages of non-conventional energy sources ?
6. Explain briefly any two of the following :
   (i) Wind power ;
   (ii) Solar energy ;
   (iii) Geothermal energy.
7. What is a hydraulic turbine ?
8. How are hydraulic turbines classified ?
9. Give the comparison between impulse and reaction turbines.
10. With the help of neat diagram explain the construction and working of a Pelton wheel turbine.

11. Draw a schematic diagram of a Francis turbine and explain briefly its construction and working.
12. State the advantages and disadvantages of a Francis turbine over a Pelton wheel.
13. What is a draft tube?
14. What is the difference between a Propeller turbine and Kaplan turbine?
15. Where is Kaplan turbine used?
16. State the advantages of a Kaplan turbine over Francis turbine.
17. Write a short note on Deriaz turbine.
18. What are tubular or bulb turbines?
19. What are the advantages and disadvantages of bulb sets compared to Kaplan turbines?
20. How is specific speed of a turbine defined?
21. What is cavitation? How can it be avoided in reaction turbines?
22. What is the principle of operation of a steam turbine?
23. How are steam turbines classified?
24. State the advantages of steam turbines over the steam engines?
25. What is the difference between an impulse and a reaction turbine?
26. Give the description of a simple impulse turbine with the help of a heat sketch.
27. Mention the drawbacks of simple impulse turbine.
28. Draw a neat sketch of a reaction turbine and explain its working.
29. What is compounding of impulse turbine?
30. What is the difference between pressure compounding and velocity compounding?
31. Give the comparison between impulse and reaction turbines.

MODULE – 5

Chapters :

8. Machine Tools

9. Manufacturing Processes

MODULE - 5

8

Machine Tools

8.1. Classification of cutting tools. 8.2. Types of chips. 8.3. Cutting tool materials. 8.4. Cutting conditions. 8.5. Machinability. 8.6. Tool life. 8.7. Forces of a single point tool. 8.8. Machining processes. 8.9. Machine tools. 8.10. **Lathe**—Introduction—Parts of a lathe—Size and specifications of lathe—Types of lathe—Lathe tools—Lathe operation—Lathe accessories—Eccentric turning—Thread rolling—Cutting speed, feed and depth of cut—Testing of lathes—Exercise. 8.11. **Drilling machines**—Introduction—Specifications of a drilling machine—Operations performed—Classification of drilling machines—Cutting speeds and feeds—Work holding devices—Drill holding devices—Drilling machine tools. 8.12. **Shaping machine** (Shaper)—Introduction—Classification of shapers—Principal parts—Specifications of a shaper—Operations performed—Tools used. 8.13. **Planing machine** (Planer)—Introduction—Comparison between planer and shaper—Types of planer—Principal parts of a planer—Size of a planer—Standard clamping devices—Planer operations. 8.14. **Milling machine**—General aspects—Specifications of a milling machine—Types of milling machines—Main parts of a horizontal milling machine—Types of milling cutters—Milling operations—Cutting speed, feed and depth of cut. 8.15. **Grinding machines**—Introduction—Types of grinding machines—The grinding wheel—Abrasives—Selection of grinding wheels—Wheel shapes—Mounting of wheels—Wheel truing—Highlights—Objective Type Questions—Theoretical Questions.

8.1. CLASSIFICATION OF CUTTING TOOLS

Cutting tools are *classified* as follows :

1. Single point cutting tools
2. Multi-point cutting tools
   (i) Solid tool
   (ii) Brazed tool
   (iii) Inserted bit tool.

The various angles of a single point tool are shown in Fig. 8.1.

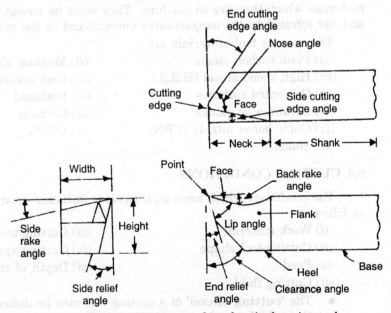

Fig. 8.1. Various angles of a single point tool.

283

8.2. TYPES OF CHIPS

The chips produced, whatever the cutting conditions be, may belong to one of the following three types. Refer Fig. 8.2.

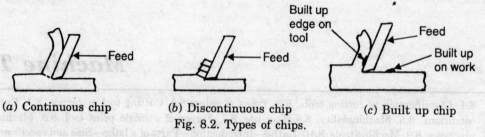

(a) Continuous chip   (b) Discontinuous chip   (c) Built up chip

Fig. 8.2. Types of chips.

1. Continuous chips
2. Discontinuous chips
3. Continuous chips with build-up edge (B.U.E.).

- *Continuous chips* are produced while machining more *ductile materials*. This type of chip is *most desirable*.
- *Discontinuous chips* are usually produced while cutting *more brittle materials* like greasy cast iron, bronze and hard brass.
- When machining ductile materials, conditions of high local temperature and extreme pressure in the cutting zone and also high friction in the tool-chip interface may cause the work material to adhere or weld to the cutting edge of the tool forming the built up edge (B.U.E.) as shown in Fig. 8.2.

8.3. CUTTING TOOL MATERIALS

The materials of which cutting tools are made must, of course, be *harder* than the materials which they are to machine. They must be *strong, tough, rigid, abrasive-resistant,* and *not affected by the temperatures* encountered in the machining.

The various tool materials are :

(i) Plain carbon steels
(ii) Medium alloy steels
(iii) High speed steels (H.S.S.)
(iv) Cast non-ferrous alloys (stellite)
(v) Cemented carbides
(vi) Diamond
(vii) Ceramics or oxides
(viii) Cermets
(ix) Cubic boron nitride (CBN)
(x) UCON
(xi) Sialon.

8.4. CUTTING CONDITIONS

The *conditions* which have an *important influence upon metal cutting in machining* are as follows :

(i) Work material
(ii) Cutting-tool material
(iii) Cutting-tool shape
(iv) Cutting speed
(v) Feed
(vi) Depth of cut
(vii) Cutting fluid.

- The **'cutting speed'** of a cutting tool may be defined as its *rate of forward travel, through the work material, or relative to the work material.*
- **'Feed'** may be defined as the *relatively small movement per cycle* of the cutting tool relative to the workpiece in a direction which is usually normal to the cutting speed direction.

MACHINE TOOLS

- **Cutting fluids** are used for the following reasons :
  (i) To cool the cutting tool and the workpiece ;
  (ii) To lubricate the chip, tool, and workpiece ;
  (iii) To help carry away the chips ;
  (iv) To lubricate some of the moving parts of the machine tool.

8.5. MACHINABILITY

Machinability may be defined as the *ease with which a material can be machined*.

The more important criteria for *measuring* both tool performance and machinability are :

(i) The rate at which the material can be removed.

(ii) The smoothness and accuracy of the machined surface obtained.

(iii) The life of the cutting tool, or how long the cutting edge cuts satisfactorily before resharpening is required.

(iv) The power required for making the cut.

The relative machinability of a metal is expressed in percentage. All machinable metals are compared to a *basic standard steel* which is given an *arbitrary machinability rating of 100%*. The standard steel is a free cutting steel containing 0.13% carbon, 0.06 to 1.1% manganese and 0.08% to 0.03% of sulphur. The machinability index of a metal is given by the following relationship (this is a way to compare machinability of metals).

$$\text{Machinability index} = \frac{\text{Cutting speed of metal for 20 min. tool life}}{\text{Cutting speed of standard steel for 20 min. tool life}} \times 100.$$

8.6. TOOL LIFE

Tool life is defined as the *time interval between the two successive regrinds*. It *depends* upon the following *factors* :

(i) Tool material

(ii) Hardness of the material

(iii) Type of material being cut

(iv) Type of the surface on the metal (scaly or smooth)

(v) Profile of the cutting tool

(vi) Type of the machining operation being performed

(vii) Microstructure of the material

(viii) Finish required on the workpiece.

Tool life of a cutting tool may be calculated by using the following relation :

$$VT^n = C$$

where V = cutting speed in m/min.,

T = tool life in min.,

C = a constant (which is numerically equal to cutting speed that gives the tool life of one min.), and

n = another constant (depending upon finish, workpiece material and tool material)

  = 0.1 for H.S.S. steel tools ; 0.2 to 0.25 for carbide tools and 0.4 to 0.55 for ceramic tools.

8.7. FORCES OF A SINGLE POINT TOOL

Refer Fig. 8.3.

Orthogonal cutting : Resultant, $R = \sqrt{F_x^2 + F_z^2}$

Oblique cutting : Resultant, $R = \sqrt{F_x^2 + F_y^2 + F_z^2}$

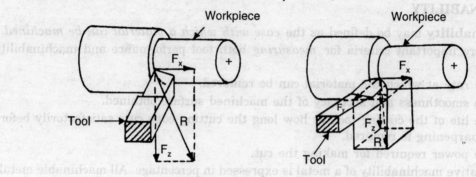

Fig. 8.3. Forces on a cutting tool.

Torque to be developed on the workpiece,

$$T = \frac{F_z \times D}{2 \times 1000} \text{ Nm (neglecting the components } F_x \text{ and } F_y\text{)}$$

(where D = diameter of the workpiece in mm)

Heat produced (= work done in cutting metal) = $\dfrac{F_z \times v}{60 \times 1000}$ kN m/s or kJ/s or kW

where v = cutting speed in m/min.

Power required = $\dfrac{F_z \times v}{60 \times 1000 \times \eta}$ kW

(where η = efficiency of the machine)

The approximate values of efficiencies of the different machines when working at full loads are :

1. Lathes .... 80 to 90%
2. Drilling machines .... 85 to 90%
3. Milling machines .... 80 to 90%
4. Shapers and planers .... 65 to 75%
5. Grinding machines .... 80 to 85%.

8.8. MACHINING PROCESSES

- *Machining* is the process of cold working the metals into different shapes by using different types of machine tools. This process is mainly used to bring the metal objects produced by means of different fabrication techniques to final dimensions.
- *Machinability* which is defined as *the ease of removing metal while maintaining dimensions and developing a satisfactory surface finish* is an important aspect affecting the metallurgical and properties stand-point of metals. Tool wear and power

MACHINE TOOLS

consumption are two factors which affect the metal removal rate. Greater effort and time is required to keep the tools sharp due to rapid tool wear and frequent machine stoppage for replacing the dull tools. Types of *metal chips* formed during machining operation also affect the different characteristics. Machinability of a metal is generally indicated by *machinability ratings* (which are dependent upon their techniques of determination as well as upon the particular metal cutting operation used for their measurement).

- *Machining* is accomplished with the use of machines known as *"machine tools"*. For production of variety of machined surfaces different types of machine tools have been developed. *The kind of surface produced depends upon the shape of cutting, the path of the tool as its passes through the material or both.* Depending on them metal cutting processes are called either turning or planing or boring or other operations performed by machine tools like lathe, shaper, planer, drill, miller, grinder etc. as illustrated schematically in Fig. 8.4.

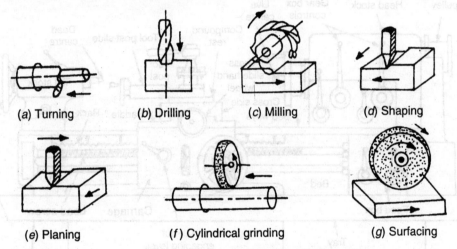

(a) Turning (b) Drilling (c) Milling (d) Shaping
(e) Planing (f) Cylindrical grinding (g) Surfacing

Fig. 8.4. Principal machining methods.

8.9. MACHINE TOOLS

Machine tools are used for machining. They employ cutting tools to remove excess material from the given job. The machine tools can be classified as follows :

1. General Purpose
 (i) Lathe   (ii) Drilling machine
 (iii) Shaping machine   (iv) Planing machine
 (v) Milling machine   (vi) Sawing machine.

2. Special Purpose
 (i) Special lathes like capstan, turret and copying lathes
 (ii) Boring machine   (iii) Broaching machine
 (iv) Production milling machine   (v) Production drilling machine.

3. Automatic Machine Tools

These machine tools, also called Automatic screw cutting machines (or simply automats), are used for mass production of essentially small parts using a set of pre-designed and job-specific cams.

4. Computer Numerical Control (CNC) Machine Tools

Under CNC machine tools, we have *CNC turning centre,* which does all the work of a lathe and *CNC machining centre,* which does milling, drilling etc., with provision for automatic tool changing and tool wear correction built into it.

8.10. LATHE

8.10.1. Introduction. Refer to Fig. 8.5.

- A lathe is one of the oldest and perhaps most important machine tools ever developed. The job to be machined is rotated (turned) and the cutting tool is moved relative to the job. That is why, the lathes are also called *"Turning machines"*. If the tool moves parallel to the axis of rotation of the workpiece, cylindrical surface is produced, while it moves perpendicular to their axis, it produces a flat surface.

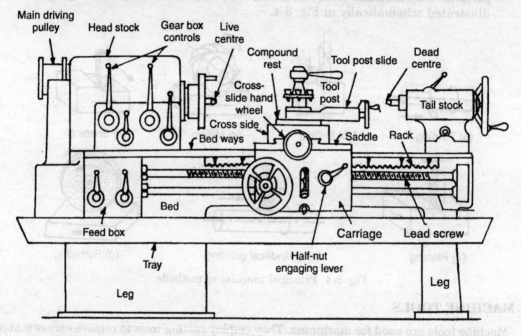

Fig. 8.5. Centre lathe.

- Fig. 8.6 shows the *working principle of a lathe.* In a lathe, the workpiece is held in a chuck or between centres and rotated about its axis at a uniform speed. The cutting tool held in tool post is fed into the workpiece for desired depth and in desired direction (*i.e.*, in linear, transverse or lateral direction). Since there exists a relative motion between the workpiece and the cutting tool, therefore the material is removed in the form of chips and the desired shape is obtained.

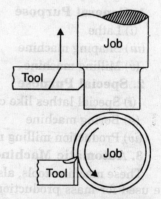

Fig. 8.6. Working principle of a lathe.

8.10.2. Parts of Lathe

1. Bed :
- It is a heavy, rigid casting made in one piece.
- It is the base or foundation of the lathe.

2. Headstock :
- It is permanently fastened to the innerways at the left hand end of the bed.
- It serves to support the spindle and driving arrangements.
- All lathes receive their power through the headstock, which may be equipped with a step-cone pulleys or a gear head drive (the modern lathes are provided with all geared type head stock to get large variations of spindle speeds).
- In order to allow the long bar or work holding devices to pass through, the head stock spindle is made hollow. A tapered sleeve fits into the tapered spindle hole.

3. Tailstock :
- It is situated at the right hand end of the bed.
- It is used for supporting the right end of the work.
- It is also used for holding and feeding the tools such as drills, reamers, taps etc.

4. Carriage :
- The carriage controls and supports the cutting tool.
- The carriage has the following five major parts :

(i) *Saddle.* It is a H-shaped casting fitted over the bed. It moves along the guide ways.

(ii) *Cross-slide.* It carries the compound slide and tool post ; can be moved by power or by hand.

(iii) *Compound rest.* It is marked in degrees ; used during taper turning to set the tool for angular cuts.

(iv) *Tool post.* The tool is clamped on the tool post.

(v) *Apron.* It is attached to the saddle and hangs in front of the bed. It has gears, levers and clutches for moving the carriage with the lead screw for thread cutting.

5. Feed mechanism :
- It is employed for imparting various feeds (longitudinal, cross and angular) to the cutting tool.
- It consists of feed reverse lever, tumbler reversing mechanism, change gears, feed gear box, quick change gear box, lead screw, feed rod, apron mechanism and half nut mechanism.

8.10.3. Size and Specifications of Lathe

Size of a lathe is *specified* in any *one* of the following ways :

1. The height of the centres measured over the lathe bed.
2. Swing or maximum diameter that can be rotated over the bed ways.
3. Swing or diameter over carriage. This is the largest diameter of work that will revolve over the lathe saddle.
4. Maximum job length in mm that may be held between the centres (headstock and tailstock centres).
5. Bed length in metres which may include the headstock length also.
6. Diameter of the hole through lathe spindle for turning bar material.

8.10.4. Types of Lathe

The following are the types of lathes :

1. Speed lathe :
- In this lathe spindle can rotate at a very high speed with the help of a variable speed motor built inside the headstock of the lathe.
- It is used mainly for wood working, centering, metal spinning, polishing etc.

2. Engine or centre lathe :
- It is the most common types of lathe and is widely used in workshop.
- The speed of the spindle can be widely varied as desired which is not possible in a speed lathe.
- The cutting tool may be fed both in cross and longitudinal directions with reference to the lathe axis with the help of a carriage.

3. Bench lathe :
- It is usually mounted on a bench.
- It is very similar to speed or centre lathe, the only difference being it is smaller in size which enables it handle small work (usually requiring considerable accuracy such as in the production of gauges, punches and beds for press tools).

4. Tool room lathe :
- It is similar to an engine lathe, designed for obtaining accuracy.
- It is used for manufacturing precision components, dies, tools, jigs etc. and hence it is called as tool room lathe.

5. Turret and capstan lathes :
- These lathes have provision to hold a number of tools and can be used for performing wider range of operations.
- These are particularly suitable for *mass production of identical parts in minimum time.*

6. Automatic lathes :
- These lathes are so designed that the tools are automatically fed to the work and withdrawn after all operations, to finish the work, are complete.
- They require little attention of the operator, since the entire operation is automatic.
- These are used for mass production of identical parts.

7. Special purpose lathes :

These lathes are primarily designed for carrying out a particular operation with utmost efficiency.

8.10.5. Lathe Tools

- In a lathe, for a general purpose work, the tool used is a *single point tool* (a tool having one cutting edge), but for special *operations multi-point tools may be used*.
- The commonly used materials are *high carbon steel, high speed steel, cemented carbides, diamond tips* and *ceramics*.

Depending upon the nature of operation done by the tool, the lathe tools are *classified* as follows :

(i) Turning tool (left hand or right hand)

(ii) Facing tool (left hand or right hand)

(iii) Chamfering tool (left hand or right hand)
(iv) Form or profile tool
(v) Parting or necking tool
(vi) External threading tool
(vii) Internal threading tool
(viii) Boring tool
(ix) Knurling tool.

The above mentioned tools are shown in Fig. 8.7.

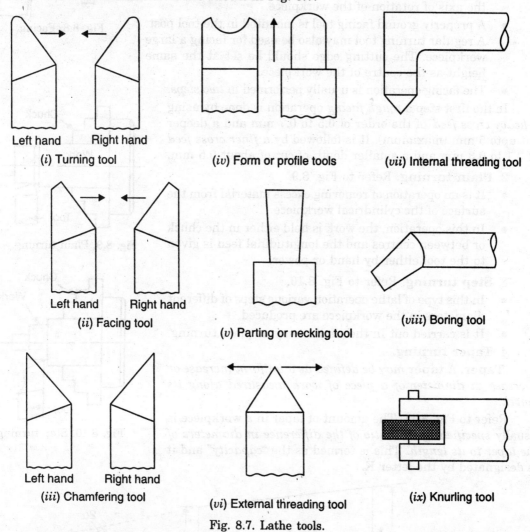

Fig. 8.7. Lathe tools.

8.10.6. Lathe Operation

Common lathe operations which can be carried out on a lathe are enumerated and briefly discussed as follows :

1. Facing
2. Plain turning
3. Step turning
4. Taper turning
5. Drilling
6. Reaming

7. Boring
8. Under cutting or grooving
9. Threading
10. Knurling
11. Forming.

1. **Facing.** Refer to Fig. 8.8.
   - *"Facing"* is an *operation of machining the ends of a workpiece to produce a flat surface square with the axis.* It is also used to cut the work to the required length.
   - The operation involves feeding the tool *perpendicular* to the axis of rotation of the workpiece.
   - A properly ground facing tool is mounted in the tool post. A regular turning tool may also be used for facing a large workpiece. The cutting edge should be set at the same height as the centre of the workpiece.
   - The facing operation is usually performed in *two steps*.

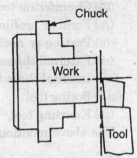

Fig. 8.8. Facing.

In the first step a *rough facing* operation is done by using a *heavy cross feed* of the order of 0.5 to 0.7 mm and a deeper cut upto 5 mm (maximum). It is followed by a *finer cross feed* of 0.1 to 0.3 mm and a smaller depth of cut of about 0.5 mm.

2. **Plain turning.** Refer to Fig. 8.9.
   - It is an operation of removing excess material from the surface of the cylindrical workpiece.
   - In this operation, the work is held either in the chuck or between centres and the longitudinal feed is given to the tool either by hand or power.

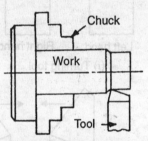

Fig. 8.9. Plain turning.

3. **Step turning.** Refer to Fig. 8.10.
   - In this type of lathe operation various steps of different diameters in the workpiece are produced.
   - It is carried out in the similar way as plain turning.

4. **Taper turning.**

Taper. A **taper** *may be defined as a uniform increase or decrease in diameter of a piece of work measured along its length.*

Refer to Fig. 8.11. The amount of taper in a workpiece is usually *specified by the ratio of the difference in diameters of the taper to its length.* This is termed as the *"capacity"* and it is designated by the letter K.

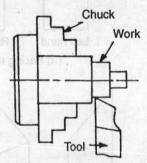

Fig. 8.10. Step turning.

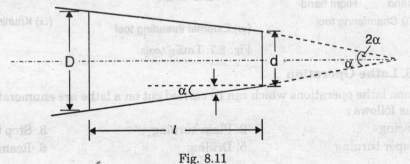

Fig. 8.11

i.e.,
$$K = \frac{D-d}{l}$$

where, D = large diameter of taper in mm,
 d = small diameter of taper in mm,
 l = length of tapered part in mm, and
 α = half of taper angle.

Taper turning. *Taper turning means to produce a conical surface by gradual reduction in diameter from a cylindrical workpiece.*

— The tapering of a part has wide applications in the construction of machines. Almost all machine spindles have taper holes which receive taper shanks of various tools and work holding devices.

Taper turning methods. Taper turning can be carried out on lathes by the following methods :

1. By setting over the tailstock centre.
2. By swivelling the compound rest.
3. By using a taper turning attachment.
4. By manipulating the transverse and longitudinal feeds of the slide tool simultaneously.
5. By using a broad nose form tool.

1. By setting over the tailstock centre :

- This method is used for *small tapers only* (the amount of setover being limited)
- It is *based upon the principle of shifting the axis of rotation of the workpiece, at an angle to the axis, and feeding the tool parallel to the lathe axis.* The angle at which the axis of rotation of the workpiece is shifted is *equal to half angle of taper*. This is done when the body of the tailstock is made to slide on its base towards or away from the operator by a setover screw as shown in Fig. 8.12.

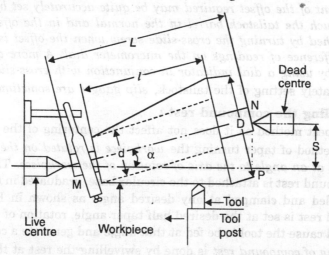

Fig. 8.12. Turning taper by tailstock set-over method.

- By setting tailstock centre to the back (away from the operator) the taper will have bigger diameter towards the tailstock. If the tailstock centre is taken in the front, bigger diameter will be on the headstock side. The reduction in diameter will be twice the offset of tailstock centre if entire length is turned.

- The major **disadvantage** of this method is that the *live and dead centres are not equally stressed and the wear is non-uniform.* Also, the lathe carrier being set at an angle, the *angular velocity of the work is not constant.*

Calculation of setover (S) :

The amount of setover required may be calculated as follows :

From the right angled triangle MNP (Fig. 8.12), we have

$$NP \ (= \text{setover}), \ S = MN \sin(\alpha) = L \sin \alpha$$

For a very small angle, (α) it can be safely considered that :

$$\sin \alpha = \tan \alpha$$

i.e., setover, $\quad S = L \tan \alpha$

or $\quad S = L \times \dfrac{D-d}{2l}$ in mm

$$= \text{Total length} \times \dfrac{\text{total taper}}{2 \times \text{taper length}}$$

where
- S = the required setover in mm,
- D = large diameter in mm,
- d = small diameter in mm,
- L = total length of work in mm, and
- l = length of tapered portion in mm.

In case the job is to be tapered over its full length, l will be equal to L. Therefore, the setover will be given by

$$S = \dfrac{D-d}{2} = \dfrac{\text{Total taper}}{2}$$

— *The amount of the offset required may be quite accurately set by allowing the tool post to touch the tailstock barrel in the normal and in the offset position This is accomplished by turning the cross-slide screw when the offset is measured directly by the difference of readings on the micrometer dial. A more accurate reading is obtained by using a dial indicator in conjunction with cross-slide.*

— For accurately setting of the tailstock, *slip gauges are sometimes used.*

2. By swivelling the compound rest :

- It is the best method as it does not affect the centering of the job or centres.
- In this method of taper turning the *workpiece is rotated on the lathe axis and the tool is fed at an angle to the axis of rotation of the workpiece.* The tool mounted on the compound rest is attached to the circular base, graduated in degrees, which may be swivelled and clamped at any desired angle as shown in Fig. 8.13. After the compound rest is set at the desired half taper angle, rotation of the compound slide screw will cause the tool to be fed at that angle and generate a corresponding taper.
- The *setting of compound rest* is done by swivelling the rest at the half taper angle, if this is already known. However, if the diameters of large (D) and small (d) ends are known, the half taper angle can be calculated as follows :

MACHINE TOOLS

$$\tan \alpha \text{ or } \alpha = \tan^{-1}\left(\frac{D-d}{2L}\right)$$

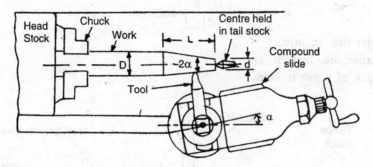

Fig. 8.13. Taper turning.

- Owing to the limited movement of the cross-slide, this method is limited to turn a *short taper* ; a *small taper* may also be turned.
- This method gives a *low production capacity and power surface finish because the movement of the tool is completely controlled by hand.*

3. By using a taper turning attachment

- This method *provides a very wide range of taper.*
- In this method of taper turning a *tool is guided in a straight path set at an angle to the axis of rotation of the workpiece, while the work is being revolved between centres or by a chuck aligned to the lathe axis.*
- As shown in Fig. 8.14, a taper turning attachment essentially consists of a *bracket or frame* which is attached to the rear end of the lathe bed and supports a *guide bar* pivoted at the centre. The bar is provided with graduations and may be swivelled on either side of the zero graduation and is set at the desired angle with the lathe axis.

The taper turning attachment is used as follows :

— The cross-slide is first made free the lead screw by removing the *binder screw*. The rear end of the cross-slide is then tightened with the guide block by means of a bolt.

— On the engagement of the longitudinal feed, the tool mounted on the cross-slide will follow the angular path, as the guide block slides on the guide bar set at an angle to the both axes.

— The required depth of cut is given by the compound slide which is placed at right angles to the axis of the lathe.

— The guide bar must be set at half taper angle and the taper on the work must be converted in degrees. The maximum angle through which the guide bar may be swivelled is 10° to 12° on either side of the centre line.

- After every cut, the feed to the tool is given by moving the compound rest which is positioned parallel to the cross-slide (*i.e.*, at 90° to the axis of the job).
- The required angle (*i.e.*, angle of swivelling the guide bar) can be found out from the following relation :

$$\tan \alpha = \frac{D-d}{2l} \quad \text{(all dimensions in mm)}$$

or
$$\alpha = \tan^{-1}\left(\frac{D-d}{2l}\right) \text{ degrees.}$$

where D = larger dia. in mm,
        d = smaller dia. in mm, and
        l = length of taper in mm.

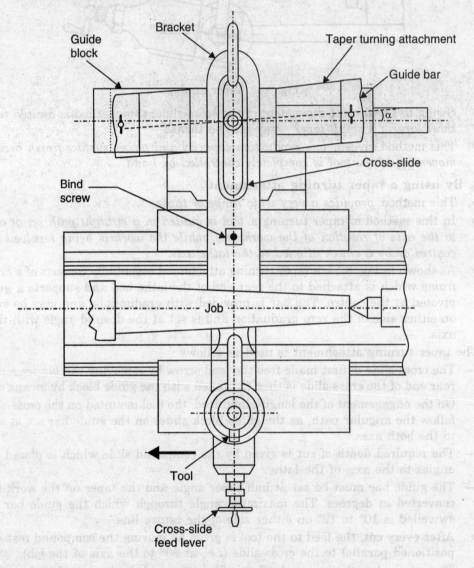

Fig. 8.14. Use of taper turning attachment.

Advantages of using a taper turning attachment :
1. Easy and quick setting.

2. The operator may not be highly skilled.
3. Accurate tapers can be easily obtained in a single setting.
4. Very steep taper on a long workpiece may be turned which is not possible with any other method
5. It is quite suitable for internal tapers as well.
6. It provides a better finish.
7. It ensures an increased rate of production because it is possible to employ longitudinal power feeds easily.
8. During the operation, normal set-up and alignment of the lathe and main parts are not disturbed (as is the case with the other methods).

4. By manipulating the transverse and longitudinal feeds of the slide tool simultaneously :
- Taper turning by manipulation of both feeds is inaccurate and requires skill on the part of the operator.
- It is used for *sharp tapers only*.

5. By using a broad nose form tool :
- In this method of taper turning (Fig. 8.15) a broad nose tool having straight cutting edge is set on to the work at half taper angle and is fed straight into the work to generate a tapered surface.
- With this method, tapers of short length only can be turned.

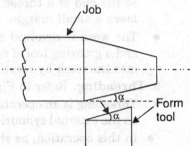

Fig. 8.15. Taper turning by a form tool.

5. Drilling. Refer to Fig. 8.16.
- It is an operation of producing a cylindrical hole in a workpiece by the rotating cutting edge of a cutter known as the *drill*.
- For this operation, the work is held in a suitable device, such as chuck or face plate, as usual, and the drill is held in the sleeve or barrel of the tail stock. The drill is fed by hand by rotating the hand-wheel of the tailstock.

6. Reaming. Refer to Fig. 8.17.
- *Reaming* is the operation which usually follows the earlier operation of drilling and boring in case of those holes in which a *very high grade of surface finish and dimensional accuracy* is needed.
- The tool used in called the *reamer*, which has multiple cutting edges. The reamer is held on the tailstock spindle, either direct or through a drill chuck and is held stationary while the work is revolved at very slow speed. The feed varies from 0.5 to 2 mm per revolution.
- For reaming tapered holes, taper reamers are used.

7. Boring. Refer to Fig. 8.18.

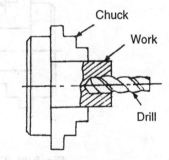

Fig. 8.16. Drilling.

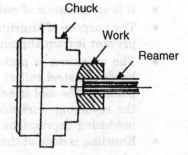

Fig. 8.17. Reaming.

- It is the operation of *enlarging and turning a hole produced by drilling, punching, casting or forging*.
- In this operation, as shown in Fig. 8.18, a *boring tool or a bit* mounted on a rigid bar is held in the tool post and fed into the work by hand or power in the similar way as for turning.
- *Boring cannot originate a hole.*

8. **Under cutting or grooving**. Refer to Fig. 8.19.
- It is the process of reducing, the diameter of a workpiece over a very narrow surface. It is often done at the end of a thread or adjacent to a shoulder to leave a small margin.
- The work is revolved at half the speed of turning and a grooving tool of required shape is fed straight into the work by rotating the cross-slide screw.

9. **Threading**. Refer to Fig. 8.20.
- Threading is an operation of cutting helical grooves on the external cylindrical surface of the workpiece.
- In this operation, as shown in Fig. 8.21, the work is held in a chuck or between centres and the threading tool is fed longitudinally to the revolving work. The longitudinal feed is equal to the pitch of the thread to be cut.

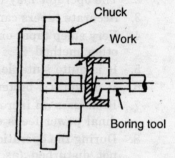

Fig. 8.18. Boring.

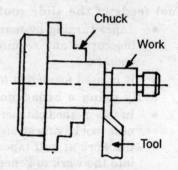

Fig. 8.19. Undercutting.

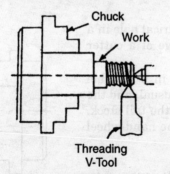

Fig. 8.20. Threading.

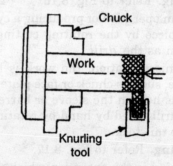

Fig. 8.21. Knurling.

10. **Knurling.** Refer to Fig. 8.21.
- It is an operation of *embossing a diamond shaped pattern on the surface of a workpiece*.
- The purpose of knurling is to provide an effective *gripping surface* on a workpiece to prevent it from slipping when operated by hand.
- The operation is performed by a special *knurling tool* which consists of 1 set of hardened steel rollers in a holder with the teeth cut on their surface in a definite pattern. *The tool is held rigidly on the tool post and the rollers are pressed* against the revolving workpiece to squeeze the metal against the multiple cutting edges, producing depressions in a regular pattern on the surface of the workpiece.
- Knurling is done at the *slowest speed* available in a lathe. Usually the speed is reduced

to $\frac{1}{4}$ th of that of turning, and plenty of oil is flowed on the tool and workpiece.

11. **Forming.** Refer Fig. 8.22.
- It is an operation of turning a convex, concave or any irregular shape.
- Form-turning may be accomplished by the following methods : (*i*) Using a forming tool, (*ii*) Combining cross land longitudinal feed, (*iii*) Tracing or copying a template.

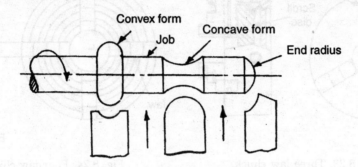

Fig. 8.22. Forming.

8.10.7. Lathe Accessories

The devices employed for handling and supporting the work and the tool on the lathe are called its accessories. The various accessories are enumerated below :

1. Chucks
   (*i*) Three jaw universal chuck
   (*ii*) Four jaw independent chuck
   (*iii*) Combination chuck
   (*iv*) Magnetic chuck
   (*v*) Air or hydraulic chuck
   (*vi*) Collet.
2. Face plate
3. Angle plate
4. Driving plate
5. Lathe carriers or dogs
6. Lath centres
7. Lathe mandrels
8. Rests
9. Jigs and fixtures
10. Lathe attachments :
    (*i*) Stops
    (*ii*) Grinding attachment
    (*iii*) Milling attachment
    (*iv*) Taper turning attachment
    (*v*) Copying attachment

(*vi*) Relieving attachment.

The figures of some accessories are given below :

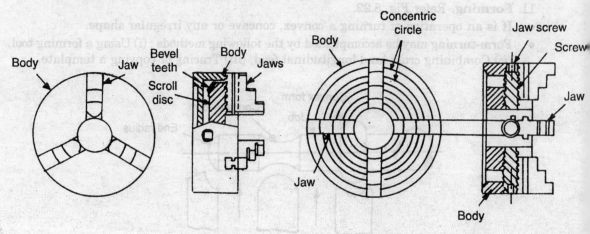

Fig. 8.23. Three jaw chuck. Fig. 8.24. Four jaw chuck.

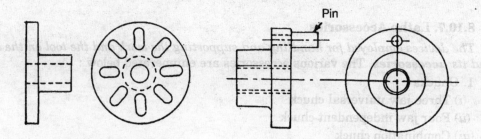

Fig. 8.25. Face plate. Fig. 8.26. Drive plate.

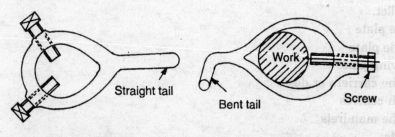

(*a*) Straight tail type. (*b*) Bent tail type.

Fig. 8.27. Lathe dog or carrier.

MACHINE TOOLS

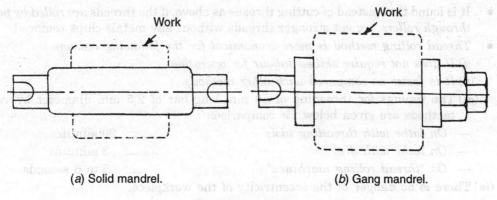

(a) Solid mandrel. (b) Gang mandrel.

Fig. 8.28. Mandrels.

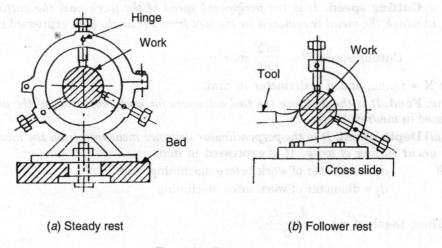

(a) Steady rest (b) Follower rest

Fig. 8.29. Rests.

8.10.8. Eccentric Turning

Although lathe does normal concentric turning, if some parts are required eccentric they can be turned so, on the lathe, as discussed below :
- One of the methods used, is to have *two centres* on the job faces countersunk. The central centre will turn out concentric turning. *Eccentric centre will produce eccentric turning.* The distance of the centres must be *half the eccentricity* required for the job.
- Another method is if the part is to be turned cylindrical around an axis other than central axis, the job can be turned centrally if it can be held in independent jaw chucks so the part to be turned remains central.

8.10.9. Thread Rolling

- Threads are normally cut on a lathe by using threading tools, by use of taps and dies or by milling. In all these methods, the metal chips are formed when the metal is removed from the grooves of the threads. *It breaks the threads at the grooves and the load carrying capacity of the threads is reduced.*

- It is found that instead of cutting threads as above, if the threads are *rolled by passing through rollers*, we get stronger threads without any metals chips removed.
- *Thread rolling method is more economical for the following reasons :*
    (i) *It does not require skilled labour for operation.*
    (ii) *It is faster as compared with other methods.*
    (iii) The timings for threading of 50 mm long bar of 2.5 mm diameter by various methods are given below for comparison :
    — On lathe with threading tools ...... 20 minutes
    — On lathe with dies ...... 3 minutes
    — On "thread rolling machines" ...... 5 to 6 seconds.
    (iv) There is no danger of the eccentricity of the workpiece.

8.10.10. Cutting Speed, Feed and Depth of Cut

(i) **Cutting speed.** *It is the peripheral speed of the work past the cutting tool* or *the speed, at which the metal is removed by the tool from the work.* It is expressed in metres/min.

$$\text{Cutting speed, } V = \frac{\pi d N}{1000} \text{ m/min,}$$

where N = r.p.m., and d = diameter in mm.

(ii) **Feed.** *It is the distance the tool advances for each revolution of the workpiece.* It is expressed in mm/revolution.

(iii) **Depth of cut.** *It is the perpendicular distance measured from the machined surface to the uncut surface of work.* It is expressed in mm.

If d_1 = diameter of work before machining, and
d_2 = diameter of work after machining.

Then, Depth of cut = $\dfrac{d_1 - d_2}{2}$.

8.10.11. Testing of Lathes

Indian standards have laid down specifications for testing lathes upto 800 mm swing over bed by Is : 1878—1961 as given below.

Test 1 : Levelling of machine-longitudinal and transverse directions.
- This is carried out by spirit level of 0.03 to 0.05 mm per metre accuracy and by gauge blocks, to suit the guide ways. The spirit levels are placed on the front and rear guide ways at intervals of 500 mm. The permissible error is 0.02 mm per metre for front guide way.
- The permissible error for rear guide way is 0.01 mm per metre convex and 0.02 mm per metre concave.
- In transverse direction, level is placed on measuring bridge and error of 0.02 mm per metre is allowed but no twist is permitted.

Test 2 : Straightness of saddle is tested by cylindrical mandrel 600 mm long and dial gauge. If dial indicator is placed on the saddle, the mandrel, should not show a variation of more than 0.02 mm in its length.

Test 3 : Alignment of both centres in vertical place is checked by dial indicator on saddle above the mandrel. It is essential that the lathe has run sufficiently to heat the main spindle bearings to normal working temperature. The permissible error is 0.02 mm.

MACHINE TOOLS 303

Test 4: *Parallelism of spindle to saddle movement in both horizontal and vertical planes* is measured by dial indicator and mandrel. The error allowed is 0.02 mm for every 300 mm.

Test 5: *Movement of upper slide paralleled with main spindle in the vertical plane.* The permissible error is 0.03 mm for every 100 mm.

Test 6: *True running of locating cylinder of main spindle* by dial gauge within 0.01 mm.

Test 7: *True running of head stock center* within 0.01 mm accuracy by dial gauge.

Test 8: *Parallelism of tail stock guide ways with movement of carriage in both planes.* The permissible error is 0.04 mm for turning lengths up to 5 metre and 0.05 mm for shorter turnings. For every 1,000 mm the error should not exceed 0.03 mm.

Test 9: *Pitch accuracy of lead screw* should be within 0.03 mm between any 2 threads at a maximum distance of 300 mm from each other.

Test 10: *Axial slip of lead screw* should be within 0.01 mm in each direction by dial gauge.

Test 11: *Alignment of lead screw bearings with respect to each other in both planes* is to be within 0.1 mm by dial gague.

Test 12: *Working accuracy of machine and cylindrical turning* is to be within 0.01 mm.

Test 13: *Working accuracy for facing* is to be within 0.02 mm over the diameter of the test piece.

Test 14: *Pitch accuracy of thread cut* is to be within 0.02 mm on 50 mm length when the work piece is held between the centres and threads are cut.

The following **additional tests** are also carried out if possible :

(a) *Alignment of lead screw bearings with split nut in both planes within* + 0.15 mm.
(b) *Parallelism of tail stock sleeve to saddle movement with in* 0.01 mm.
(c) *Parallelism of tail stock sleeve taper socket to saddle movement.*
(d) *Axial slip of main spindle and true running of shoulder face of spindle nose.*

8.10.12. Exercise : *Prepare the job as per diagram given below* :

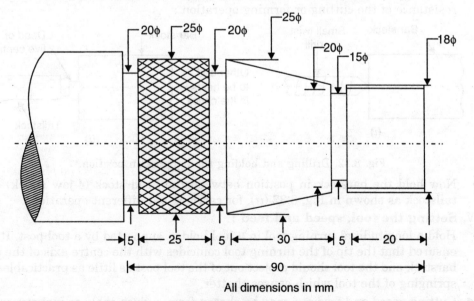

All dimensions in mm

Fig. 8.30

Solution :

Equipment required :

1. *Machine tool* ...... centre lathe
2. *Cutting tools* ...... longitudinal turning tool, grooving tool, knurling tool, drilling tool, threading tool etc.
3. Measuring tools ...... steel rule, outside caliper vernier caliper, micro-meter, etc.
4. Dead or live centres
5. Bar stock ...... 30 mm dia × 150 mm

Procedure :

The following *procedural steps* may be followed to make the given job :

I. Holding and centering :

- Using a four jaw lathe chuck, hold the barstock in such a way that at least 110 mm of bar stock is projected outside the chuck, rotating about the axis of its centre.
- Centre the workpiece (using a set procedure) so as it rotates exactly around its centre axis.

II. Facing or squaring :

- Turn the projected surface and make it square using a single point side turning tool. The feed is given from centre axis towards the circumference as shown in Fig. 8.31.

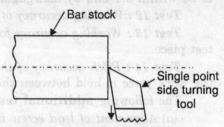

Fig. 8.31. Facing operation.

III. Drilling and holding the bar stock in position :

- Mark the centre position with a small twist drill. Hold the twist drill in the tailstock and feed it into the rotating bar's squared face. The depth and diameter of the hole is to support the dead or live centre and should be large enough to withstand the resistance of the cutting or forming operation :

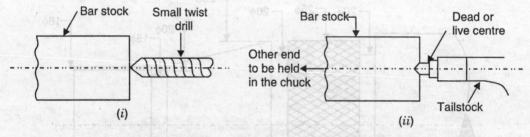

Fig. 8.32. Drilling and holding the barstock in position.

- Now hold the barstock in position between the head stock (4-jaw chuck) and the tailstock as shown in Fig. 8.32 (*ii*), for carrying out different operations.

IV. Setting the tool, speed and feed :

- Hold a longitudinal turning tool in a tool holder, supported by a tool post. It may be ensured that the tip of the turning tool coincides with the centre axis of the rotating barstock and the tool should project out of the tool post as little as practicable to avoid springing of the tool which causes chatter.
- Cutting speed and feed etc. may be chosen from a given table or instructions of the shop instructor may be followed.

V. Turning:

- Obtain the required size by first giving one or more rough cuts and then a finishing cut may be made (It should not be more than 0.75 mm). *First the larger diameter i.e., 25 mm should be obtained.* The direction of feed should be towards head stock as illustrated in Fig. 8.33.

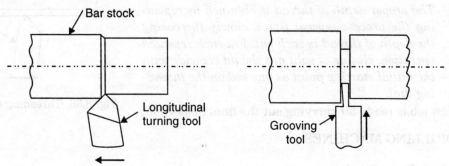

Fig. 8.33. Turning operation.   Fig. 8.34. Grooving operation.

VI. Grooving:

Mark the position of various grooves on the work and obtain the grooves using a grooving tool as illustrated in Fig. 8.34.

VII. Knurling:

Obtain knurling at the required portion, using a knurling tool.

VIII. Turn *the next lower diameter i.e., 18 mm.*

IX. Taper turning: Refer Fig. 8.35.

In order to obtain the required taper the following methods may be used:

(i) By off-setting the tailstock
(ii) By using compound rest
(iii) By using taper turning attachment.

The method used depends on *length the taper, the angle of the taper and the number of pieces to be machined.*

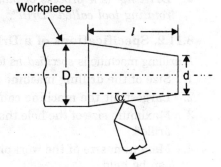

Fig. 8.35. Taper turning operation.

The best method to obtain the required taper for the class exercise work is the use of *compound rest*. In this method the tapering angle is calculated as follows:

$$\tan \alpha = \frac{D-d}{2l} = \frac{25-20}{2 \times 30} = 0.08333$$

or
$$\alpha = \tan^{-1}(0.08333) = 4.76° \text{ or } 4° \ 45°.$$

X. Threading. Refer Fig. 8.36.

Lastly, the threads are cut on the workpiece.

- In order to obtain the required threads, the set the gear mechanism of the lathe for the threading operation.
- The thread cutting in a lathe is obtained by causing the lathe carriage to move, positively, a certain distance for each revolution of main spindle. The positive movement

is obtained by connecting the main spindle to the lead screw by gears and then closing the nut tightly upon the load screw, thereby ensuring a positive movement of the carriage for each revolution of the lead screw.

- *The proper depth of thread is obtained by repeating the process several times, slowly increasing the depth of thread is each cut. For each repeated operation, closing of split nut should coincide with the initial starting point as marked on the threading dial.*

Now job is ready for carrying out the final inspection.

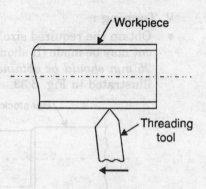

Fig. 8.36. Threading operation.

8.11. DRILLING MACHINES

8.11.1. Introduction

- Drilling machine is one of the simplest, moderate and accurate machine tool used in production shop and tool room. It consists of a *spindle* which imparts rotary motion to the drilling tool, or mechanism for feeding the tool into the work, a *table* on which the work rests and a *frame*. It is considered as *a single purpose machine tool since its chief function is to make holes*. However, it can and does perform operations *other than drilling also.*

- *Drilling* is a *process of making hole or enlarging a hole in an object by forcing a rotating tool called "Drill".*

8.11.2. Specifications of a Drilling Machine

A drilling machine is *specified* as follows (Refer Fig. 8.37).

1. Size of the drilling machine table.
2. Largest bit the machine can hold.
3. Maximum size of the hole that can be drilled.
4. Maximum size of the workpiece that can be held.
5. Power of the motor, spindle speed or feed.

Specifically, the various types of drilling machines are specified as follows :

- **Portable drilling machine.** *Maximum diameter of drill which can be held.*
- **Sensitive and upright drilling machines.** *The diameter of the largest workpiece that can be drilled.*
- **Radial drilling machine**. *The length of the arm and column diameter.*

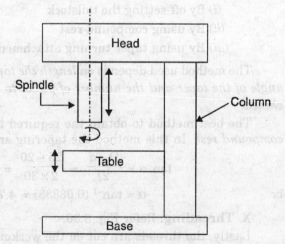

Fig. 8.37. Block diagram of a drill press.

MACHINE TOOLS

- **Multiple sprindle drilling machine.** *The drilling area, the size and number of holes a machine can drill.*

8.11.3. Operations Performed

Although drill press is mainly meant for drilling operation, it can also be used for performing the following operations : Refer Fig. 8.38.

1. **Reaming.** It is an *operation of finishing an existing drilling hole.* The tool used is reamer.

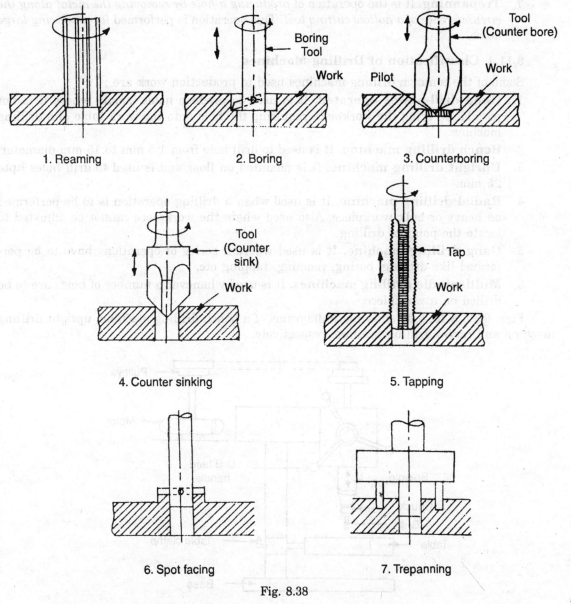

Fig. 8.38

2. **Boring.** It is an *operation of enlarging an existing hole.*

3. **Counter boring.** It is an *operation of enlarging a drilled hole partially, that is for a specific length*.
4. **Countersinking.** It is an *operation of forming a conical shape at the end of a drilled hole*.
5. **Tapping.** It is an *operation in which external threads are cut in the existing hole*.
6. **Spot facing.** It is the *operation of smoothing and squaring the surface around a hole for the seat for the nut or the head of a screw*. A counter bore or a special spot facing tool may be employed for this purpose.
7. **Trepanning.** It is the operation of *producing a hole by removing the metal along the circumference of a hollow cutting tool*. This operation is performed for *producing large holes*.

8.11.4. Classification of Drilling Machines

Some of the common drilling machines used in production work are :
1. **Hand drill-power operated.** It is used to produce holes (small) where it is not possible to bring the workpiece requiring the hole onto the work table of a drilling machine.
2. **Bench drilling machine.** It is used to drill hole from 1.5 mm to 15 mm diameter.
3. **Upright drilling machine.** It is mounted on floor and is used to drill holes upto 25 mm.
4. **Radial drilling machine.** It is used when a drilling operation is to be performed on heavy or bulk workpiece. Also used where the workpiece cannot be adjusted to locate the point of drilling.
5. **Gang drilling machine.** It is used where a series of operations have to be performed like drilling, boring, reaming, tapping etc.
6. **Multispindle drilling machines.** It is used whenever a number of holes are to be drilled on a workpiece.

Figs. 8.39, 8.40, 8.41 show block diagrams of a bench drilling machine, upright drilling machine and radial drilling machine respectively.

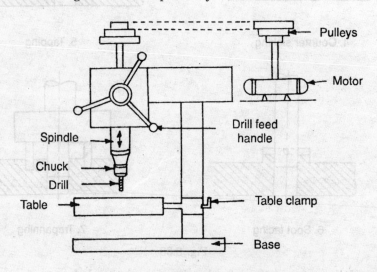

Fig. 8.39. Bench drilling machine.

MACHINE TOOLS 309

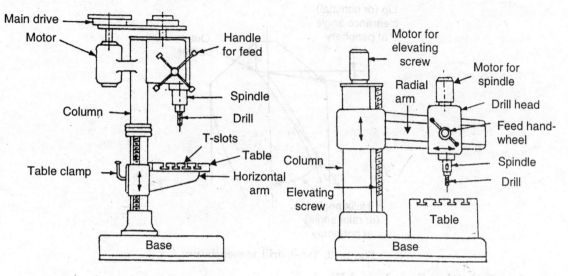

Fig. 8.40. Upright drilling machine.　　　Fig. 8.41. Radial drilling machine.

Figs. 8.42, 8.43 show a twist drill and its nomenclature respectively.

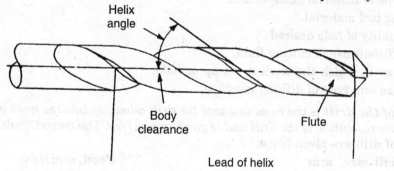

Fig. 8.42. Twist drill.

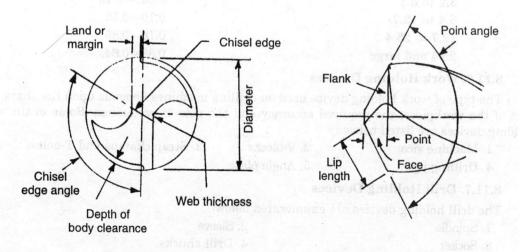

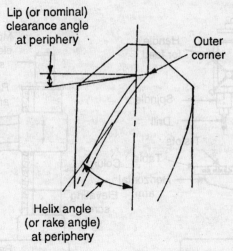

Fig. 8.43. Twist drill nomenclature.

8.11.5. Cutting Speeds and Feeds

- The *cutting speed* depends upon the following factors :
  (*i*) The type of material being drilled.
  (*ii*) Cutting tool material.
  (*iii*) The quality of hole desired.
  (*iv*) The efficient use of cutting fluid.
  (*v*) The way in which the work is set up or held.
  (*vi*) The size and type of drilling machine.

- **Feed of the drill** is the *axial distance the drill advances into the work piece for each complete revolution of the drill and is given in mm/rev*. The correct feeds for different sizes of drill are given below :

Drill size, mm	Feed, mm/rev
3.2 and less	0.025—0.050
3.2 to 6.4	0.050—0.10
6.4 to 12.7	0.10—0.18
12.7 to 25.4	0.18—0.38
25.4 and large	0.38—0.64

8.11.6. Work Holding Devices

The type of work holding device used on drilling machines depends upon the shape and size of the workpiece, the required accuracy and the rate of production. Some of the work holding devices are listed below :

1. Machine vice
2. V-blocks
3. Strap champs and T-bolts
4. Drilling jigs
5. Angle plate.

8.11.7. Drill Holding Devices

The drill holding devices are enumerated below :

1. Spindle
2. Sleeve
3. Socket
4. Drill chucks.

MACHINE TOOLS

8.11.8. Drilling Machine Tools

Drilling machine tools include the following :
1. Flat drill
2. Straight drill
3. Twist drill
4. Taper shank core drill
5. Oil tube drill
6. Centre drill
7. Reamer
8. Centre punch
9. Drift
10. Hammer.

8.12. SHAPING MACHINE (SHAPER)

8.12.1. Introduction

Refer to Fig. 8.44.

- A *shaper is a reciprocating type of machine tool intended primarily to produce horizontal, vertical or inclined flat surfaces (upto 1000 mm long).*
- In the shaper, the cutting tool has a reciprocating motion, and *it cuts only during the forward stroke only.*
- The work is held in a vice bolted to the work table. The *regular feed is obtained by moving the work table automatically at right angles to the direction of the cutting tool* and the *tool head gives downward feed at right angles to the regular feed or at any other angle as desired.*

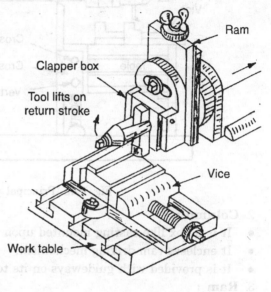

Fig. 8.44. Shaper.

8.12.2. Classification of Shapers

The shapers are classified as follows :
1. *According to the type of mechanism used for giving reciprocating motion to the ram* :
    (*i*) Crank type
    (*ii*) Geared type
    (*iii*) Hydraulic type.
2. *According to the position and travel of ram* :
    (*i*) Horizontal type
    (*ii*) Vertical type
    (*iii*) Travelling head type.
3. *According to the type of cutting stroke* :
    (*i*) Push type
    (*ii*) Draw type.
4. *According to the type of design of the table* :
    (*i*) Standard shaper
    (*ii*) Universal shaper.

8.12.3. Principal Parts

Refer Fig. 8.45. The principal parts of a shaper are described briefly below :
1. **Base :**
- It is made of cast iron to resist vibration and takes up high compressive load.
- It is so designed that it can take up the entire load of the machine and the forces set up by the cutting tool over the work.

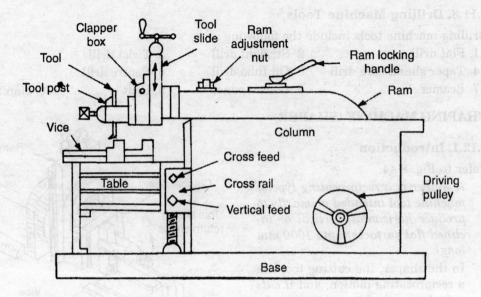

Fig. 8.45. Principal parts of a shaper.

2. **Column :**
- It is a box like casting mounted upon the base.
- It encloses ram driving mechanism.
- It is provided with guideways on its top to enable the ram to slide on it.

3. **Ram :**
- It is a reciprocating member which reciprocates on the guideways provided above the column.
- It carries a tool-slide on its head and a mechanism for adjusting the stroke length.

4. **Cross-rail :**
- It is mounted on the front vertical guideways of the column.
- It has two parallel guideways on its top in the vertical plane that are perpendicular to the ram axis. The table may be raised or lowered to accommodate different sizes of jobs by rotating an elevating screw which causes the cross-rail to slide up and down on the vertical face of the column.

5. **Table :**
- It is made of cast iron and is rectangular in shape.
- It has T-slots on its top surface.
- The table can be moved upward, downward or sideward with the help of elevating screws and other feed handle.

8.12.4. Specifications of a Shaper

The shaper is *specified* as follows :
1. Maximum length of the stroke (in mm).
2. Size of the table *i.e.*, length, width and depth of the table.
3. Maximum horizontal and vertical travel of the table.

4. Maximum number of strokes per minute.
5. Type of quick return mechanism.
6. Power of the drive motor.

8.12.5. Operations Performed

On a standard shaper the following operations can be performed :

(i) Machining of *vertical surfaces* (ii) Machining of *horizontal surfaces*
(iii) Machining of *angular surfaces* (iv) Machining of *curved surfaces*
(v) Machining of *irregular surfaces* (vi) Machining of *slots and keyways*.

8.12.6. Tools Used

Following tools are used :

(i) Try square and square head of combination set
(ii) Micrometer (iii) Surface gauge
(iv) Sine bar (v) Dial test indicator.

- The cutting tools used in shapers are similar to those used in lathe work *except for side and front clearance*.
- Shaper tools have *less* side and front clearance because the work feeds into it on the return stroke, whereas the lathe tool is constantly feeding into the work. For best results these angles should not exceed 4° and be less than 2°.

8.13. PLANING MACHINE (PLANER)

8.13.1. Introduction

Refer to Fig. 8.46.

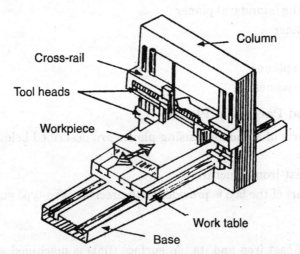

Fig. 8.46. Planer.

- The planing machine **(planer)** is a machine tool *used in the production of flat surfaces on workpieces too large or too heavy to hold in a shaper.*

- In this machine, the table called *PLATEN, on which the work is securely fastened, has a reciprocating motion.*
- The *tool head is automatically* fed horizontal in either direction along the heavily supported cross-rail over the work, and automatic downward feed is also provided.

8.13.2. Comparison between Planer and Shaper

Comparison between a planer and a shaper is given in the Table 8.1 below :

Table 8.1. Comparison between Planer and Shaper

S.No.	Planer	Shaper
1.	Heavier, more rigid and costlier machine.	A comparatively lighter and cheaper machine.
2.	Requires more floor area.	Requires less floor area.
3.	Work reciprocates horizontally.	Tool reciprocates horizontally.
4.	Tool is stationary during cutting.	Work is stationary during cutting.
5.	Heavier cuts and coarse feeds can be employed.	Very heavy cuts and coarse feeds cannot be employed.
6.	Work setting requires much of skill and takes a longer time.	Clamping of work is simple and easy.
7.	Several tools can be mounted and employed simultaneously, usually four as a maximum, facilitating a faster rate of production.	Usually one tool is used on a shaper.
8.	Used for machining large size workpieces.	Used for machining small size workpieces comparatively.

8.13.3. Types of Planer

The various types of planers commonly used are :
1. Double housing (standard) planer
2. Open side planer
3. Pit planer
4. Edge or plate planer
5. Divided table planer.

8.13.4. Principal Parts of a Planer

The principal parts of a double housing planer are described below :

1. Bed :
- It is a big cast iron structure.
- The upper part of the bed is provided with precision Vee-type guide ways on which the table slides.

2. Table :
- It is made of cast iron and its top surface (flat) is machined accurately.
- It reciprocates along the ways of the bed and supports the work.
- Its top surface is provided with slots to clamp the workpieces.
- It may be driven by rack and gear, by rack and double helical gear or by hydraulic system.

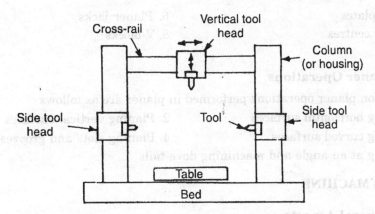

Fig. 8.47. Double housing or standard planer.

3. **Column or Housing :**
- The columns or housings are rigid column-like castings placed on each side of the bed.
- The front vertical surface of the column has guide ways to enable movement of the cross-rail vertically up and down.

4. **Cross-rail :**
- It is mounted on the precision machined ways of the two housings.
- It may be raised or lowered on the housings to accommodate work of differents heights on the table and to allow for the adjustment of the tools.

5. **Tool heads :**

These are mounted on the cross-rail or housings by means of a saddle which slides along the rail or housing ways. The saddle may be made to move transversely on the cross-rail to give cross feed.

8.13.5. Size of a Planer

- The size of a standard planer is specified by the *size of the largest rectangular solid that can reciprocate under the tool.*
- Double housing planers range from 750 mm × 750 mm × 2.5 m as the smallest and upto 3000 mm × 3000 mm × 18.25 m as the largest size.
- In addition to the basic dimensions, other particulars given below also need to be stated for specifying the planer completely :
    — Number of speeds and feeds available
    — Power input
    — Floor space required
    — Net weight of the machine
    — Type of drive etc.

8.13.6. Standard Clamping Devices

The following are the standard clamping devices used for holding most of the work on a planer table :

1. Heavy duty vices
2. T-bolts and clamps
3. Step blocks, clamps, T-bolts
4. Poppets or stop pins and the dogs

5. Angle plates
6. Planer jacks
7. Planer centres
8. V-blocks
9. Stops.

8.13.7. Planer Operations

The common planer operations performed in planer are as follows :
1. Planing horizontal surfaces
2. Planing vertical surfaces
3. Planing curved surfaces
4. Planing slots and grooves
5. Planing at an angle and machining dove-tails.

8.14. MILLING MACHINE

8.14.1. General Aspects

- The **milling machine** is a *machine tool in which metal is removed by means of a revolving cutter with many teeth, each tooth having a cutting edge which removes metal from a workpiece.*
- The work is supported by various methods on the work table, and may be fed to the cutter, longitudinally, transversely or vertically.
- A great variety of work may be done on a milling machine.
- This machine is perhaps next to the lathe in importance.

Generally there are two types of milling process, namely :

(*i*) Upmilling (or conventional) process
(*ii*) Downmilling (or climb) process.

— In *upmilling process* (Refer Fig. 8.48), the workpiece is fed *opposite* to the cutter's tangential velocity.

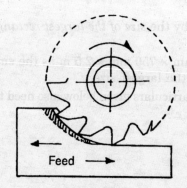
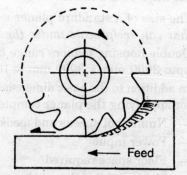

Fig. 8.48. Upmilling process.   Fig. 8.49. Downmilling process.

— In *downmilling process* (Refer Fig. 8.49), the workpiece is fed in the same direction as that of the cutter's tangential velocity.

Fig. 8.50 shows the nomenclature of a milling cutter.

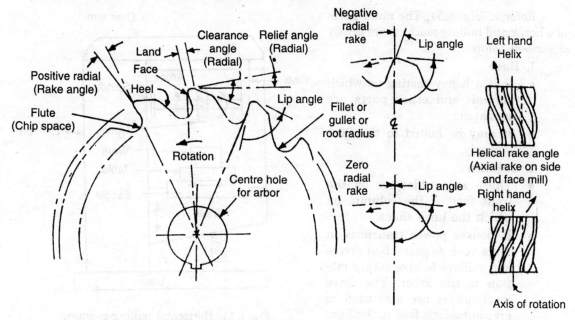

Fig. 8.50

8.14.2. Specifications of a Milling Machine

The following are the specifications of a *column and knee type milling machine* :
1. Width and length of the table.
2. Maximum distance the knee can travel.
3. Maximum longitudinal movement and cross feed of the table.
4. Number of spindle speeds.
5. Power of the main drive motor.

8.14.3. Types of Milling Machines

According to the general design of the milling machine, the usual classifications of the milling machine are :
1. *Column and knee type* :
   (i) Hand milling machine   (ii) Plain milling machine
   (iii) Universal milling machine (iv) Omniversal milling machine
   (v) Vertical machine.
2. *Manufacturing of fixed bed type* :
   (i) Simplex milling machine   (ii) Duplex milling machine
   (iii) Triplex milling machine.
3. *Planer type*.
4. *Special type* :
   (i) Rotary table milling machine (ii) Drum milling machine
   (iii) Planetary milling machine
   (iv) Pantograph, profiling and tracer controlled milling machine.

8.14.4. Main Parts of a Horizontal Milling Machine

Refer to Fig. 8.51. The main parts of a horizontal milling machine are briefly described below :

1. **Base**
   - It is a heavy casting on which column and other parts are mounted.
   - It may be bolted to the floor strongly.

2. **Column**
   - There are guideways on the front face of the column, on which the knee slides.
   - It houses power transmission units such as gears, belt drives and pulleys to give rotary motion to the arbor. The drive mechanisms are also used to give automatic feed to the handle and table.

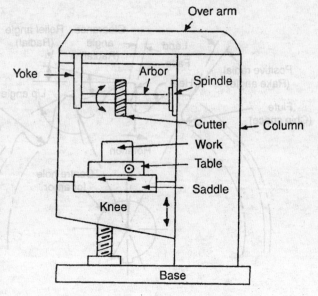

Fig. 8.51. Horizontal milling machine.

3. **Knee**
   - It supports the saddle, table, workpiece and other clamping devices.
   - It moves on guideways of the column.
   - It resists the deflection caused by the cutting forces on the workpiece.

4. **Saddle**
   - It is mounted on the knee and can be moved by a handwheel or by power.
   - The direction of travel of the saddle is restricted to be towards or away from the column face.

5. **Table**
   - It is mounted on the saddle and can be moved by a handwheel or power.
   - Its top surface is machined accurately to hold the workpiece and other holding devices.
   - It moves perpendicular to the direction of saddle movement.

6. **Arbor**
   - Its one end is attached to the column and the other end is supported by an over arm.
   - It *holds and drives different types of milling cutters.*

7. **Spindle**
   - It gets power from gears, belt drives, to drive the motor.
   - It has provision to add or remove milling cutters onto the arbor.

8.14.5. Types of Milling Cutters

Common types of milling cutters are enumerated below :

1. Plain milling cutters
2. Side milling cutters
3. End milling cutters
4. Face milling cutters

MACHINE TOOLS 319

5. Metal slitting cutters
7. Formed milling cutters
9. T-slot milling cutter

6. Angle milling cutters
8. Woodruff-key milling cutters
10. Fly cutter.

8.14.6. Milling Operations

The milling operations are classified as follows :
1. Plain or slab milling
3. Angular milling
5. Straddle milling
7. End milling
9. Dove-tail milling
11. Involute gear cutting.

2. Face milling
4. Form milling
6. Gang milling
8. T-slot milling
10. Saw milling

1. **Plain or slab milling.** Plain milling is used to *machine flat and horizontal surfaces* (Fig. 8.52). Here plain milling cutter is used, which is held in the arbor and rotated. The table is moved upwards to give the required depth of cut.

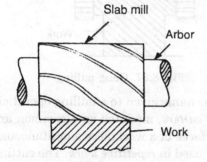

Fig. 8.52. Plain or slab milling.

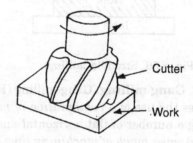

Fig. 8.53. Face milling.

2. **Face milling.** This milling process (Fig. 8.53) is used for machining a flat surface which is at right angles to the axis of the rotating cutter. The cutter used in this operation is the *face milling cutter*.

3. **Angular milling.** In angular milling, an angle milling cutter is used (Fig. 8.54). The cutter used may be a *single* or *double* angle cutter, depending upon whether a single surface is to be machined or two mutually inclined surfaces simultaneously.

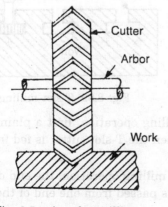

Fig. 8.54. Angular milling.

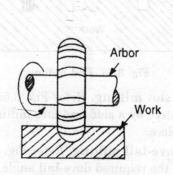

Fig. 8.55. Form milling.

4. **Form milling.** This milling process (Fig. 8.55) is used for machining those surfaces which are of *irregular shapes*. The form milling cutter used has the shape of its cutting teeth conforming to the profile of the surfaces to be produced.

5. **Straddle milling.** Refer Fig. 8.56. Straddling milling is an operation in which a pair of *side milling cutters* is used for machining two parallel vertical surfaces of a workpiece simultaneously. The distance between the cutters is adjusted by the spacers. This process is used to mill *square and hexagonal surfaces*.

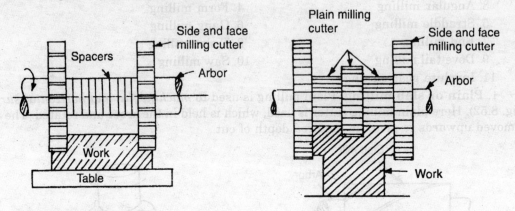

Fig. 8.56. Straddle milling.   Fig. 8.57. Gang milling.

6. **Gang milling.** Gang milling (Fig. 8.57) is the name given to a milling operation which involves the use of a *combination of more than two cutters*, mounted on a common arbor, for milling a number of flat horizontal and vertical surfaces of a workpiece simultaneously. This method *saves much of machining time* and is widely used in *repetitive work*. The cutting speed of a gang of cutters is calculated from the cutter of the largest diameter.

7. **End milling.** Refer Fig. 8.58. It is an operation of producing narrow slots, grooves and keyways using an end mill cutter. The mill tool may be attached to the vertical spindle for milling the slot. Depth of cut is given by raising the machine table.

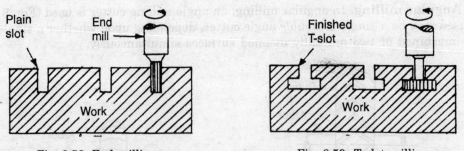

Fig. 8.58. End milling.   Fig. 8.59. T-slot milling.

8. **T-slot milling.** Refer Fig. 8.59. In this milling operation, first a plain slot is cut on the workpiece by a side and face milling cutter. Then the T-slot cutter is fed from the end of the workpiece.

9. **Dove-tail milling.** Refer Fig. 8.60. In this milling operation, the end of the cutter is shaped to the required dove-tail angle. The cutter is passed from one end of the workpiece to the other end.

MACHINE TOOLS

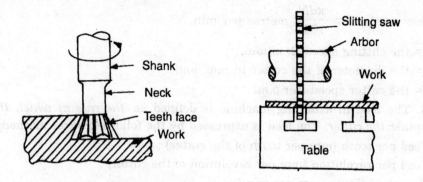

Fig. 8.60. Dove-tail milling.   Fig. 8.61. Saw milling.

10. Saw milling. Refer to Fig. 8.61. It is an operation of producing narrow grooves and slots on the workpiece. A slitting saw is used for saw milling.

11. Involute gear cutting. Gear milling operation, often referred as *gear cutting*, involves cutting of different types of gears on a milling machine. For this, either an *end mill cutter* or a *form relieved cutter* is used, which carries the profile on its cutting teeth corresponding to the required profile of the gap between gear teeth.

Fig. 8.62 shows involute gear cutting operation. Shape of the cutter teeth resembles the involute profile. Gear blank is indexed after cutting each tooth.

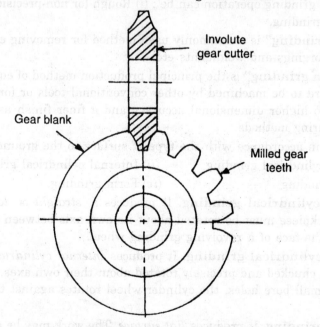

Fig. 8.62. Involute gear cutting.

8.14.7. Cutting Speed, Feed and Depth of Cut

Cutting speed. The cutting speed of a milling cutter is the *distance travelled per minute by the cutting edge of the cutter*. It is expressed in metres per minute.

In other words, $V = \dfrac{\pi d N}{1000}$ metres per min.

where, V = the cutting speed in m/min,
d = the diameter of the cutter in mm, and
N = the cutter speed in r.p.m.

Feed. The feed in a milling machine is defined as *the rate at which the workpiece advances under the cutter*. The feed is expressed by the following *three* methods :

(i) Feed per tooth (mm per tooth of the cutter) ;

(ii) Feed per revolution (mm per revolution of the cutter) ;

(iii) Feed per minute (mm per minute).

8.15. GRINDING MACHINES

8.15.1. Introduction

- *Grinding* is a *metal cutting operation performed by means of a rotating abrasive tool, called "grinding wheel"*. Such wheels are made of fine grains of abrasive materials held together by a bonding material, called a *"bond"*. Each individual and irregularly shaped grain acts as a cutting element (a single point cutting tool).

- Grinding is done on surfaces of almost all conceivable shapes and materials of all kinds. The grinding operation can be : (i) Rough (or non-precision) grinding and (ii) Precision grinding.

   (i) **"Rough grinding"** is a commonly used method for removing excess material from castings, forgings and weldments etc.

   (ii) **"Precision grinding"** is the principal production method of cutting materials that are too hard to be machined by other conventional tools or for producing surfaces on parts to higher dimensional accuracy and a finer finish as compared to other manufacturing methods.

- Grinding, in accordance with the type of surface to the ground, is classified as :

   (i) External cylindrical grinding   (ii) Internal cylindrical grinding

   (iii) Surface grinding   (iv) Form grinding.

(i) **External cylindrical grinding.** It produces a *straight* or *tapered surface* on a workpiece. The workpiece must be rotated about its own axis between centres as it passes lengthwise across the face of a revolving grinding wheel.

(ii) **Internal cylindrical grinding.** It produces *internal cylindrical holes and tapers*. The workpieces are chucked and precisely rotated about their own axes. The grinding wheel or, in the case of small bore holes, the cylinder wheel rotates against the sense of rotation of the workpiece.

(iii) **Surface grinding.** It produces *flat surface*. The work may be ground by either the periphery or by the end face of the grinding wheel. The *workpiece is reciprocated at a constant speed below or on the end face of the grinding wheel.*

(iv) **Form grinding.** This operation is done with *specially shaped grinding wheels* that grind the formed surfaces as in grinding *gear teeth, threads, splined shafts, holes* etc.

MACHINE TOOLS

Fig. 8.63 shows *three basic kinds of precision grinding*.

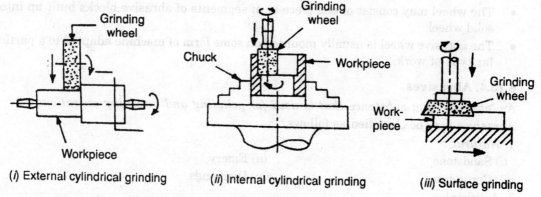

(i) External cylindrical grinding   (ii) Internal cylindrical grinding   (iii) Surface grinding

Fig. 8.63. Basic kinds of precision grinding.

8.15.2. Types of Grinding Machines

The grinding machines are *classified* as follows :

I. According to the quality of surface finish
1. *Roughing or non-precision grinders* :
   (i) Bench, pedestal or floor grinders.
   (ii) Swing frame grinders
   (iii) Portable and flexible shaft grinders.
   (iv) Belt grinders.
2. *Precision grinders*.

II. According to the type of the surface generated or work done
1. *Cylindrical grinders* :
   (i) Plain cylindrical grinders (ii) Universal cylindrical grinders
   (iii) Centreless grinders.
2. *Internal grinders* :
   (i) Plain internal grinders (ii) Universal internal grinders
   (iii) Chucking internal grinders (iv) Planetary internal grinders
   (v) Centreless internal grinders.
3. *Surface grinders* :
   (i) Reciprocating table
      (a) Horizontal spindle (b) Vertical spindle.
   (ii) *Rotating table* :
      (a) Horizontal spindle (b) Vertical spindle.
4. *Tool and cutter grinders* :
   (a) Universal (b) Special.
5. *Special grinding machines*.

8.15.3. The Grinding Wheel

- A **grinding wheel** is a *multi-tooth cutter made up of many hard particles known as 'abrasive'* which have been crushed to leave sharp edges which do the cutting. The

abrasive grains are mixed with a suitable bond, which acts as a matrix or holder when the wheel is in use.
- The wheel may consist of one piece or of segments of abrasive blocks built up into a solid wheel.
- The abrasive wheel is usually mounted on some form of machine adapted to a particular type of work.

8.15.4. Abrasives

An *'abrasive'* is a substance that is used for grinding and polishing operations. Abrasives may be classified as follows :

1. *Natural :*
   (*i*) Sandstone (*ii*) Emery
   (*iii*) Corundum (*iv*) Diamonds.
2. *Artificial :*
   (*a*) Silicon carbide (*b*) Aluminium oxide.

8.15.5. Selection of Grinding Wheels

The following factors need to be considered while selecting a grinding wheel :
1. The material to be ground   2. Amount of shock to be removed
3. Area of contact
4. Type of grinding machine :
   (*i*) Wheel speed (*ii*) Work speed
   (*iii*) Condition of the grinding machine (*iv*) Personal factor.

8.15.6. Wheel Shapes. Refer to Fig. 8.64.

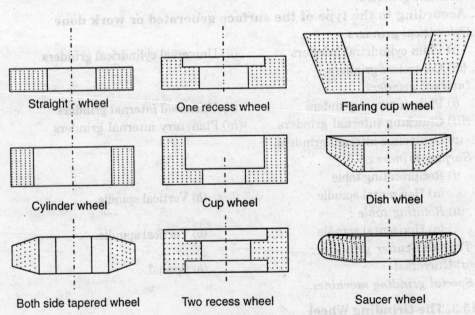

Fig. 8.64. Wheel shapes.

MACHINE TOOLS

- Grinding wheels are used in almost all shapes. *Most common is disc wheel*. It has a central hole for mounting and both edges or the sides of the wheels are used.
- For grinding piston rods or pins, *rod shape wheel* is used.
- For grinding in restricted spaces like wheel teeth, etc. *tapered edges dish shape or wheel disc shape wheels are used.*
- Some grinding wheels have shapes like saucer, cylinder, flaring cup.

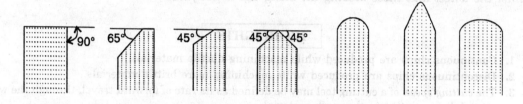

Fig. 8.65. Standard wheel edges.

Some of the grinding wheel shapes are shown in Fig. 8.64.

Maximum speeds for flexible shaft grinders :

  3 × 100 mm shaft size ..... 40,000 r.p.m.
  4 × 1250 mm shaft size ..... 30,000 r.p.m.
  7 × 1500 mm shaft size ..... 18,000 r.p.m.
  10 × 1500 mm shaft size ..... 15,000 r.p.m.
  12 × 2000 mm shaft size ..... 8,000 r.p.m.
  20 × 2500 mm shaft size ..... 5,000 r.p.m.

Recommended wheel speeds for grinding :

	Metres/min.
Internal Grinding	650—1950
Surface Grinding	1300—1630
Knife Grinding	1150—1450
Cutter Grinding	1630—1950
Tool Grinding	1630—1950

8.15.7. Mounting of Wheels

- Normally grinding wheels are supplied with a central hole fitted with a lead bush. This is then mounted on the spindle. The wheels are clamped axially with clamping collar. To distribute the clamping pressure, soft washers are put between wheels and collars. *In absence of the soft washers, there is a danger of breaking of wheels.*
- The soft washers can be of the *blotting paper or any other thick paper. Small washer increases the tightening pressure or if the lead bush is out of square, the pressure of tightening may not be equalised. Large washers are therefore essential.* To test the soundness of wheels, these are subjected to over speeds for testing. A good wheel should not run out of centre and should not have hollows at the testing speed. Defective wheels will **crack at over speed test.**

8.15.8. Wheel Truing

In the market, two types of wheel dressers are available. 'Huting done' type and 'Diamond dresser'. Diamond dresser can be used as a hand dresser or can be mounted. It contains a diamond point. 'Diam carbo' dresser is a substitute for 'Diamond dresser'. The wheel has to be correctly mounted and *truer can traverse wheel backwards and forwards on the saddle. The purpose of the dressing is to expose newly made sharp wheel faces for grinding. Glazed wheels do not grind and require dressing.* Dressing can be carried out by pressing abrasive sticks against the wheel and these moving all along the wheel face.

HIGHLIGHTS

1. Continuous chips are produced while machining ductile materials.
2. Discontinuous chips are produced while machining more brittle materials.
3. The *cutting speed* of a cutting tool may be defined as its rate of forward travel, through the work material, or relative to the work material.
4. *Feed* may be defined as the relatively small movement per cycle of the cutting tool relative to the workpiece in a direction which is usually normal to the cutting speed direction.
5. *Machinability* may be defined as the ease with which a material can be machined.
6. *Tool life* is defined as the time interval between its two successive regrinds.
7. Machine tools are used for machining. They employ cutting tools to remove excess material from the given job.
8. A lathe is one of the oldest and perhaps most important tools ever developed.
   Types of lathe. Speed lathe, engine or central lathe, bench lathe, tool room lathe, turret and capstan lathes, automatic lathes, special purpose lathes.
   Lathe operations : Facing, plain turning, step turning, taper turning, drilling, reaming, boring, undercutting or grooving, threading, knurling and forming.
9. *Drilling* is a process of making hole or enlarging hole in an object by forcing a rotating tool called *Drill. Operations performed on a drilling machine* : Reaming, boring, counterboring, countersinking, tapping, spot facing, trepanning.
10. A *shaper* is a reciprocating type of machine tool intended primarily to produce horizontal, vertical or inclined flat surfaces (upto 1000 mm long).
11. The *planing machine* (planer) is a machine tool used in the production of flat surfaces on pieces too large or too heavy to hold in a shaper.
12. The *milling machine* is a machine tool in which metal is removed by means of a revolving cutter with many teeth, each tooth having a cutting edge which removes metal from a workpiece.
13. *Grinding* is a metal cutting operation performed by means of a rotating abrasive tool, called "grinding wheel".
14. In sawing, the individual tooth of the saw *tracks* through the work, in each tooth deepening the cut made by the proceeding teeth in the direction of feed.
15. *Numerical control*. NC can be defined simply as control by numbers.
    CNC machine coil. A machine tool having a dedicated computer to help prepare the program and control some or all of the operations of the machine tool is called computer numerical control (CNC) machine tool.

OBJECTIVE TYPE QUESTIONS

Fill in the Blanks :

1. Discontinuous chips are produced while cutting more .... materials.
2. .... may be defined as the ease with which a material can be machined.

MACHINE TOOLS

3. The devices employed for handling and supporting the work and the tool on the lathe are called its ....
4. Portable drilling machine is specified by maximum .... to the drill.
5. .... is a metal cutting operation performed by means of a rotating abrasive tool.
6. Surface grinding produces .... surface.
7. A grinding wheel is a .... cutter made up of many hard particles (abrasives).
8. An .... is a substance that is used for grinding and polishing operation.
9. A .... saw carries an endless steel blade, having the teeth cut on its one edge.
10. A .... saw carries a disc type rotating blade which provides a continuous action.
11. Numerical control can be defined simply as control by ....
12. A machine tool having a dedicated computer to help prepare the program and control some or all of the operations of the machine tool is called .... machine tool.
13. In NC machines the input information for controlling the machine tool motion is provided by means of punched tapes or magnetic tapes in a coded language.
14. In NC machine tool the operator is replaced by the data processing unit of the system and .... unit.
15. The magnetic box is one of the main elements of a NC machine tool.
16. Closed loop system is less expensive than open loop system.
17. NC machines cannot be used for manufacturing complex parts.
18. The use of NC machines cuts down the lead time in manufacturing.
19. In case of NC machines, the production rates are low.
20. In a CNC machine, a .... is used to control machine tool functions from stored in information or punched tape input or computer terminal input.
21. CNC machines are unsuitable for long run applications.
22. CNC machines entail lower maintenance cost.

Answers

1. brittle	2. Machinability	3. accessories	4. diameter	5. grinding
6. flat	7. multitooth	8. abrasive	9. band	10. circular
11. number	12. CNC	13. Yes	14. control	15. Yes
16. No	17. No	18. Yes	19. No	
20. minicomputer	21. Yes	22. No.		

THEORETICAL QUESTIONS

1. How are cutting tools classified ?
2. Describe briefly, with neat sketches, various types of chips.
3. Enumerate various type of tool materials.
4. Define 'cutting speed' and 'feed'.
5. Define machinability.
6. What is tool life ? On what factors does it depend ?
7. How is the size of a lathe specified ?
8. Enumerate common lathe operations which can be carried on a lathe.
9. Enumerate various lathe accessories.
10. What is a drill press ? Draw the block diagram of a drill press.
11. How are drilling machines specified ?
12. Give the classification of drilling machines.

13. List the various types of grinding machines.
14. What is a grinding wheel ?
15. What is an abrasive ?
16. Give the classification of sawing machines.
17. Enumerate main parts of a band saw and show them on a diagram.
18. How are shapers classified ?
19. Enumerate different parts of a shaper and explain them briefly.
20. How is the size of a shaper specified ?
21. Name and describe the various work holding devices in shapers.
22. Give the fundamental difference between a planar and a shaper.
23. How is the size of a planer specified ?
24. List and describe in brief the main parts of a planer.
25. Classify milling machines.
26. Classify milling cutters.
27. Name and describe the principal parts of a milling machine.
28. Describe various milling processes with neat sketches.
29. What do you mean by *"Numerical control"* ?
30. What are the areas where *"Numerical control"* can be used ?
31. Describe briefly working of NC machine tool.
32. Explain with a neat diagram the main elements of a NC machine tool.
33. How are NC machines classified ?
34. Enumerate various applications of NC machines.
35. List the advantages of NC machines.
36. Define CNC.
37. What are the function of CNC ?
38. State the advantages of CNC machines over NC machines.
39. What are the disadvantages of CNC machines ?
40. What are the applications of CNC machines ?

9
Manufacturing Processes

9.1. Moulding and casting : Introduction—*Mould making*—Core—Moulding sand—Foundry hand tools—Melting equipment—*Casting*—Advantages of casting process—Preparation of a casting—Design of a casting—Casting processes—Defects in castings—Cleaning of castings—Inspection of castings—Exercise. 9.2. **Forging :** Introduction—Advantages and disadvantages of forging—Applications of forging—Classification of forging—Hand forging—Machine forging—Defects in forging—Heat treatment of forgings—exercise 9.3. **Rolling.** 9.4. **Welding :** Pressure welding—Fusion welding—Highlights—Objective Type Questions—Theoretical Questions.

9.1. MOULDING AND CASTING

9.1.1. Introduction

- In any manufacturing workshop foundry section occupies an important place. In foundry the jobs are manufactured by *pouring molten metals* such as cast iron, brass, cast steel, while metal, gun metal, aluminium into *moulds* prepared in moulding boxes by wooden or metal patterns.
- The *process of shaping metals by pouring is very fast and much cheaper than fabrication methods*. This is *cheaper than welding*. Hence designs are many times changed to get the components cast rather than fabricated.
  — The possibility of blow holes can be avoided by centrifugal method of pouring castings.
  — With modern techniques, it is possible to produce such castings which do not require any machining operation at all except some grinding.
  — With improved heat treatment techniques it is possible to modify the properties of metals also by casting (*e.g.,* malleable castings) etc.
- The products in foundry are poured metals cast in moulds, hence the production of foundry is known as '*castings*'.

9.1.2. Mould Making

9.1.2.1. General aspects

A **mould** *may be defined as the negative print of the part to be cast and is obtained by the pattern in the moulding sand container (boxes) into which molten metal is poured and allowed to solidify.* Sand moulds are destroyed as the casting is removed from the moulds.

Moulding is *an art of making sound mould out of sand by means of pattern and cores so that metal can be poured into the moulds to produce castings.*

— *Moulding is done both by hands and by machines.* Hand moulds are restored to *odd castings* generally less than 50 pieces at a time or so. Here ramming in done by hand which takes more time than machine moulding. However the quality is better for odd castings.

For mass manufacture, machine moulding is suitable. Moulding machines are prominently used in big foundries.

The **moulding machines** *perform are the following basic operations :*
1. Ramming or sand in the mould,
2. Lifting or drawing of pattern from the mould, and
3. Rolling over mould section.

Following are the two main classes of moulding machines :
(a) Hand moulding machines.
(b) Power moulding machines :
  (i) Jolt-machine
  (ii) Squeezing machine
  (iii) Jolt-squaeeze machine
  (iv) Sand slinger
  (v) Diaphragm moulding machine
  (vi) Stripper-plate machine.

- Moulding is carried out in moulding boxes called *flasks*, which are open at the top and bottom ; the top part is called the *cope* and the lower part as the *drag*. The moulding boxes are usually either of fabricated mild steel or cast iron which can be damped together. To avoid misfitting of two halves there are two pins on one side and one pin on other side in the top half which go into corresponding holes of the bottom half. This avoids the possible misfit. In case of very big castings, the moulds may be made on ground without moulding boxes. In some cases moulding boxes are put in 3 pieces (the intermediate part is called a *cheek*) to facilitate moulding. The section of moulding box is shown in Fig. 9.1.

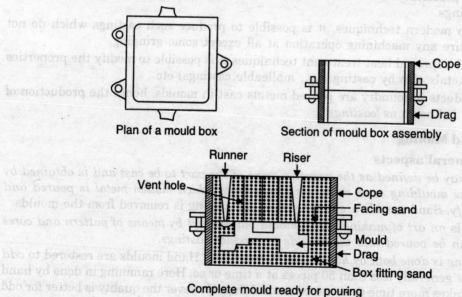

Fig. 9.1. Mould making.

MANUFACTURING PROCESSES

9.1.2.2. Types of moulds
The moulds are of the following two types :

1. Temporary moulds :
These moulds are destroyed at the time of removing castings from them.
Example : *Sand moulds.*

2. Permanent moulds :
These moulds are used in die casting. These moulds are used time and again.
Example : *Metallic moulds.*

9.1.2.3. Moulding processes
The moulding processes may be classified as follows :

1. Bench moulding :
- The moulding done on a bench of convenient height to the moulder is called *bench moulding*.
- It is used for small work.

2. Floor moulding :
- The moulding done on the foundry floor is called *floor moulding*.
- It is used for all medium sized and large castings.

3. Pit moulding :
- Very large moulds made in a pit or cavity cut in the floor to accommodate very large castings is called *pit moulding*. The pit acts as a drag.
- Since pit moulds can resist pressures developed by hot gases, therefore, it greatly saves pattern expenses.

4. Machine moulding :
- The mouldings done by a machine is called *machine moulding*.
- Small, medium and large moulds may be made with the help of a variety of machines.
- Machine moulding is usually faster and more uniform than bench moulding.
- Machine moulding generally requires mounted patterns.

9.1.2.4. Types of sand moulding
Sand moulding methods may be classified as follows :
1. Green sand moulds
2. Dry-sand moulds
3. Skin-dried moulds
4. Loam moulds
5. Metal moulds.

1. Green sand moulding (moulds)

Among the sand-casting processes, moulding is often done with green sand. *Green moulding sand* may be defined as a *plastic mixture of sand grains, clay, water, and other materials, which can be used for moulding and casting processes*. The sand is called *"green"* because of moisture present and is thus distinguished from dry sand.

The basic steps in green-sand moulding are as follows :

(*i*) **Preparation of the pattern.** Most green-sand moulding is done with match plate or cope and drag pattern. Loose patterns are used when relatively few castings of a type are to be made. In simple hand moulding the loose pattern is placed on a mould board and surrounded with a suitable-sized flask as illustrated in Figs. 9.2 and 9.3 respectively.

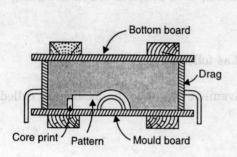

Fig. 9.2. Drag half of mould made by hand. Drag is ready to be rolled over in preparation for making the cope.

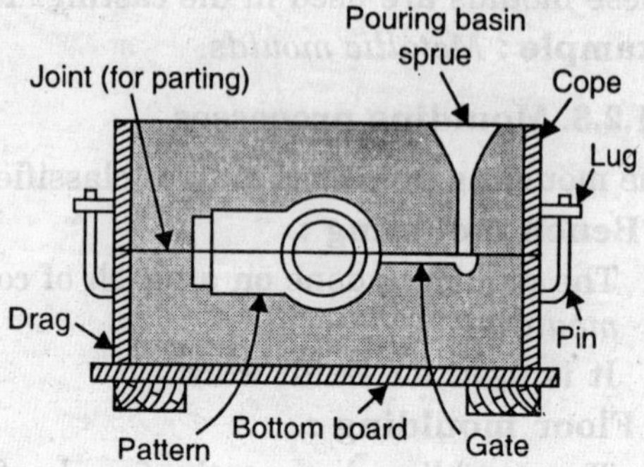

Fig. 9.3. Cope mould rammed up.

(*ii*) **Making the mould.** Moulding requires ramming of sand around the pattern. As the sand is packed, it develops strength and becomes rigid within the flask. Ramming may be done by hand, as in simple set up illustrated in Fig. 9.2. Both cope and drag are moulded in the same way, but the cope must provide for the sprue. The gating-system parts of the mould cavity are simply channels for the entry of the molten metal, and can be moulded as illustrated in Fig. 9.3.

(*iii*) **Core setting.** With cope and drag halves of the mould made and the pattern withdrawn, cores are set into the mould cavity to form the internal surfaces of the casting.

(*iv*) **Closing and weighing.** With cores set, the cope and drag are closed. The cope must usually be weighted down or clamped to the drag to prevent it from floating when the metal is poured :

Advantages :
1. Great flexibility as a production process.
2. The least costly method of moulding.
3. Less time consuming since no backing operations or equipment is required.
4. Green sand moulds can be used for all of the ferrous and non-ferrous alloys.

Disadvantages/Limitations :
1. Certain metals and some castings develop defects if poured into moulds containing moisture.
2. More intricate castings cannot be made.
3. The dimensional accuracy and surface finish of green-sand castings may not be adequate.

4. Green sand moulds are not very strong and may be damaged during handling or by metal corrosion.
5. Storage of green sand moulds for longer periods is not possible.

2. Dry-sand moulds :
- The sand mould made with a sand that does *not* require moisture to develop strength (the binder provides strength) are called *dry sand moulds*.
- These moulds are used for *steel castings*.
- The dry-sand moulds are stronger and may be handled more easily with less damage.
- Dry sand eliminates the possibilities of moisture related defects in casting.
- These are *more expensive* comparatively.

3. Skin-dried moulds :
- The sand moulds with a dry sand facing and a green sand backing are called *skin-dried moulds*.
- They can be employed for casting all ferrous and non-ferrous alloys.
- They are more commonly used for *large moulds*.
- As compared to dry-sand moulds they are less expensive to construct, but more expensive than green sand moulds.
- They are less stronger than dry-sand moulds.

4. Loam moulds :
- These moulds are made with loam sand (a mixture of sand and clay). The loam sand also contains fire clay or gainsters.
- They are used for large work.
- A loam mould is constructed of porous bricks cemented together with loam mortar. The inside of the brick structure forms the rough contour of the casting and it is faced with a 6 to 12 mm layer of loam sand to give the required shape.
- A loam mould requires enough area and space, difficult to give proper contour and shape and is *suitable only for a single casting*.

5. Metal moulds :
- The metal moulds are permanent type of moulds. These are used in die casting where molten metal is introduced into the metallic mould cavity by means of pressure. Sometimes even gravitational force is sufficient to feed the metal into the mould cavity.
- These moulds are used *in the casting of low-melting temperature alloys*.

Advantages :
   (*i*) Improved surface furnish.
   (*ii*) Since the castings produced by metal moulds have a smooth finish, therefore much of the machine work is eliminated.
   (*iii*) High production rate.
   (*iv*) Thin sections can be cast.
   (*v*) Castings produced are less defective.

Disadvantages :
  (*i*) High cost of moulds and equipment.
  (*ii*) For maintenance of moulds/equipment, special skill is required.

9.1.3. Core

- A *'core' is a body made of refractory material* (sand or metal, metal cores being less frequently used), *which is set into the prepared mould before closing and pouring it,* for *forming the holes, recesses, projections, undercuts and internal cavities.*
- The cores are subject to much more severe thermal and mechanical effects than the moulds, because they are surrounded on all sides (except for the ends) by molten metal. Consequently core sands should meet more stringent requirements.
- *Refractiveness or thermal stability of core can be increased by giving a thin coating of graphite or similar material to the surface of the core.*

9.1.3.1. Core making

- Cores are made in simple wooden, metallic or plastic core boxes. These core boxes are part of the pattern equipment for the castings. The complicated shapes may require support on sand or metal formers until these are baked.
- The *simple method of core making is similar to that of mould making.* The sand mixture is rammed into the core box with a wooden rammer. Sometimes the cores may need reinforcement with wire or nails in order to provide internal support so that they may not collapse while handling. The core-sand mixture is rammed by hand or pneumatic rammers. Venting and other necessary operations are performed during construction of the core.
- For *production work, machines are used for core makings* where core-sand mixture is rammed by *jolting, squeezing or blowing* by *means of suitable machines.* The most common core making machine is the *core blower.* Venting, reinforcing and other necessary operations are performed by hand during core construction.

Sometimes cores may be made by *extruding a core-sand mixture through a suitable die opening* and called **"stock cores"** which are of symmetrical cross-section.

The cores are removed from the core box placed in metal trays and are baked in an oven at a suitable temperature varying from 150°C to 400°C for the required duration of time. The source of heat may be the burning of gas, oil, coke, or electric heating.

9.1.3.2. Types of cores

A *core is a specially designed shape employed to take the place of metal in a mould.*
The cores may be classified as follows :

I. According to the state of the core :
  (*i*) Green sand cores
  (*ii*) Dry sand cores
  (*iii*) Oil sand cores
  (*iv*) Loam cores
  (*v*) Metal cores.

(*i*) **Green sand cores :**
- These are made from ordinary moulding sand, mixed with floor-sand, thoroughly vented and not too damp.

- They are restricted to *simple shapes,* reinforced with substantial core irons for handling, and *not dried before using.*

(*ii*) **Dry sand cores :**
- These cores consist of moulding sand with controlled conditions of such opening materials as horse manure, sawdust or chopped straw, blended together in a mill with either water, clay water, molasses or any other suitable bindings agent. Ample core irons are used for strengthening.
- The interiors of large cores are filled with coke to facilitate venting, and to overcome contraction strains.
- They are baked until perfectly dry.

(*iii*) **Oil sand cores :**
- They consist chiefly of sea shore or other silica sands, to which has been added a binding medium ; a few typical binders being linseed oil, resin, molasses and cereal flour.
- All cores of this type are baked before handling.

(*iv*) **Loam cores :**
- Loam cores are perforated cast-iron or steel barrels on to which has been wound straw or a hayband and then coated with loam. The whole is then baked perfectly dry.

(*v*) **Metal cores :**
- These cores are usually made of steel (See Fig. 9.4) and are mostly used in the making of non-ferrous castings, acting as densers, and they also impart a fine finish.

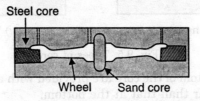

Fig. 9.4. Steel cores in a sand casting.

- They are also used for the production of iron castings, they produce a *white hard skin.*

Note. Owing to contraction stresses, means should be provided whereby the cores can be released at a reasonable time after the metal in the casting has set.

II. According to the position of the core in the mould :

(*i*) Horizontal core (*ii*) Vertical core
(*iii*) Balanced core (*iv*) Cover core
(*v*) Hangings core (*vi*) Wing core.

(*i*) **Horizontal core :** Refer Fig. 9.5. (*i*).
- The core is usually cylindrical in form and is laid horizontally at the parting line of the mould.
- The ends of the core rest in the seats provided by the core prints on the pattern.

(*ii*) **Vertical core :** Refer Fig. 9.5 (*ii*).
- The core is placed vertically in the mould.

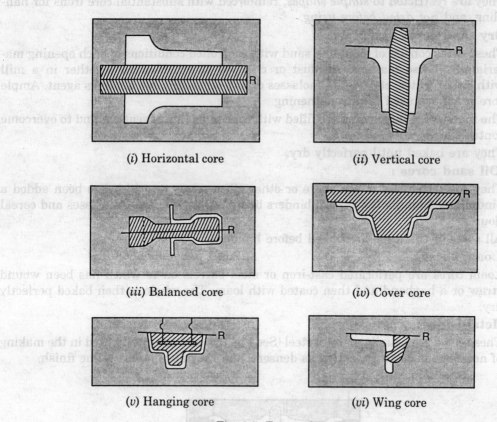

Fig. 9.5. Types of cores.

- Usually top and bottom of the core are provided with a taper, but the amount of taper on the top is greater than that at the bottom.

(*iii*) **Balanced core :** Refer Fig. 9.5 (*iii*)

- This core is similar to horizontal core, but it is supported at one end only.
- The core print in such cases should be large enough to give proper bearing to the core.

(*iv*), (*v*) **Cover core and hanging core :** Refer Fig. 9.5 (*iv, v*)

- The cover core, as shown in Fig. 9.5 (*iv*), is used when the entire pattern is rammed in the drag and the core is required to be supported from the top of the mould. This type of core usually requires a hole through the upper part to permit the metal to reach the mould.
- If the core hangs from the cope and does not have any support at the bottom in the drag, then it is called *hanging core* [Fig. 9.5 (*v*)].

(*vi*) **Wing core :** Refer Fig. 9.5 (*vi*)

- A wing core is used when a hole or recess is to be obtained in casting either above or below the parting line. In this case, the side of the core point is given sufficient amount of taper so that the core can be placed readily in the mould.
- The core is sometimes designated by other names such as *tail core, drop core, chair core* and *saddle core* according to its shape and position in the mould.

Core sand :

- Core sand is a variety of silica sand. Rock sand, river sand and sea shore sand, commonly known as sharp sand, are generally used for making of cores, chiefly because they are *capable of withstanding high temperatures, and resisting the penetrating action of the molten metal.* They have in addition, *high porosity, together with good permeability.*
- Having no natural bond, these sands are mixed with a suitable binder of which there are several kinds in the form of creams, oils and resins. These binders are burnt out by the time the casting is cold making the core friable and easy to remove.

9.1.3.3. Core prints

For supporting the cores in the mould cavity an impression in the form of a recess is made in the mould with the help of a projection on pattern. This projection is known as **core print**. Core prints are of the following types :

1. Horizontal core print

It produces seats for horizontal core in the mould,

2. Vertical core print

It produces seats to support a vertical core in the mould.

3. Balanced core print :

It produces a single seat on one side of the mould and the core remains partly in this formed seat and partly in the mould cavities, the two portions balancing each other. The hanging portion of the core may be supported on chaplets.

4. Cover core print :

It forms seat to support a cover core.

5. Wing core print :

It is used to form a seat for a wing core.

9.1.3.4. Core box

- A core box is a *type of box used for the production of sand cores.* Core boxes are used in foundry work to form shapes in sand, called cores, which are used in connection with moulds, when *holes or internal shapes are required.*

Various methods of construction are used depending on the shape or size of the core, and its removal from the box after it is made. The inside of a core box must be cut out to form the exact shape of the hollow part required in the casting, plus the extensions for locating in the prints.

- The core boxes are filled with sand which is made firm with ramming and removed by opening the box at a centre joint, or taking it away from the core in various directions by a number of joints.
- If a plain hole is required in a casting, the core is a cylindrical piece of sand, the length of the hole plus the print portion ; this core can be made in a **simple core box** cut from *two pieces of wood held together with dowels on the centre joint.*Fig. 9.6 (*i*).
- Another method of buildings up core boxes is to make the joints in such positions as will enable straight cuts to be made, *e.g.*, **piston core box**, Fig. 9.6 (*ii*). Each piece of the box is made to thickness of the sections at which the diameter of the piston

alters, and screwed together. After making out of the joint on the box, the pieces are taken apart and cut through, glued and screwed back in position.

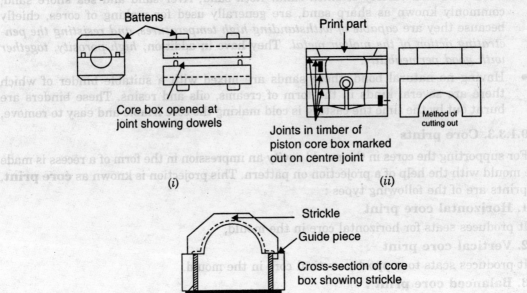

Fig. 9.6. Core boxes : (*i*) Simple box for cylindrical core,
(*ii*) Piston core box, arranged for straight cuts ;
(*iii*) Strickle used to shape large cores.

- **Stickles** are *frequently used with core boxes to obtain shapes of large works*. They work from a *guide held against a parallel side of the core box*. Fig. 9.6 (*iii*). A considerable amount of time in building up and shaping is saved by their use.

9.1.4. Moulding sand

In a foundry shop sand is the principal moulding material and is used for all types of castings. The moulding sand possesses all the properties which are vital for foundry purpose and is used time and again.

9.1.4.1. Properties of moulding sand

1. Permeability :
- It is the property to allow gases to escape easily from the mould.
- Higher the silt content of sand, the lower is gas permeability. If the mould is rammed too hard, its permeability will decrease and *vice versa*.
- It is measured in number such as 60, 80, 100, 120 etc.

2. Strength or cohesiveness :
- It is defined as the *property of holding together of sand grains*.
- A moulding sand should have ample strength so that mould does not collapse or get partially destroyed during conveying, turning over or closing.

- The strength of the moulding sand *grows with density, clay content of the mix* and *decreased size of sand grains*. Thus as the *strength of the moulding sand increases, its porosity decreases*.

3. Refractiveness :
- It is *the ability of the moulding sand mixture to withstand* the *heat of melt without showing any sign of softening* or *fusion*.
- It *increases with the grain size of sand and its content and with the diminished amount of impurities and slit*.

4. Plasticity or flowability :
It should be of plastic nature so that it can easily take any desired shapes.

5. Collapsibility :
- *This is the ability of the moulding sand mixture to decrease in volume to some extent under the compressive forces developed by the shrinkage of metal during freezing and subsequent cooling.*
- This property permits the moulding sand to collapse easily during shake out and permits the core to collapse easily during its knock out from the cooled casting.
- Lack of collapsibility in the moulding sand and core may result in the formation of cracks in the casting.
- This property depends on the amount of quartz sand and binders and their type.

6. Adhesiveness :
- This is the property of sand mixture to *adhere* to another body.
- The moulding sand should cling to the sides of the moulding boxes so that it does not fall out when the flasks are lifted and turned over.
- This property depends on the type and amount of binder used in sand mix.

7. Coefficient of expansion :
The sand should have low co-efficient of expansion.

8. Chemical resistivity :
The sand should not chemically react or combine with molten metal.

9.1.4.2. Types of moulding sand
The moulding sands are classified as follows :

I. *According to their clay bonding material* :

1. **Natural sand** : It contains sufficient amount of binding clay and, therefore, no more binder is required to be added.

2. **Synthetic sand** : It is one which is artificially compounded by mixing sand, and selected type of clay binders etc. These sands have the following *advantages* :
- Lower cost in large volume.
- Widespread availability.
- The possibility of sand reclamation and reuse.

II. *According to their use* :

1. **Green sand** :
- The sand in its natural or moist state is called *green sand*.
- It is a mixture of silica sand with 20 to 30% clay, having total amount of water from 6 to 10%.

- The green sand moulds are used for small size castings of ferrous and non-ferrous metals.

2. Dry sand :
- When the moisture from the green sand is evaporated by drying or baking, after the mould is made is called *dry sand* mould.
- The dry sand moulds have greater strength, rigidity and thermal stability. The dry sand moulds are used for large and heavy castings.

3. Loam sand :
- The loam sand consists of as high as 50% of clay contents.
- It is used for loam moulding of *large grey-iron castings*.

4. Facing sand :
- A sand used for facing of the mould is called *facing sand*.
- Since it comes in contact with molten metal when poured, therefore it must possess high strength and refractoriness.

5. Parting sand :
- Parting sand is purely clay-free silica sand which is sprinkled on the pattern and the parting surfaces of the mould so that the sand mass of cope and drag separate without clinging and do not stick to the pattern.

6. Backing or flour sand :
- A sand used to back up the facing sand not used next to the pattern, is called *backing sand*.
- Because of its black colour, it is sometimes called *black sand*.

7. Core sand :
- A sand used for the preparation of the cores is called *core sand*.
- It is sometimes called *oil sand*.

9.1.4.3. Composition of the Green Sand

- (i) Silica .... up to 75 per cent
- (ii) Clay .... 8 to 15 per cent
- (iii) Bentonite .... 2 to 5 per cent
- (iv) Coal dust .... 5 to 10 per cent
- (v) Water .... 7 to 8 per cent.

9.1.5. Foundry Hand Tools

The various hand tools commonly used in foundry are described below :

1. Shovel : Refer to Fig. 9.7.
- It consists of a square pan fitted with a wooden handle.

Fig. 9.7. Shovel.   Fig. 9.8. Riddle.

- It is used for *mixing and for moving sand from one place to another in the foundry*.

2. Riddle : Refer to Fig. 9.9.
- It consists of a wooden frame fitted with a screen of standard wire mesh at its bottom.
- It is used for *hand riddling of sand to remove foreign material from it*.

3. Rammers :
- A rammer is a tool used by foundry workers for consolidating sand. Unless the sand is rounded evenly, swelling may occur in the metal in the neighbourhood of any soft spots, so that the resulting casting will not be true.
- Fig. 9.9 shows floor rammer and bench rammer, the only difference being the size of the shaft and weight of head.

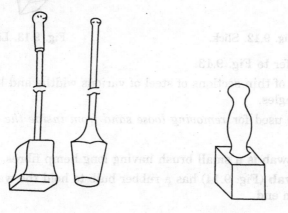

(i) Floor rammers   (ii) Bench rammers

Fig. 9.9. Rammers.

4. Strike off bar : Refer to Fig 9.10.
- It is a flat bar, made of wood or iron, to *strike off the excess sand from the top of a box after ramming*.
- Its one edge is made bevelled and the surface perfectly smooth and plane.

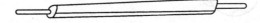

Fig. 9.10. A strike off bar.

5. Vent wire : Refer to Fig. 9.11.

After ramming and striking off the excess sand, vent wire is *used to make small holes in the sand mould to allow the exit of gases and steam during casting*.

Fig. 9.11. Vent wire.

6. Slick : Refer to Fig 9.12.
- It is a small double ended tool having a flat on one end and a spoon on the other. It is also made in a variety of other shapes.

- A slick is used for *repairing and finishing the mould*.

Fig. 9.12. Slick.   Fig. 9.13. Lifter.

7. Lifter : Refer to Fig. 9.13.
- It is made of thin sections of steel of various widths and lengths with one end ben at right angles.
- Lifters are used for *removing loose sand from inside the mould cavity*.

8. Swab :
- A simple swab is a small brush having long hemp fibres.
- A **bulb swab** (Fig. 9.14) has a rubber bulb to hold the water and a soft hair brus at the open end.

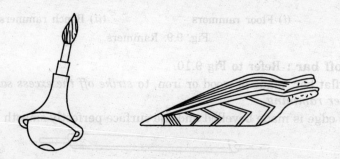

Fig. 9.14. Swab bulb.   Fig. 9.15. Bellow.

- It is used for *moistening the sand around the edge before the pattern is remove*

9. Bellow : Refer to Fig. 9.15.

It is *used to blow out the loose or unwanted sand from the surface and cavity of t mould*.

10. Trowels : Refer to Fig. 9.16.
- Trowels are used for *finishing flat surfaces and joints in a mould*.
- They are *made of iron and are provided with a wooden handle*.

MANUFACTURING PROCESSES

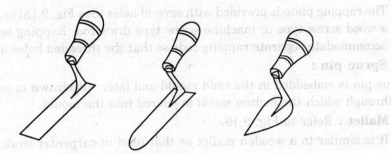

Fig. 9.16. Trowels.

11. Gate cutter : Refer to Fig. 9.17.
- A gate cutter is a *"U-shaped"* piece of thin sheet.

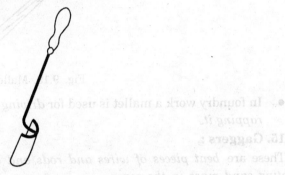

Fig. 9.17. Gate cutter.

- It is *used for cutting shallow trough in the mould to act as a passage for the hot metal.*

12. Draw screws and rapping plate : Refer to Fig. 9.18.
- Draw screws are straight mild steel rods carrying a loop or ring at one end and a wood or machine screw at the other. They are always used in conjunction with a *rapping plate* for rapping and withdrawing the pattern from sand.

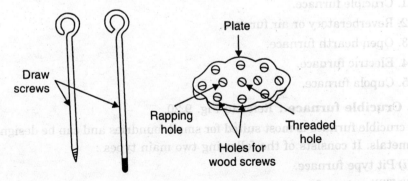

Fig. 9.18. Draw screws and rapping plate.

- The rapping plate is provided with several holes (See Fig. 9.18) to accommodate either a wood screw type or machine screw type draw rod. Rapping holes are provided to accommodate separate rapping rods so that the threaded holes are not spoiled.

13. Sprue pin :

Sprue pin is embedded in the sand mould and later withdrawn to produce a hole called runner, through which the molten metal is poured into the mould.

14. Mallet : Refer to Fig. 9.19.

- It is similar to a wooden mallet as that used in carpentry work.

Fig. 9.19. Mallet.

- In foundry work a mallet is used for *driving the draw spike into the pattern and then rapping it.*

15. Gaggers :

These are *bent pieces of wires and rods and are used for reinforcing the downward projecting sand mass in the cope.*

16. Clamps, cotters and wedges :

They are *made of steel* and are used *for clamping the moulding boxes firmly together during pouring.*

9.1.6. Melting Equipment

The main types of furnaces used in foundries for melting various varieties of ferrous and non-ferrous metals and alloys are enumerated and described below :

1. Crucible furnace.
2. Reverberatory or air furnace.
3. Open hearth furnace.
4. Electric furnace.
5. Cupola furnace.

1. Crucible furnace : Refer to Fig. 9.20.

A crucible furnace is most suited for small foundries and can be designed for melting any of the metals. It consists of the following two main types :

(*i*) Pit type furnace.

(*ii*) Tilling type furnace.

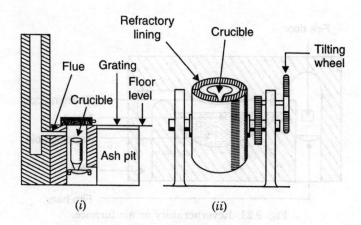

Fig. 9.20. Crucible furnace.

(a) **Pit type crucible furnace :** Refer to Fig. 9.20 (i).
- It is built to suit the type of metal to be melted.
- These are fixed wholly or party in the ground from which the crucible must be lifted when the metal is ready.
- Here a crucible (a heat resisting pot for metal melting and made of fire clay mixed with coke dust or graphite) is placed in a pit in the floor. The furnace is usually fired with sufficient coke being packed round and above the crucible pots to melt and superheat the charge without re-coking. The natural draught provided by tall chimney is controlled by means of loose brick or damper at the foot of the stack.

(b) **Tilting type crucible furnace :** Refer to Fig. 9.20 (ii).
- This type of furnace is built above the ground level, and contains a firmly fixed crucible.
- The furnace is fired with coke, oil or gas and the forced draught is used.
- When the metal charge is ready for pouring, the whole furnace is tilted and the crucible emptied by operating a geared trunnion.
- For the metals of high melting points, *clay or plumbago crucibles* are used ; for the low-melting-point metals, such as zinc-base or aluminium, *cast iron or steel cucibles* are suitable.

2. **Reverberatory or air furnace :** Refer to Fig. 9.21.
- This is used for *melting in one heat large quantities of metal,* those most suited being all grades of cast-iron and the alloys of brasses and bronzes.

This type of furnace is also used for the production of wrought iron, and is then situated not in a foundry, but near forge or rolling mill, and is known as a puddling furnace.
- It may have either a sloping roof, or a double arched roof which forms a dip in the centre. A chimney is provided at one end and a fire grate or burners at the other end. A hearth or well is provided in the centre for holding the metal.
- It employs natural draught, which is controlled by dampers.

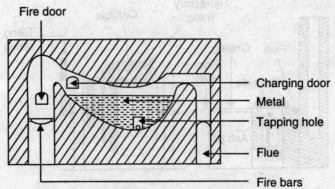

Fig. 9.21. Reverberatory or air furnace.

- The fuel can be either small lumpy coal, which is used on the fire grate, or powdered fuel, which is supplied through burners. The object is to create a long flame which reverberates or strikes back from the furnace roof on to the metal to be melted in the hearth.

3. Open hearth furnace : Refer to Fig. 9.22.

- An open hearth furnace is used chiefly for the *production of steel and for refining purposes.*

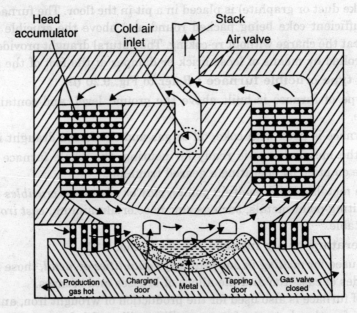

Fig. 9.22. Open hearth furnace.

- Gas and heated air, admitted through the ports on the left, burn above the hearth. The hot, spent gases heat a brickwork chamber, before reaching the chimney stack. After about twenty minutes the direction of air and gas flow is reversed, so that the cold air passes through the newly-heated brick chamber, while that on the left is re-heated in preparation for the next cycle.

4. Electric furnace

This is used especially where rigid control over temperature and analysis is required. It is *suitable for all types of metals and alloys*.

The electric furnaces are classified as follows :

I. *Arc type furnaces* :
   (*i*) Direct arc.
   (*ii*) Indirect arc.

II. *Induction furnaces*.

I. *Arc type furnace* :

(*i*) **Direct-arc furnace** : Refer to Fig. 9.23.

- It consists of a round, bowl-shaped carbon hearth with a domeshaped roof supporting one or more carbon electrodes through which passes the current which strikes arcs with the metal in the hearth, thus giving heat direct to the metal.
- This type of furnace can be either stationary or made to tilt. The roof is usually so made that it can be removed for charging purposes.

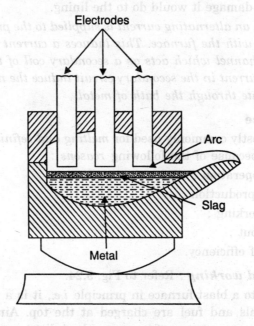

Fig. 9.23. Direct arc furnace.

- The capacity of these furnaces for production work varies from 3 to 10 tonnes. These are best suited for laboratory work where very small quantity of a few kg is needed for research work.
- These furnaces give *high melting rate, high pouring temperature and excellent-control of metal analysis and temperature*.

(ii) Indirect-arc furnace

- This furnace is used for melting *all types of metallic alloys but especially useful in the production of copper-base alloys.*
- It consists of a horizontal cylinder lined with a refractory material with two electrodes on the horizontal axis. An arc is struck between the electrodes in the centre of the furnace. The arc does not come in contact with the metal to be melted, the heat being given to the charge by radiation from the arc and reflection from the walls of the furnace. The furnace is designed to give a rocking motion as the melting proceeds, thus quickening up the melt by distributing the heat more rapidly. The charging, tapping and slagging are done through an opening in the side of the furnace.

II. *Induction furnace* :

- An induction furnace is a tilting furnace used chiefly *for the melting of non-ferrous metals. Heat is generated by the resistance offered to an induced current set up within the metal in the furnace.*
- The design of the furnace is such that a small channel is formed inside at its base ; this channel in filled with metal, which should never be allowed to solidify owing to the amount of damage it would do to the lining.
- *When working, an alternating current is supplied to the primary coil of a transformer which is built with the furnace. This induces a current to pass through the liquid metal in the channel which acts as a secondary coil of the transformer. The forces set up by the current in the secondary circuit induce the metal in the channel to heat up and circulate through the bath of metal.*

5. Cupola furnace

This furnace is mostly commonly used for *melting and refining pig iron* (alongwith cast iron and steel scarps) because of the following *reasons* :

    (*i*) Simplicity of operation ;
   (*ii*) Continuity of production ;
  (*iii*) Economy of working ;
  (*iv*) Increased output ;
   (*v*) High degree of efficiency.

Construction and working : Refer to Fig. 9.24.

It is very similar to a blast-furnace in principle *i.e.,* it is a vertical shaft furnace, into which the raw materials and fuel are charged at the top. Air for combustion of fuel is introduced through one or more rows of tuyeres a short distance above the bottom. *Since the cupola is only concerned with the melting of the metal and not with the reduction of ores as in the blast furnace, it is considerably smaller than a blast furnace of the same output.* Its diameter varies from 1 to 2 metres with a height of 4 to 5 times diameter.

MANUFACTURING PROCESSES

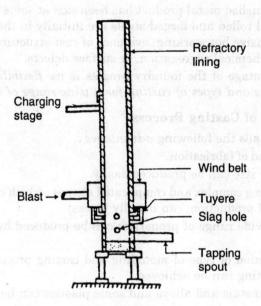

Fig. 9.24. Cupola.

- In a cupola, the first operation is to light the fire at the bottom. When the fire is burning strongly, coke is added gradually till the level above the tuyeres is about 0.6 metres. This coke serves as a bed for the alternate charges of metal and coke which follow. When the shaft of the cupola is filled level with the charging door the blast is put on and the combustion of the coke near the tuyeres increases rapidly until a very intense heat is attained. "The gases of combustion move upwards and pass on a portion of the heat to the metal and coke waiting to descend. In 5 to 10 minutes the first charge of metal starts melting and trickles down through the coke and finally collects at the bottom of the cupola. When an adequate quantity (say 1 or 2 tonnes) has accumulated the plug of clay called *'bout'* is removed from the tap hole and metal allowed to run into the ladle. The temperature of tapping metal is 1200–1400°C. After melting a number of charges as per requirements the bed coke is removed through a drop-bottom door and quenched with water so as to be available for use the next day.
- Although it is usual practice to operate a cupola with cold blast (since no reduction of ores is required) a few cupolas have recently been equipped for hot blast. It may be noted *whereas a blast furnace operates continuously, a cupola works intermittently*.

9.1.7. Casting

- *'Casting' means the pouring of molten metal into a mould, where solidification occurs.*

Metal casting may also be *defined as a process of production of objects of desired shapes and sizes by introduction of molten metal into a predesigned mould cavity created commonly in a compact sand mass, with the help of a pattern, or in a metallic mould (as in die casting); and allowing it to solidify.*

- Almost every finished metal product has been cast at some stage of its manufacture. For example, all rolled and forged steels are initially in the form of cast ingots, and even after extensive hot working, evidence of cast structure may still remain in the form of solids, chemical segregation, or surface defects.
- The main advantage of the foundry process is its *flexibility* and the *possibility of making all sorts and types of casting for a wide range of applications*.

9.1.8. Advantages of Casting Process

Casting process entails the following *advantages* :
1. Cheapest method of fabrication.
2. Objects of large size can be produced easily.
3. The objects having complex and complicated shapes ; which cannot be produced by any other method of production, can usually be cast.
4. Castings with wide range of properties can be produced by adding various alloying elements.
5. By proper selection of type of moulding and casting process, required dimensional accuracy in casting can be achieved.
6. Almost all the metals and alloys and some plastics can be cast.
7. The number of castings can vary from very few to several thousands.

9.1.9. Preparation of a Casting

Preparation of a casting involves the following steps :
1. Preparation of a pattern.
2. Preparation of moulding sand.
3. Preparation of mould and core(s).
4. Melting the metal.
5. Pouring of metal into the mould.
6. Cooling and solidification.
7. Removing the casting from the mould.
8. Fettling (*i.e.,* cutting off the unwanted projection in the from of gates, risers etc.)
9. Heat treatment
10. Testing and inspection.

9.1.10. Design of a Casting

As castings are produced by pouring hot metal into sand moulds at one time, the casting designed should consider the following *points* :
1. No section of a casting is less than 3 mm thick, as thinner sections are liable to break.
2. The joint between, the thinner and thicker sections is not suddenly changed in sections leading to stress concentration and ultimate weak sections.
3. As metal flows along smooth curves and does not flow in sharp corners, the smoothening of sharp edges by rounded fillets is necessary.
4. As cooling rates for thin sections is faster than those for thick surfaces care should be taken to see that the stress concentrations are not left in thin sections nor the metals (such as cast iron) get handened due to faster cooling.

MANUFACTURING PROCESSES

9.1.11. Casting Processes

The various casting processes in use are enumerated and described below :
1. Sand casting.
2. Shell moulding.
3. Permanent mould casting.
4. Die casting.
5. Centrifugal casting.
6. Investment casting.
7. Plaster casting.
8. Slush casting.

1. Sand casting :
- A commonly used method involves pouring molten metal into a cavity in a mass of packed sand.
- Fig. 9.25 (*i*) shows a typical mould in cross-section. It illustrates the use of *chills* to produce a local hard surface, a *core* to form a shaft opening, and a *sprue* for running the molten metal into the cavity. A wood or metal *pattern* approximately the shape of the final casting is used to produce the cavity in the sand mould. So that the pattern may be removed from either the cope (upper) or the drag (lower) section of the mould without disturbing the sand that has been packed around it, a taper or *draft* of a few degrees must be allowed on the metal faces of the pattern.
- Since casting alloys decrease in volume as they solidify and cool to room temperature, it is necessary to make the pattern larger than the final casting by an amount known as the *shrinkage allowance*. Shrinkage depends on such factors as the kind of alloy being cast, the design of the casting, the pouring temperature, and the size of the casting. In making castings such as those of a U-shape it is necessary to "*fake*" or distort the pattern in order to obtain the desired from in the final casting. This is called *distortion allowance*. An additional *machine finish allowance* of 1.6 mm or more must be allowed on surfaces that are to machined. Finally even with the use of the best available information on shrinkage allowance, it is unlikely that final dimensions of the casting can be corrected exactly. Therefore, *size tolerance* equal to half the shrinkage allowance are suggested for use with castings of new design.

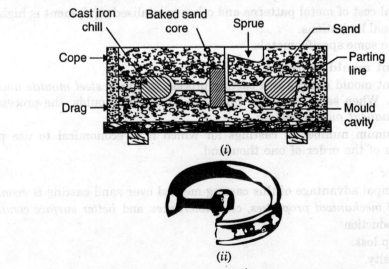

Fig. 9.25 (*i*) and (*ii*). Typical mould in cross-section.

Advantages :
Any metal with no limit on size and shape can be cast with low tool cost :

Limitations :
  (i) Product gives rough surface.
  (ii) Dimensional accuracy difficult.
  (iii) Thin projections not practical.
  (iv) Machining always necessary.

2. Shell moulding :
- Shell moulding is modification of sand casting in which a *relatively thin shell forms the mould cavity* into which the molten metal is poured.
- Typically, each of the two halves of a shell moulding *is made by placing a mixture of fine sand and a resin binder in contact with a heated metal pattern.* Melting of the resin occurs in thin layer of the sand-resin mixture at the surface of the pattern, and this thin shell remains attached to the pattern when the excess mixture is allowed to fall off. The shell is then baked at a high temperature, removed from the pattern, and finally assembled with matching half to form the completed shell moulding.
- Castings produced by this process have *better surface finish and closer dimensional tolerances than sand castings.*

Advantages :
  (i) A very smooth surface is generally obtained.
  (ii) The shell cast parts can be produced with dimensional tolerance of ± 0.2 mm.
  (iii) Reduced cleaning and machining costs.
  (iv) Gives rapid production rate.
  (v) Uniform grain structure.
  (vi) Minimum finishing operations.

Disadvantages :
  (i) The resin binder is more expensive than other binders.
  (ii) The initial cost of metal patterns and other specialised equipment is high.
  (iii) Dimensional limitations.
  (iv) Limited to some specific metals.

3. Permanent mould casting :
- Permanent mould casting is a casting process in which *steel moulds and cores are employed*. When sand cores are employed with steel moulds, the process is called *semi-permanent mould casting.*
- The minimum number of castings for which it is economical to use permanent moulds is of the order of one thousand.

Advantages :
  (i) The principal advantage of this casting method over sand casting is *economy*.
  (ii) *Improved mechanical properties, close tolerances* and *better surface conditions.*
  (iii) Rapid production.
  (iv) Low scrap loss.
  (v) Low porosity.
  (vi) Very high mould life.

MANUFACTURING PROCESSES

Disadvantages :
 (i) High cost of mould,
 (ii) Dimensional limitations,
 (iii) Limited to low melting point metal castings.

4. Die casting : Refer to Fig. 9.26.
- Die casting is essentially permanent mould casting *in which pressure forces the molten metal into the mould cavity.* However, the *mould used is much more expensive* (it is called a *"die"*) and a complex machine is employed to produce castings at a very high rate.

Advantages :
 (i) Large quantities of identical parts can be produced rapidly and economically.
 (ii) Very little machining is required on the parts produced.
 (iii) The parts having thin and complex shapes can be casted accurately and easily.
 (iv) The die casting requires less floor area than is required by other casting processes.
 (v) The castings produced by die-casting process are *less defective, owing to increased casting soundness.*
 (vi) The rapid cooling rate produces high strength and quality in many alloys.

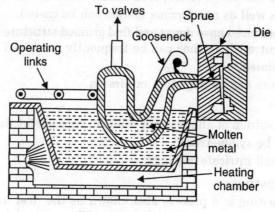

Fig. 9.26. Die casting.

Disadvantages :
 (i) The cost of equipment and die is high.
 (ii) There is a limited range of non-ferrous alloys which can be used for die castings.
 (iii) The die castings are limited in size.
 (iv) It requires special skill in maintenance.

5. Centrifugal casting :
- Castings that have rotational symmetry, such as long *cylinders,* are conveniently made by pouring the casting alloy into a metal, graphite or sand mould rotating above its axis of symmetry.
- A machine for producing such centrifugal castings is shown in Fig. 9.27. Non-metallic inclusions and slag particles, being less dense than the liquid metal, are forced to the inner surface of the casting and are removed in a latter machining operation.

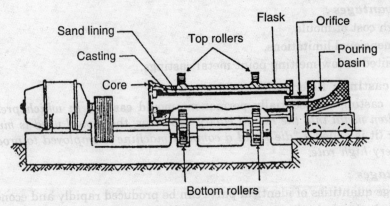

Fig. 9.27. Centrifugal casting.

- *The mechanical properties of these castings are superior to those of sand castings, but gravity segregation is encountered in some alloys.*

Advantages :
   (i) Quick and economical than other methods.
   (ii) In this process the use of risers, feed heads, cores etc. is eliminated.
   (iii) The ferrous as well as non-ferrous metals can be casted.
   (iv) The castings produced have dense and find grained structure with all impurities forced back to the centre where they can be frequently machined out.
   (v) Good surface finish.
   (vi) Gates and risers can be kept to a minimum.

Disadvantages :
   (i) Metallic composition of alloys is not uniform throughout the casting.
   (ii) Casting must be symmetrical.
   (iii) Limited to small intricate castings.

6. Investment casting :

- Investment casting is a process also known as the *"lost wax"* process or *"precision casting."* The term investment refers to a clock or special covering apparel, in this case a refractory mould, surrounding a refractory-covered wax pattern.
- In this method, a wax pattern of the part to be made is embedded (invested) in a fluid ceramic material that subsequently becomes solid. This mould is heated, causing the wax to melt and flow out, leaving a cavity of the desired shape. Molten metal is poured into the mould cavity, and after the metal has solidified the mould material is broken away, leaving the final casting.
- Investment castings have *excellent surfaces* and *dimensional accuracy,* and for this reason, they are used for parts made of *non-machinable and non-forgetable alloys. All extremely complex sections can be produced by this method, since there are no problems of draft, parting lines* and so on (as in sand casting).

Advantages :
   (i) In average work the close tolerance (± 0.05 mm) are easily maintained.
   (ii) Extremely smooth surfaces are produced.

MANUFACTURING PROCESSES

(*iii*) Most machining operations including thread cutting and gear tooth forming are eliminated.

(*iv*) Adaptable to the metallic alloys.

Disadvantages :

(*i*) The large size objects are impractical for investment casting due to equipment size limits.

(*ii*) The investment moulds as well as the materials from which they are made are single purpose, therefore they can not be reused ; this increases the production cost.

7. Plaster casting

If the sand-casting process is changed so that the Plaster of Paris is substituted for sand as the moulding material the method is called *plaster casting*.

Advantages :

(*i*) High dimensional accuracy.

(*ii*) Smooth surface.

(*iii*) Low porosity.

(*iv*) Mould easily repairable.

Disadvantages :

(*i*) Limited to non-ferrous metallic castings.

(*ii*) Dimensional limitations.

(*iii*) Time consuming.

8. Slush casting :

Hollow castings, such as *statues*, can be made by pouring a low-melting point alloy into a bronze or a plaster mould and quickly pouring out the excess molten metal after a thin solid shell has formed. The resulting slush casting can be finished by *electroplating* and *laquering*.

9.1.12. Defects in Castings

A large number of defects occur in sand castings produced through various methods. The *factors* which are normally responsible for the *production of these defects are* :

- Design of casting ;
- Design of pattern equipment ;
- Moulding and core-making equipment ;
- Mould and core materials ;
- Gating and risering ;
- Melting and core-making techniques ;
- Melting and pouring ;
- Composition of the metal.

Some of the common defects in casting are described below :

1. Blow holes : Refer to Fig. 9.28.

- They appear as *cavities in a casting*. When they are visible on the upper surface of the casting, they are called *"open blows"*. When they are concealed in casting and are not visible from outside, they are known as **blow holes**. They are due to the *entrapped bubbles of gases* in the metal and are exposed only after machining.

Fig. 9.28. Blow or gas holes.

- They are caused mainly by *hard ramming, excessive moisture, low permeability, excessive fine grains and incomplete or improper venting*.

2. Misrun : Refer to Fig. 9.29.

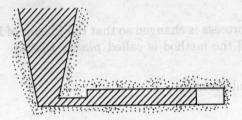

Fig. 9.29. Misrun.

- This defect is *incomplete cavity filling*.
- It is caused mainly by *inadequate* metal supply, too low mould or melt temperature and improperly designed gates.
- This defect determines the minimum thickness that can be cast for a given metal, superheat, and type of mould.

3. Cold shut : Refer to Fig. 9.30.

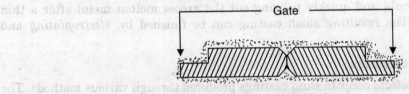

Fig. 9.30. Cold shut.

A *cold shut is an interface within a casting that is formed when two metal streams meet without complete fusion*. The causes are the same as for mistrun.

4. Mismatch : Refer to Fig. 9.31.

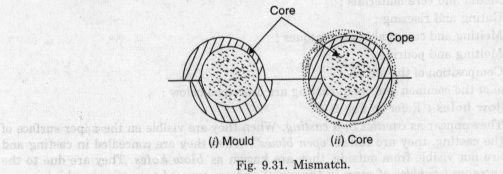

Fig. 9.31. Mismatch.

MANUFACTURING PROCESSES

- It is *shift of the individual parts of a casting with respect to each other.*
- It is caused by an inexpert assembling of the two halves of the mould and dimensional discrepancy between the core prints of the pattern and the core prints of the core.

5. Drop : Refer to Fig. 9.32.

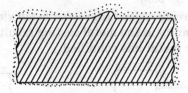

Fig. 9.32. Drop.

- This defect appears as *an irregular deformation of a casting.*
- It occurs on account of a portion of the sand breaking away from the mould and dropping into the molten metal.
- *Increase in green strength of the sand by suitable modification in its composition, hard ramming and adequate reinforcing of cope and other sand projections by means of bars, nails and gaggers etc. are the principal remedies of this defect.*

6. Flashes or fins :
- These are *thin projections of metal not intended as a part of casting*. These usually occur at the parting line of the mould or core sections.
- These are caused by loose clamping of the mould, insufficient weight on the top part of the mould and excessive rapping of the pattern before it is withdrawn from the mould.

7. Fusion :
- This defect appears as a *rough glassy surface over the casting.*
- It is caused due to lack of enough refractoriness in sand, faulty gating, too high pouring temperature of the metal and poor facing sand.

8. Metal penetration :
- This defects occurs as a *rough and uneven external surface on the casting.*
- The principal causes for the promotion of this defect are the use of coarse sand, having high permeability and low strength, and soft ramming.

9. Cut or wash : Refer to Fig. 9.33.
- It is a low projection on the drag face of a casting that extends along the surface, decreasing in height as it extends from one side of the casting to the other end.
- It usually occurs in bottom gating castings in which the moulding sand has insufficient hot strength, and when too much metal is made to flow through one gate into the mould cavity.

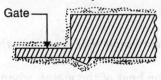

Fig. 9.33. Wash.

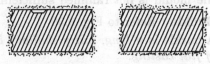

Scar   Blister
Fig. 9.34. Scar and Blister.

10. **Scars and blisters** : Refer to Fig. 9.34.
- A *scar* is a shallow blow. It generally occurs on a flat surface, where as a blow occurs on a convex casting surface.
- A *blister* is a shallow blow like a scar with a thin layer of metal covering it.

11. **Hot tears** : Refer to Fig. 9.35.
- These are the cracks having *ragged edges due to tensile stresses during solidification*. It is due to the discontinuity in the metal casting resulting from hindered contraction, occurring just after the metal has solidified.

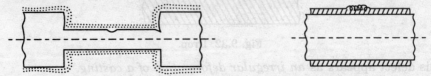

Fig. 9.35. Hot tears.   Fig. 9.36. Sponginess.

- This defect is caused by excessive mould hardness by ramming, high dry and hot strength and improper metallurgical and pouring temperature controls.

12. **Sponginess** : Refer to Fig. 9.36.
- Sponginess or honeycombing is an external defect, *consisting of a number of small cavities in close proximity*.
- It is caused by '*dirt*' or '*inclusions*' held mechanically in suspension in molten metal.

13. **Scab** : Refer to Fig. 9.37.
- This defect occurs when a portion of the face of a mould lifts or breaks down and the recess thus made is filled by metal.
- It is caused by too fine a sand, low permeability of sand and uneven ramming of the mould.

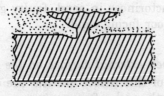 

Fig. 9.37. Scab.   Fig. 9.38. Swell.

14. **Swell** : Refer to Fig. 9.38.
- A swell is a slight, smooth bulge usually found on vertical faces of castings, resulting from liquid metal pressure.
- It is caused due to low strength of mould because of too high water content or when the mould is not rammed sufficiently.

15. **Buckle** : Refer to Fig. 9.39.
- A buckle is a *long, fairly shallow, broad, vee depression that occurs in the surface of flat casting*.
- It occurs due to the sand expansion caused by the heat of the metal, when the sand has insufficient hot deformation. It is also caused due to poor casting design.

Fig. 9.39. Buckle.

Fig. 9.40. Rat tail.

16. Rat tail : Refer to Fig. 9.40.

A rat tail is a *long, shallow, angular depression in the surface of a flat casting and resembles a buckle except that it is not shaped like broad vee*. The reasons for this defect are the same as for buckle.

17. Slag holes :
- These are smooth depressions on the upper surfaces of the casting. These usually occur near the ingates.
- This defect is due to imperfect skimming of the metal or due to poor metal.

18. Pour short :
- It occurs when the mould cavity is completely filled because of insufficient metal.
- This defect occurs due to interruptions during pouring operation, and insufficient metal in the ladles being used to pour the metal.

9.1.13. Cleaning of Castings

Generally, the cleaning of casting refers to all operations related to the *removal of adhering sand, gates, risers or other metal not a part of the casting*. The cleaning operations may also include a certain amount of metal finishing or machining for obtaining the required casting dimensions.

The various cleaning operations usually performed on a casting are enumerated and discussed below :

1. Rough cleaning
2. Surface cleaning
3. Trimming
4. Finishing

1. Rough cleaning. Rough cleaning includes the *removal of gates of risers*. The following points are worth-noting :

- In case of a ductile material casting, rough cleaning way be done with mechanical cut-off machines (using abrasive cut-off wheels, band saws and metal shears).
- The gating system of a brittle material casting way be broken off by impact when the castings are dumped and vibrated in shake-out or knock-out devices.
- In case of steel castings, very large risers and sprues way be removed by cutting torches
- In case of risers being large and cast of oxidation-resisting alloys, *powder cutting* (in which a stream of iron powder in introduced into the oxygen torch flame) is employed.

2. Surface cleaning. Surface cleaning includes cleaning of interior and exterior surfaces when sand, scale and other adhering materials are involved. This type of cleaning involves the following procedures :

(*i*) **Tumbling.** This operation is carried out with a barrel-like machine called *tumbling mill*, which removes sand, scale and some fins and wires.

(*ii*) **Blasting.** The *sand blasting* is performed by using coarse sand as abrasive and air as the carrying medium. The grit or short blasting is carried out by throwing the metallic particles by centrifugal force from a rapidly rotating wheel.

(*iii*) *Other surface cleaning methods* :
The following methods aid in surface cleaning :
— Wire brushing ;
— Buffing ;
— Pickling ;
— Various polishing procedures.

3. Trimming. Trimming involves the removal of fins, gate and riser pads, chaplets, wires and other similar *unwanted* appendages to the casting which are *not* a part of its final dimensions.

It involves the following *procedures* :

(*i*) **Chipping.** It is used to remove pins, gates and riser pads, wires etc. It may be carried out by hammer and chisel or by pneumatic chipping hammers.

(*ii*) **Grinding.** It is employed to remove excess metal and is carried out through portable grinders, stand grinders and swing-frame grinders.

4. Finishing. It is the later stage of cleaning. In certain cases cleaning is complete after trimming operations, but others may require additional surface finishing *e.g.*, machining, polishing, buffing etc.

9.1.14. Inspection of Castings

In order to determine the presence of any defects (not readily visible) it becomes necessary to inspect the casting. Following methods are employed to inspect the casting.

1. Destructive inspection method. In this type of inspection the casting sample is destroyed inspection. This method is used to test mechanical properties *e.g.*, tensile strength, hardness etc. These tests are performed on the test bars or pieces cut from the casting sample.

2. Non-destructive inspection method. Following are the various methods of non-destructive inspection :

(*i*) **Visual inspection.** The main aim of this type of inspection is to ensure that the outward appearance of the casting looks good. Through this inspection the defects like cracks, tears, run outs, swells etc. may be detected.

(*ii*) **Dimensional inspection.** The dimensional inspection may be carried out by surface plates, height and depth gauges, swap and plug gauges etc. Through this inspection it can be ascertained whether certain details are within tolerances or not.

(*iii*) **Pressure testing.** It is employed to locate leaks in a casting or to check the overall strength of a casting in resistance to bursting under hydraulic pressure. It is carried out on tubes and pipes.

(*iv*) **Radiographic inspection.** This type of inspection is employed to inspect *internal defects of a casting*, by the use of X-ray or gamma ray technique.

(*v*) **Magnetic particle inspection.** This inspection method is employed on magnetic ferrous castings for detecting invisible surface or slightly subsurface defects.

MANUFACTURING PROCESSES

(vi) **Fluorescent penetrant.**

- This type of inspection is employed to find minute pores and cracks on the surface of castings that may be missed even under magnification.
- In this method a fluorescent penetrating oil mixed with whiting powder is applied to the casting surface by dipping, spraying or brushing. The cracks or other defects become visible after the surface has been wiped dry (the oil creeping out of cracks).

9.1.15. Exercise. *To prepare casting of a pipe bend from a pattern provided.*

Solution.

Material and equipment required :

(i) Moulding box
(ii) Core box
(iii) Pattern
(iv) Well prepared moulding sand
(v) Core sand
(vi) Melting furnace
(vii) Cast metal (*e.g.,* cast iron or aluminium)
(viii) Ladle
(ix) Crucible handling tong, corrier
(x) Moulding tools
(xi) Fettling tools (*e.g.,* hammer, chisel, hacksaw, wire brush, file, grinder, etc.)

Procedure of preparing the casting :

The following steps (I to III) may be followed to prepare to casting of a pipe bend : Refer to Fig. 9.41.

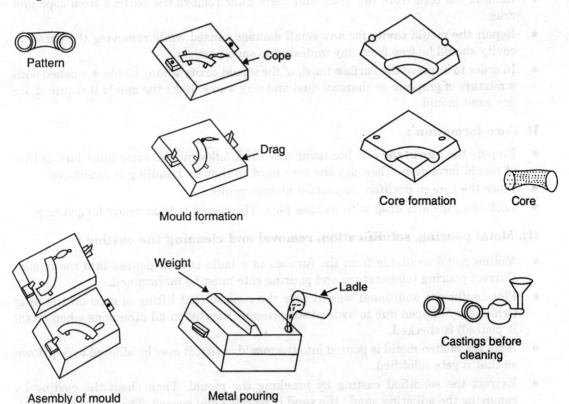

Fig. 9.41. Procedural steps for casting of a pipe bend.

I. Mould formation :

- Depending upon the size of the pattern select a suitable moulding box such that adequate space around the pattern for ramming the sand is available.
- Place the drag part of the moulding box upside down on the moulding bench and then place lower part of the pattern in the centre of drag on the moulding bench. Fill the drag with well prepared green sand and ram it properly. Using a strike-off bar, remove the excess sand so that now sand is in level with the edges of the drag. Sprinkle small amount of parting sand over the tap surface to avoid sticking. Now turn the drag upside down with lower half of the pattern in it.
- Place the cope over the drag in its proper position in alignment with locking pins. Then assemble top part of the pattern in position.
- Sprinkle parting sand over the surface of the drag and the pattern.
- Place the runner and riser in position and fill the cope with green sand and ram it properly. Cut off the excess sand to bring it in level with the edges of the cope.
- Remove the runner and riser and form the pour basin. Sprinkle parting sand on the top surface.
- Using a venting wire perform the venting operation. It is done to allow exit of gases and steam generated during pouring.
- Remove the cope from the drag, and there after remove the pattern from cope and drag.
- Repair the mould cavity for any small damage caused while removing the pattern ; cavity should be free from any undesirable sand particle.
- In order to improve the surface finish of the mould cavity it may be black washed with a mixture of graphite or charcoal dust and clay water. Bake the mould if required, for dry sand mould.

II. Core formation :

- Prepare the core in the core box using core sand, following the same procedure as that of mould formation. Then dry the core hard so that its handling is facilitated.
- Place the core in position, supported by core prints.
- Lock the cope and drag with locking pins. The mould is thus ready for pouring.

III. Metal pouring, solidification, removal and cleaning the casting :

- Molten metal available from the furnace in a ladle is then poured into the mould. Currect pouring temperature and pouring rate must be maintained.
- Place sufficient additional weight over the cope so that lifting of cope off the drag (which may happen due to hydrostatic pressure exerted in all directions when metal is poured) is checked.
- After the molten metal is poured into the mould cavity, it may be allowed to cool down so that it gets solidified.
- Extract the solidified casting by breaking the mould. Then clean the casting by removing the adhering sand ; the sand is recycled and reused. The unwanted projections fins and nails etc. may also be removed.

9.2. FORGING

9.2.1. Introduction

- **Forging** is *the process by which heated metal is shaped by the application of sudden blows or steady pressure and characteristics of plasticity of material is made use of.* This is one of the oldest manufacturing process.
- The working of small objects heated in an apex furnace, operated by manual labour, is known as **"Smithing"** whereas, the large objects heated in close furnace and worked by hammers or forging machines are called **"Forgings"**.
- In any machine building industry *e.g.,* manufacture of tractors, automobiles, agricultural machinery etc. forging processes are extremely important. Under forging operation, the *machine parts become stronger as the quality of steel improves and forged parts are especially made to take up heavy duty load without much wear and tear.* Moreover the material wastage in the form of chipping is minimum under forging operation as compared to machining work.
- The 'forgings' may **finished forgings** or **"Blanks"**. In *finished forging no machining is required* whereas **"Blanks"** *need machining operations for finishing the product.*

Kneading action. *It is the action under which the forged metal becomes tough and strong because of the directional qualities of the fibrous grain flow of hot metal as it is shaped between the dies.*

By repeated blows, the hot metal is made to fill the die cavities. In this process of kneading, the grain structure obtained is in the form of unbroken lines and is the exact repetition of the part contour, thereby imparting strength and toughness to the forged parts.

9.2.2. Advantages and Disadvantages of Forging

Following are the advantages and disadvantages of forging :

Advantages :

1. The forged machine part can withstand heavy duty load conditions (which may not be possible under machining or casting processes).
2. It is a time saving operation.
3. The metal parts can be easily shaped in required form without damaging the structure.
4. Defects such as porosity get eliminated.
5. Ductility and resistance to impact loading is increased.
6. The structure of the metal and internal strength are improved.
7. Under forging operation 50 to 60 per cent of the material, wasted in the form of chipping in machining operation, is saved.
8. The grain structure continuity is maintained in the direction of the shape.

Disadvantages :

1. Initial set up cost is quite high.
2. It is difficult to maintain close tolerances.
3. Poor surface finish.
4. Some metals get cracked under the impact of hammer when in hot state.
5. The metal gets cracked or distorted if worked at temperature below a specified limit, and gets burnt if it is heated above the required range.

9.2.3. Applications of Forging

- The forging processes are of significant importance in the *machine building industry*.
- When the machine parts are manufactured by adopting forging processes, *considerable time and material are saved*. Forging *improves the quality of steel. The forging product becomes tough and strong and are capable of taking up heavy and impact loading.*
- The use of forging is particularly widespread in the manufacture of :

    (i) Tractor parts
    (ii) Automobile parts
    (iii) Agriculture machinery
    (iv) Ship building
    (v) Locomotive building
    (vi) Cutting tools
    (vii) Cams and camshafts
    (viii) Connecting rods
    (ix) Axles
    (x) Levers
    (xi) Helical and laminated springs
    (xii) Arms and weapons etc.

9.2.4. Classification of Forging

Forging can be classified in two ways :
1. Hand forging
2. Machine forging.

9.2.5. Hand Forging

- Hand forging or blacksmithing is *employed for small quantity production and for special work*.

    Generally speaking, the *accuracy obtained is less than that of drop forging*.

- In hand forging the metal is heated in a **Smith's forge or hearth** (Fig. 9.42). It consists of a *hearth* for holding the fuel, a *cast iron tuyere* for supplying air blast to the fire, a *centrifugal blower* driven by a power preferably electric motor, to produce the blast, a *chimney* to carry the smoke and poisonous gases to air, a *water tank* behind the hearth to water cool the tuyere, a *coal bunker* to stock coal or coke, a *water trough* in front for quenching cutting tools and an *air valve* to control the blast.

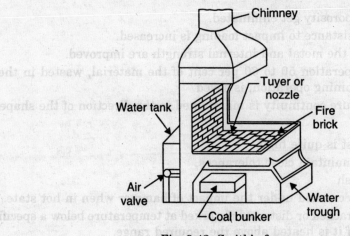

Fig. 9.42. Smith's forge.

MANUFACTURING PROCESSES

In operation, the work is paced in the fire pot and heated to the proper temperature for forging.

9.2.5.1. Smithy tools

Fig. 9.43 shows the various tools used in smithy. The list of important smithy tools and their uses are given below :

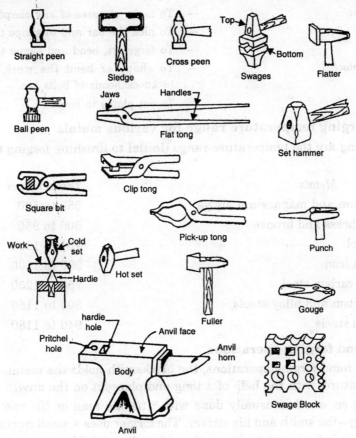

Fig. 9.43. Tools used in smithy.

Tools used in smithy

Name of tool	Use
1. *Sledge hammers, straight, flat and cross peen*	— To forge big jobs (heavy work).
2. *Smith's ball peen hammer*	— To forge light and medium work.
3. *Tongs, flat or square bit pick up tong.*	— To hold the hot work.
4. *Chisel long cold set*	— To cut cold metal.
5. *Hot set*	— To cut hot metal.
6. *Fullers, top and bottom*	— To shape inside curves. To form corrugations for elongating metal.

7. *Swages, top and bottom*	—	To shape convex surfaces and to give finish to round, square, hexagonal or octagonal shaped sections.
8. *Flatter or flattener*	—	To give smooth finish to flat surfaces.
9. *Set hammer*	—	To form square shoulders and to clean the rounding in corners.
10. *Punches*	—	To make recesses of any shape in hot metal.
11. *Hardie*	—	To nick the bar and to shape the cold work.
12. *Anvil*	—	To forge art, bend and shape the work.
13. *Swage block*	—	To shape or bend the work to any form and to knock heads of bolts etc.
14. *Gouge*	—	To cut plates to curves.

9.2.5.2. Forging temperature range for various metals

The following are the temperature-range (initial to finishing forging temperatures) for various metals :

Metals	*Temperature °C*
1. Aluminium and magnesium alloys	350 to 500
2. Copper, brass and bronze	600 to 950
3. Mild steel	750 to 1300
4. Wrought iron	900 to 1300
5. Medium carbon steel	750 to 1250
6. High carbon and alloy steels	800 to 1150
7. Stainless steels	940 to 1180

9.2.5.3. Hand forging operations

- During hand forging operations, the blacksmith holds the metal, heated to forging temperature, with the help of a tong and places it on the anvil.
- Forging on *anvil* is usually done with : (*i*) one man or (*ii*) two men-two handed working—the smith and his striker. The former uses a small hammer, the latter the sledge. To indicate, where he requires his mate to strike a blow the smith lightly taps the work with the small hammer ; the striker's job is to hit the spot with the sledge. If working three handed the same procedure is followed, a ligh tap from the smith preceding a heavy one from the striker. To indicate when to finish, the smith taps the anvil with his hammer.

The principal hand forging operations carried out in a smithy shop are :

1. Upsetting
2. Drawing down and fullering
3. Cutting out
4. Bending
5. Piercing, punching and drifting
6. Twisting
7. Welding.

MANUFACTURING PROCESSES

1. Upsetting. Refer to Fig. 9.44.

- *It is the process of increasing cross-sectional dimensions when forging.* The process implies that cross-section is increased and the length decreases.
- It may be done in a number of ways, each varying according to the details of the article required and the equipment in the shop. The simplest is to place the heated article on the anvil and hammer directly on the upper end. This increases the cross-section and reduces the length of the metal being worked.

2. Drawing down and fullering

Drawing down. Refer to Fig. 9.45.

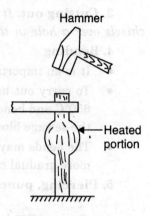

Fig. 9.44. Upsetting.

- It is the process of *increasing the length of a bar at the expense of its cross-sectional area.*
- In this operation the work-piece is heated upto the plastic stage and placed on the anvil and the metal is beaten down at the edges with a hand hammer or a sledge hammer. The work is turned frequently and struck an approximately equal number of times on each side, to prevent the metal spreading out side ways as the length increases and thickness decreases.
- **Setting down** is *a process of local thinning down effected by the set hammer.* The work is usually fullered at the place where the setting down commences.

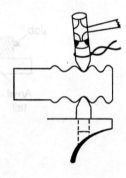

Fig. 9.45. Drawing down.

Fullering. Refer to Fig. 9.46.

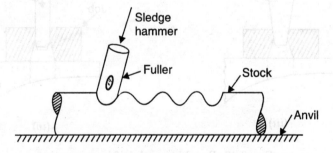

Fig. 9.46. Fullering operation.

- *This type of forging operation is carried out to increase the width of the work-piece rather than its length*
- The operation is carried out by different type of fullering tools held in position with a tang and struck with a sledge hammer.
  - *Fullers* are employed for drawing down metals and also for finishing grooves and concave surfaces. *Flatters* are then used for levelling and finishing flat surfaces.
  - *Swages* are also used for drawing down or changing the sectional shape of round stock.

3. Cutting out. *It is the process of cutting large holes of various shapes by using a hot chisels over a hole in the sewage block.*

4. Bending
- It is an important operation in forging and is very frequently used.
- To carry out bending operation the bar stock is heated to a dark-red heat *i.e.,* at 650°C and bending operation is carried out over the anvil edge or on the beak or on the sewage block.
- The bends may be either sharp cornered angle bends or they may be composed of a more gradual curve.

5. Piercing, punching and drifting

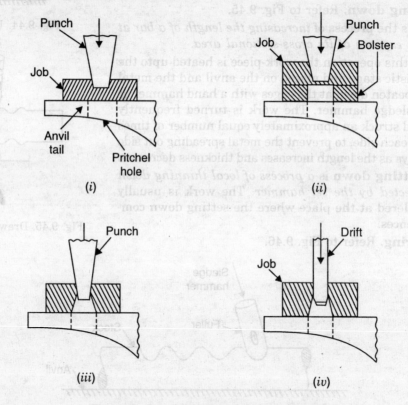

Fig. 9.47. Punching and drifting operation.

Piercing
- It is *process of making holes in 'thin sheets'*.
- The workpiece is heated to temperature of 900°C to 1000°C and placed on the anvil in such a way that the portion where the hole is to be made, coincides with the hardie-hole. The punch is then placed over the spot to be pierced and struck with a sledge hammer. When the punch passes through the metal, it forces the piece removed out through the hardie hole.

Punching

- In this operation the holes are punched in thicker workpiece.
- The job is first heated to nearly white heat and then placed flat on the anvil face. The punch is then forced into upto about half its thickness. The job is then turned upside down and placed over a tool *bolster*. The punch is again forced into the job and made to pass through by hammering.

Drifting. Punching, without using a die, (a *die* is used to produce the hole in a single operation) is usually followed by *drifting*. In this, a tool, known as a *drift* is made to pass through the punched hole to produce a finished hole of the required size.

Fig. 9.47 shows the punching and drifting operations

6. Twisting. The flats are sometimes required to be twisted at a point for connecting two pieces placed at different planes. For this purpose the workpiece is heated to cherry red temperature at the point where the twist is required. Holding one of end of the flat, from near the twist, in a suitable vice, the *other end is twisted to a required angle with the help of a suitable tong*.

7. Welding

- The welding operation carried out in a smithy shop is called *forge welding*.
- The steel when heated to a white heat (*i.e.*, at 1300°C) becomes highly plastic or pasty ; if two pieces in this condition are placed together and hammered, they will adhere and form a single piece of steel. This process is called *forge welding*. It may be noted that if the temperature is low, it will not cause the weld to take place. On the other hand, if the temperature is too high it will ruin the metal by burning it.

While welding the following points must be taken care of :

(*i*) The contact surface must be perfectly clean.

(*ii*) The point should be free from iron oxide. The presence of iron oxide results in defective welding.

(*iii*) In order to remove air, scale and drift, the surface to be welded should be made slightly convex. Under such conditions the contact is first made at a point and the impurities are squeezed out as the surfaces are joined under pressure of the hammer.

(*iv*) The fuel used for heating the steel should be free from sulphur (since sulphur is detrimental to good weld).

- In order to carry out welding operation smoothly, *flux is used to dissolve the oxide at a lower temperature*. The flux slag formed is easily expelled under pressure of hammer blows. '**sand**' is the suitable flux for wrought iron while '*calcined borax*' is useful for mild steel.

Types of welded joints

The following three types of welded joints are in common use :
1. Butt weld
2. Scarf weld
3. 'V' weld.

Fig. 9.48 shows the welded joints and end preparations.

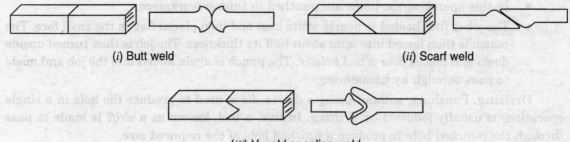

Fig. 9.48. Types of welded joints.

1. Butt weld. Refer to Fig. 9.48 (*i*).

This type of joint is usually made by a welding machine (suice this joint is difficult to be made by hammering because in this case faces to be joined and the hammering is always done perpendicular to the joint line) in which parts are gripped and their ends are forced together by hand or power operation. When they are so held, the sides of the weld may be smoothened by hammering.

2. Scarf weld. Refer to Fig. 9.48 (*ii*).
- It is also known as *lap weld*.
- It is the most reliable and straight forward weld.
- Before welding, the ends of the workpiece are rounded as shown in the diagram.

3. V-weld. Refer to Fig. 9.48 (*iii*).
- It is also known as *split, splice* or *fork weld* etc.
- It is employed where a highly strong welded joint is needed, particularly in heavy work where the greater thickness of the job enables the formation of 'V' easily.
- To ensure perfect joining of metals the scarf of one piece should be made rough by providing steps on it, as shown in the diagram.

9.2.6. Machine Forging

A forging machine is one which is designed to shape a metal article while the material is in hot plastic state.

The term forging machine in its widest sense includes :
1. Drop stamp (whether of rope, belt or board type).
2. Steam hammer.
3. Pneumatic hammer.
4. Hydraulic hammer.

1. Drop stamp
- The *drop stamp of board type* (Fig. 9.49) is a widely employed when shaping hot bars, and finally to bring the work to size and shape between a set of drop stamping dies.
- In board hammer the tup is attached to a board which passes between two rollers. The latter run in an overhead attachment, are belt-driven and run in opposite directions. The tup is lifted by means of eccentric (foot, or hand operated or self acting) and they

MANUFACTURING PROCESSES

(eccentrics) cause the rollers to grip or release the board, when the board is gripped by the rollers their direction of rotation is such as to lift it (board) and the attached tup, when the board is released the tup falls with it. The height of lift depends upon the timing of release, which is instantaneous.

- When producing small drop forgings or hot pressings the drop stamps in its various forms is a very effective method of obtaining the desired results. For shallow sheet metal work drop stamp is first class production machine as it permits a solid blow to be struck without any fear of bending a crank or breaking a press frame.

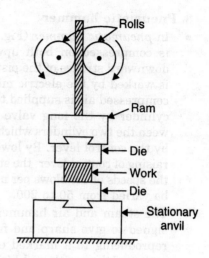

Fig. 9.49. Drop stamp of board type.

Advantages of drop forging :
1. Little wastage of material.
2. Enhancement in the strength of material.
3. Operations relatively faster.
4. No sand-casting mould are required.

Disadvantages :
1. Dies are costly.
2. The maintenance of the dies is expensive.
3. The removal of flash entails extra cost.

2. Steam hammer

- A *steam hammer* (Fig. 9.50) operates on the principle of the steam engine.
- The main parts are *frame, a steam chest or cylinder, piston, piston rod and the anvil*. The hammer head is attached to the piston rod and is raised by admitting steam in the cylinder through the valve beneath the piston. The downward stroke of the hammer is obtained by exhausting the steam from beneath the piston and admitting from above the piston. The hammer descends by gravity and steam pressure is 5.5 to 8.5 bar. For varying the intensity of the hammer blows, light to heavy, steam is admitted below the piston while the hammer is descending to create cushioning to the falling hammer. The steam inlet and outlet are controlled by a special slide valve. For generating steam a boiler is required. A wide range of work is done on this class of forging machine.

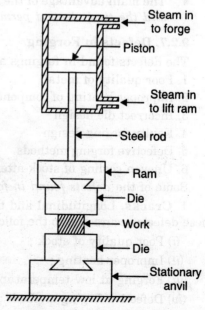

Fig. 9.50. Steam hammer.

3. Pneumatic hammer

- In *pneumatic hammer* (Fig. 9.51) air is compressed on both upward and downward strokes of the piston which is worked by the electric motor. This compressed air is supplied to the ram cylinder by the long valve kept between the two cylinders which is moved by the control lever. By lowering and raising of control lever, the strokes and the speeds of the blows per minute can be varied from 50 to 200.
- The steam and air hammers are designed to give sharp and fast blows, reproducing to a marked extent the action of the smith and his hammer.

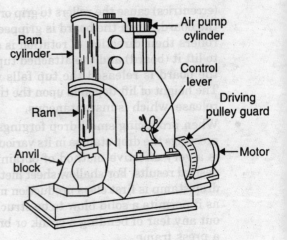

Fig. 9.51. Pneumatic hammer.

They may be used with a standard pair of anvils or with a set of dies, the latter often being so designed that the metal can be drawn out to the approximate length and width, and then placed in the dies for the final shaping stage. The flash which is formed is clipped off as the last operation.

4. Hydraulic hammer

- For the large castings and in cases where a heavy pressure is required the use of hydraulic hammer is restored to.
- The hydraulic forging machine being sluggish in action cannot usually compare with the steam or air hammer which operate more quickly, for small and medium sized forgings.
- The main advantage of the hydraulic forging press is that *it gives a definite squeeze and the time element permits the material to flow*.

9.2.7. Defects in Forging

The defects found in forgings are due to one or more of the following reasons :

1. Poor quality of metal
2. Improper heating of component for forging
3. Incorrect die design
4. Faulty forging design
5. Defective forging methods
6. Uneven cooling of stock after forging.

Some of the *defects found in forging* are as follows :

1. **Cracks.** Longitudinal and transverse cracks are the most common forging defects. These defects occurs due to the following *reasons* :

   (*i*) Poor quality of stock ;
   (*ii*) Improper heating ;
   (*iii*) Forging at low temperature ;
   (*iv*) Defective forging methods ;
   (*v*) Incorrect cooling of forge.

2. **Hair cracks.** These may occur due *to rapid cooling and defective materials.*
3. **Incomplete component.** It may be due to :
   (*i*) Less metal used ;
   (*ii*) Inadequate heating of metal ;
   (*iii*) Improper forging design ;
   (*iv*) Faulty die design ;
   (*v*) Metal not placed properly in the die ;
   (*vi*) Inadequate metal flow.
4. **Mismatched forgings.** This defect results when the upper and lower die impressions are *out of alignment* as the final blow is struck.
5. **Scale pits.** These are formed by *squeezing of scale into the metal surface* during forging.
6. **Fibre flow lines discontinued.** This defect is caused due to *very rapid flow of metal.*
7. **Oversize components.** This type of forging defects is caused due to following reasons :
   (*i*) Incorrect dies ;
   (*ii*) Worn out dies ;
   (*iii*) Misalignment of die halves.
8. **Burnt and overheated metal.** This is caused by heating the metal at high temperature or for a too long time.

9.2.8. Heat Treatment of Forgings

The forged parts are generally heat treated :
1. To relieve internal stresses set up during forging and cooling.
2. To normalise the internal structure of the forged metal.
3. To improve the machinability.
4. To improve hardness, strength and other mechanical properties.

The forged components are normally given the following *heat treatments* :
1. Annealing ;
2. Normalising ;
3. Tempering.

9.2.9. Exercise : *To make a hexagonal headed M.S. (Mild steel) bolt as per drawing given in Fig. 9.52 out of M.S. stock 25 mm in diameter and 270 mm long.*

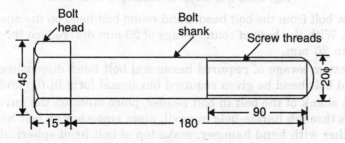

All dimensions in mm

Fig. 9.52. Hexagonal headed M.S. bolt.

Solution. *Procedure of making a bolt* : Refer to Fig. 9.53.
The following steps (I to VI) may be followed to make the hexagonal headed bolt :

I. Take a bar of 25 mm diameter and 270 mm long. In case the length is more than required, it may be cut-off using a hot set.

II. Heat one end of the bar to a length of about 80 mm in a forging furnace. After it attains forging temperatures of 1260°C grip the heated bar at one end with tongs, place it vertically on anvil and *upset* heated part with a *sledge hammer*.

III. Place cylindrical part of the upset bar in a bolt header and place the bar on the anvil in such a way that the shank of the bolt passes through the hardie-hole in the anvil and further upset head of bolt with *hand hammer*.

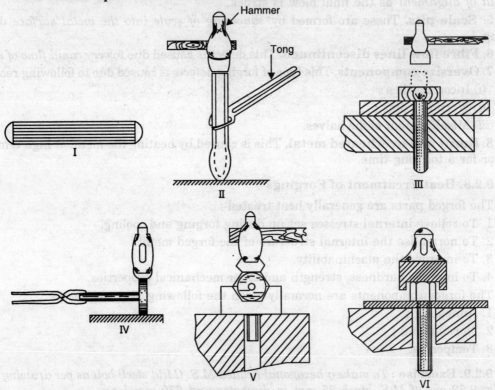

Fig. 9.53. Six steps in making a M.S. bolt.

IV. Withdraw bolt from the bolt header and round bolt head on the anvil face with a *hand hammer*. With the help of round swage of 20 mm dia. reduce the dia. of shank from 25 mm to 20 mm.

V. Insert bottom swage of required hexagonal bolt head dimensions in hardie hole of anvil and bolt head be given required hexagonal form in top and bottom swage.

VI. • Insert shank of the bolt in bolt header, place stock on the anvil so that its shank passes through hardie-hole in anvil, place smoother on bolt head and by striking smoother with hand hammer, make top of bolt head spherical in shape.

• Straighten shank of bolt on face of anvil. Check its length. If it is more than 180 mm long cut off excess length with the help of a hot set.

- Using a suitable die cut screw threads.

The bold is now ready as per drawing given.

9.3. ROLLING

Rolling *is a forming operation on cylindrical rolls where in cross-sectional area of a bar or plate is reduced with a corresponding increase in length.* The metal is thinned and elongated by compression and shear forces but increased in width only slightly. Because of the high surface finish maintained on the rolls, the surface of the stock is burnished by the rolling action and attains a smooth bright finish.

A rolling may be two high, three high, four high, or six high depending upon the number of rolls stacked above each other as illustrated in Figs. 9.54 to 9.57. The two high rolls being least expensive are most common for both hot and cold rolling.

Hot rolling consists of taking the hot ingot from a soaking pit, where it has been kept at an elevated temperature, and rolling it first into blooms (large oblong squares) and then through a series of other rollers into structural shapes, pipe, and tubing, steel is nearly always rolled hot except for finishing passes on sheet. Copper is also rolled to rod, for mking wire. Brass and nickel silver are usually cold rolled with many intermediate annealings. Fig. 9.58 illustrates grain structure during hot rolling. It is desirable that in all hot working the metal be heated throughout to the proper temperature before processing. If the temperature is non-uniform work hardening and crack may result.

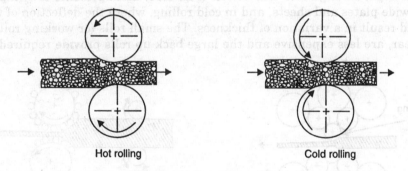

Fig. 9.54. Two high rolls.

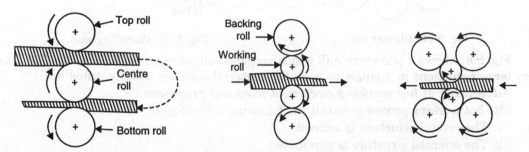

Fig. 9.55. Three high rolls.   Fig. 9.56. Four high rolls.   Fig. 9.57. Six high rolls.

Cold rolling is widely employed to produce a finish for hot-rolled metals. Sheets, strip, steel bar stocks such as shafting, flat wire etc. are produced by cold rolling. Most hot rolled material is cleaned (by acid cleaning solution) before it enters the cold rolling mill. Cold rolling is continued until the rolled section becomes too hard to continue the process, or until it reaches its final size.

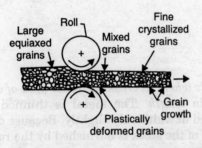
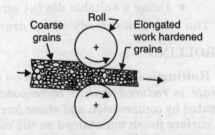

Fig. 9.58. Grain structure during hot rolling.　　Fig. 9.59. Grain structure during cold rolling.

Fig. 9.59 shows the grain structure during cold working.

The cold rolling claims the following *advantages over hot rolling* :
  (i) It gives improved physical properties by combining the cold work rolling with subsequent heat treatment.
  (ii) It gives improved surface finish.
  (iii) It produces thickness dimensions.

The two high reversing mill and three high mill are used to roll bars and plates that are up to 12 m length. For rolling strip, coils or sheets that may be thousands of metre in length, continuous mills are used.

The four high, six high and cluster roll (Fig. 9.60) arrangements are employed in hot rolling very wide plates and sheets, and in cold rolling, where the deflection of the centre of the roll would result in a variation of thickness. The small rolls (or working rolls) which are subject to wear, are less expensive and the large back up rolls provide required rigidity.

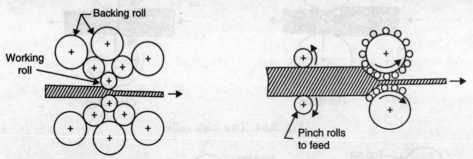

Fig. 9.60. Cluster roll.　　Fig. 9.61. Planetary mill.

Fig. 9.61 shows a planetary mill which employs rolls of very small diameter and effects very large reductions in a single pass on the material which is usually rolled hot.

Advantage of hot working over cold working processes :

The hot working processes entail the following *advantages* :
  1. The crystal structure is refined.
  2. The oriented structure is eliminated.
  3. Porosity in the metal is largely eliminated.
  4. The impurities in the form of inclusions are broken up and distributed throughout the metal.
  5. Mechanical properties, especially elongation, reduction of area and Izod values are improved.
  6. Greater homogeneity is developed in the metal.

9.4. WELDING

- *Welding is the method of joining metals by application of heat, without the use of solder or any other metal or alloy having a lower melting point than the metals being joined.*
- **Welding Processes.** These may be divided into two main groups :
1. Pressure welding
2. Fusion welding.

9.4.1. Pressure Welding

The characteristics of a pressure weld is that the metal joined is *never brought to a molten stage,* it is heated to a welding temperature and the actual union is brought about *by application of pressure.*

(I) Forge welding.
(II) Resistance electric welding.

(I) Forge Welding

- In this method of welding the surfaces to be joined are heated in an open hearth until they reach the welding temperature of metal, *which is below its melting point.* The blacksmith will judge this temperature by the colour of the metal, which may be between red-hot and white-hot. The parts are then placed on an anvil and hammered together.
- In this welding process there is a risk of oxide and other inclusion when the metal is heated in an open fire and accurate judgement of the temperature is called for if the structure of metal is not to be changed. Modern alloy steels can be ruined by injudicious heating. When the wide range of light alloys is considered, it becomes imperative to use a more scientific method. However this process is still widely used for heavier classes of work, such as manufacture of anchor chains, while controlled heating furnaces and automatic forging machines have been designed to replace the open-forge fire and the blacksmith's anvil.

(II) Resistance Electric Welding

It is the method of uniting two pieces of metal by the passage of a heavy electric current while the surfaces are pressed together. The fusing temperature is obtained by placing the surfaces to be joined in contact with one another, and passing a current of two to eight volts, at a high amperage through them. The *heat is developed around the point to which they touch, forcing them together* (by pressure mechanically applied), and at the same time *switching off the current, completes the weld.*

The important resistance welding processes are discussed below :

(*a*) **Butt welding.** Refer Fig. 9.62. In this type of welding which is employed to join bars and plates together end to end, one bar is held in a fixed clamp in the butt welding machine and the other bar in a movable clamp, the clamp being electrically insulated, the one from the other, and being connected to a source of

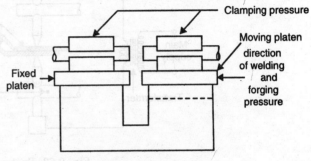

Fig. 9.62. Butt welding.

current. When the two ends to be joined are brought into contact and current is switched on, the resistance at the joint causes the ends to heat up to welding temperature. Current is then switched off and the movable clamp forced up, so that a weld is made. The voltage applied across the clamps is a low one, from 2 to 6 volts, and the current is usually alternating. If the bars being joined are different in cross-section the amounts they project from their clamps may have to be adjusted so as to modify the heat losses and ensure both bars being brought to the welding temperature simultaneously.

This process is being used for welding such things as steel rails whose cross-sectional area is as much as 6.25 cm^2.

(b) Flash welding. In this process, the parts to be welded are clamped to the electrode fixtures, as in butt welding *but the voltage is applied before the parts are butted together.* As the *parts touch each other, an arc is established which continues as long as the parts advance at the correct speed.* This arc bursts away a portion of the material from each piece. When the welding temperature is reached, the speed of travel is increased, the power switched off and weld is upset.

Flash butt welding claims the following *advantages* over upset method of welding, (*i*) Power consumed is less once the arc creates more heat with a given current. (*ii*) The weld is made in clean virgin metal as the surfaces are burned away. (*iii*) More quicker.

It is widely used in automobile construction on the body, axles, wheels, frames and other parts. It is also employed in welding motor frames, transformer tanks and many types of sheet steel containers such as at barrels and floats.

(c) Spot welding. Steel, brass, copper and light alloys can be joined by this method, which forms a cheap and satisfactory substitute for riveting. The area of fusion at each spot weld, in fact, is approximately equal to the cross-sectional area of the rivet which would be employed for a similar gauge of material.

Refer to Fig. 9.63. Spot welding, as the name implies, is carried out by overlapping the edges of two sheets of metal and fusing them together between copper electrode tips at suitably spaced intervals by means of a heavy electrical current. The resistance offered to the current as it passes through the metal raises the temperature of the metal between the electrodes to welding heat. The current is cut off and mechanical pressure is then applied by the electrodes to forge the weld. Finally the electrodes open.

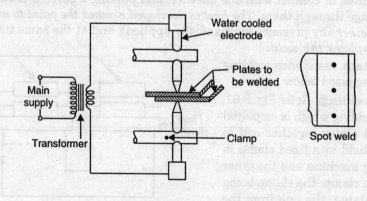

Fig. 9.63. Spot welding.

MANUFACTURING PROCESSES

When sheets of unequal thickness are joined, the current and pressure setting for the thinner sheets are used. Similarly four thickness may be welded, using the same settings as for two thickness.

(*d*) **Seam welding.** Refer to Fig. 9.64. Seam welding is analogous to spot welding with the difference the electrodes are in the form of rollers; and the work moves in direction perpendicular to roller axis. The current is interrupted 300 to 1500 times a minute to give a series of overlapping spot welds. The welding is usually done under water to keep the heating of the welding rollers and the work to a minimum, and thus to give lower roller maintenance and less distortion of the work.

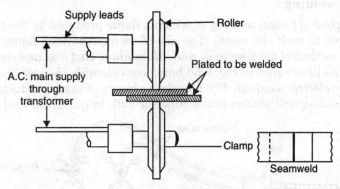

Fig. 9.64. Seam welding.

It is employed on many type of pressure (light or leak proof) tanks, for oil switches, transformers, refrigerators, evaporators and condenser, aircraft tanks, paint and varnish containers etc.

(*e*) **Projection welding.** Refer to Fig. 9.65. It is in effect, a form of multi-spot welding in which a number of welds are made simultaneously. The pieces to be welded are arranged between two flat electrodes which exert pressure as the current flows. The projections, and the areas with which they make contact, are raised to welding heat and are joined by the pressure exerted by the electrodes. The projections are flattened during the welding.

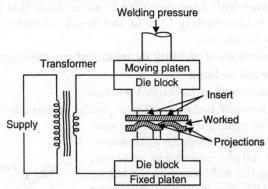

Fig. 9.65. Projection welding.

The process is used chiefly to join pressings together since it is relatively simple to make the press-tools so that the projections are produced during the main forming operation in the press.

The same principle is used in the cross welding of a number of wires or rods to make a mesh.

The materials like brass and aluminium cannot be projection welded satisfactorily.

9.4.2. Fusion Welding

The characteristic of a fusion weld is that the metal being joined *is actually melted* and the union is produced on *subsequent solidification*. The group includes :

1. Gas welding   2. Electric arc welding   3. Thermit welding.

9.4.2.1. Gas welding

- It is a method of fusion welding in which a flame produced by the combustion of gases is employed to melt the metal. The use of an oxyacetylene flame is the most widely employed method of *welding iron, steel, aluminium, cast iron and copper,* the equipment required, as illustrated in Fig. 9.66 being *considerably cheaper and simpler than that needed for electric welding.* For a certain classes of mass production work, however, electric welding will always prove superior both in quickness and cheapness.

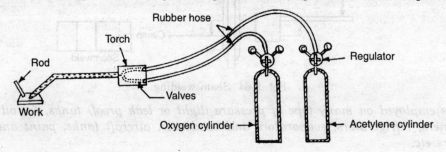

Fig. 9.66. Gas welding.

- The principle of acetylene welding is the ignition of oxygen and acetylene gases, mixed in a blow pipe fitted with a nozzle of suitable diameter ; this flame is applied to the edges of the joint and to a wire filler of the appropriate metal, which is thereby melted and run into the joint. When the acetylene is burned in an atmosphere of oxygen an intensely hot flame with a temperature of about 3300°C is produced. As the melting point of steel is approximately 1300°C, the metal is fused very rapidly at the point at which the flame is applied.

- There are two methods of welding by means of the oxyacetylene blow pipe :

  (i) Leftward or forward welding.

  (ii) Rightward or backward or backhand welding.

  — In *leftward or forward welding* after suitable preparation of the joint the weld is commenced at the right-hand side of the joint and blow pipe is given a steady forward movement, with a slight sideways motion, zigzagging along the weld towards the left as shown in Fig. 9.67. The blow pipe is kept at an angle of 60° to 70° to the surface of the work so that the flame plays ahead of it, and the filler rod held at an angle of 30° to 40°, is held just ahead of the flame and progressively

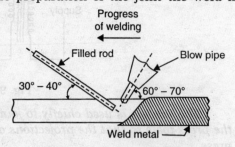

Fig. 9.67. Leftward welding.

fed into it. The leftward technique is most suitable for thin material (up to about 0.5 cm) and for all non-ferrous metals.

— In *rightward welding* the flame is directed towards the completed part of the joint and welding proceeds from left to right as shown in Fig. 9.68. The filler rod is given a circular movement as it is fed into the flame. The rightward technique is used *for thicker materials, chiefly steel. It uses less gas and filler rod and causes less distortion than leftward welding.*

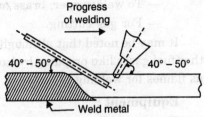

Fig. 9.68. Rightward welding.

Types of Flames :

Following are the *three* types of flames of oxygen and acetylene mixture :

1. Neutral flame
2. Carburising flame
3. Oxidising flame.

The brief description of these flames is given below :

1. Neutral flame. Refer to Fig. 9.69.

- When the ratio of oxygen and acetylene is *equal*, a neutral flame is obtained.
- This type of flame has a temperature of 3250°C, is white in colour and has a sharply defined central cone with a reddish purple envelope.
- It does not react chemically with the parent metal and protects it (the metal) from oxidation.
- The neutral flame is used to weld carbon steels, cast iron, copper, aluminium etc.

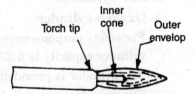

Fig. 9.69. Neutral flame.

2. Carburising flame. Refer to Fig. 9.70.

- The ratio of oxygen to acetylene is 0.9 to 1. It consists of the following three zones,
  — Luminous zone,
  — Feather or intermediate cone of white colour, and
  — Outer envelope.
- It is also called as *reducing flame* and has a temperature of 3040°C.
- *The carburising flame is used* for the following purposes :
  To join those materials which are *readily oxidised*. Thus it is used to weld aluminium since it prevents the formation of aluminium oxide at the time of welding.
  — To weld monel metal, high carbon steel and alloy steel.
  — To give a hard facing material in some cases.

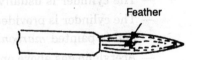

Fig. 9.70. Carbourising flame (Excess acetylene).

3. Oxidising flame. Refer to Fig. 9.71.

- The ratio of oxygen to acetylene varies from about 1.2 to 1.5.

Fig. 9.71. Oxidising flame (Excess oxygen).

- It is used in the following cases :
    - To weld copper, brass and bronze and zinc-bearing alloys.
    - For gas cutting.

It may be noted that although, in gas welding, oxygen and acetylene mixture is popular, other fuel gases like propane, hydrogen and coal gas may also be used along oxygen to produce gas flames for welding.

Equipment :

For gas welding following equipments are used :
1. Gas cylinders
2. Pressure regulators
3. Pressure gauges
4. Welding torch
5. Hoses and hose fittings
6. Safety devices etc.

The brief description of the above equipments is given below :

1. Gas cylinders

A. Oxygen cylinder

— For safety purposes oxygen cylinders are filled at a pressure 12500 to 14000 kN/m^2 and cylinder capacity is 6.23 m^3.

— The cylinder is provided with a right *hand thread valve* and is painted *black*.

— The cylinders are usually provided with fragile disc and fusible plug to relieve the cylinder of its contents if subjected to overheating or excessive pressure.

B. Acetylene cylinder :

— The cylinder is usually filled to pressure of 1600 to 2100 kN/m^2.

— The cylinder is provided with *left hand threads* for accommodating pressure regulator and is painted *maroon*.

— Acetylene gas above one atmospheric pressure is highly explosive. Hence acetylene is stored with calcium silicate saturated with acetone. Acetone can absorb 25 times its own volume of acetylene for each atmospheric pressure.

2. Pressure regulators. The cylinders are provided with pressure regulators to control the working pressure of oxygen and acetylene to the welding torch. The pressure of oxygen and acetylene depends on the thickness of the metal to be welded/cut.

3. Pressure gauges. Two pressure gauges are fitted on each pressure regulator. While one pressure gauge shows the pressure inside the cylinder, the other one shows the working pressure of the fuel gas and oxygen.

4. Welding torch/blow pipe. It is a device for mixing oxygen and acetylene in the required volume and igniting it at the mouth of its tip. Generally following two types of torches are available :

MANUFACTURING PROCESSES

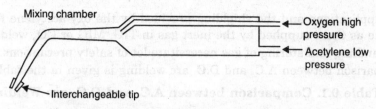

Fig. 9.72. Low pressure blow pipe.

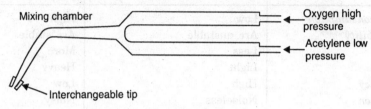

Fig. 9.73. High pressure blow pipe.

(1) Low pressure blow pipe (Injector type)
(2) High pressure blow pipe.

5. Hose and hose fittings :
- Hoses are the rubber and fabric pipes used to connect gas cylinder to blow pipe and are painted black or green for oxygen and red or maroon for acetylene. It should be strong, durable, non-porous and light.
- Special fittings are used for connecting hoses to equipment.

6. Safety devices :
- *Goggles* fitted with coloured glasses should be used to protect the eyes from harmful heat ultraviolet rays.
- *Gloves* made of leather, convas and asbestos should be worn to protect hands from any injury. Gloves should be light so that the manipulation of the torch may be done easily.

Other requirements include *spark-lighter, apron, trolley, wire brush, spindle key, spanner set, filter rods and fluxes and welding tips.*

Advantages and disadvantages of Gas Welding :

Advantages :
1. The oxy-acetylene torch is *versatile*. It can be used for brozing, bronze welding, soldering, heating, heat treatment, metal cutting, metal cleaning etc.
2. It is portable and can be moved almost everywhere for repair of fabrication work.
3. The oxy-acetylene flame is easily controlled and not as piercing as metallic arc welding, hence, extensively used for sheet metal fabrication work.

Disadvantages :
1. As compared to arc welding, it takes considerably longer time for the metal to heat up.
2. Owing to prolonged heating harmful thermal effects are aggravated which results in a larger heat affected area, increased grain growth, distortion and less of corrosion resistance.
3. Oxygen and acetylene gases are expensive.

4. Flux applications and the shielding provided by the oxy-acetylene flame are not so positive as those supplied by the inert gas in TIG, MIG or CO_2 welding.
5. The handling and storing of gas necessitate lot of safety precautions.

The comparison between A.C. and D.C. arc welding is given in the table below :

Table 9.1. Comparison between A.C. and D.C. arc Welding

S.No.	Aspect	A.C. Welding	D.C. Welding
1.	Power consumption	Low	High
2.	Arc stability	Arc unstable	Arc stable
3.	Cost	Less	More
4.	Weight	Light	Heavy
5.	Efficiency	High	Low
6.	Operation	Noiseless	Noisy
7.	Suitability	Non-ferrous metals cannot be joined	Suitable for both ferrous and non-ferrous metals
8.	Electrode used	Only coated	Bare electrodes are also used
9.	Welding of thin sections	Not preferred	Preferred
10.	Miscellaneous	Work can act as cathode while electrode acts as anode and *vice versa*	Electrode is always negative and the work is positive

Specifications :

A.C. transformer : Step down, oil cooled = 3 phase, 50 Hz ; Current range = 50 to 400 A ; Open circuit voltage = 50 to 90 V ; Energy consumption = 4 kWh per kg of metal deposit; Power factor = 0.4 ; Efficiency = 85%.

D.C. generator : Motor generator – 3 phase, 50 Hz ; Current range = 125 to 600 A ; Open circuit voltage = 30 to 80 V ; Arc voltage = 20 to 40 V ; Energy consumption = 6 to 10 kWh/kg of deposit ; Power factor = 0.4 ; Efficiency = 60%.

Electrodes :

The electrodes may be of the following two types :

1. *Consumable electrode* :
   (i) Base electrode
   (ii) Flux coated electrode.
2. *Non-consumable electrode* :

1. Consumable electrode :

(i) *Base electrode* :
- These electrodes do not prevent oxidation of the weld and hence the joint is weak. They are used for minor repairs where strength of the joint is weak.
- Employed in automatic and semi-automatic welding.

(ii) *Flux coated electrode* :
- The flux is provided to serve the following purposes :
  — To prevent oxidation of the weld bead by creating a gaseous shield around the arc.
  — To make the formation of the slag easy.
  — To facilitate the stability of the arc.

MANUFACTURING PROCESSES

2. *Non-consumable electrode* :
- These electrodes are 12 mm in diameter and 450 mm long.
- These are not consumed during the welding process.

Examples of these electrodes are : *Carbon, graphite and tungsten.*

Types of Welded Joints :

The type of joint is determined by the relative positions of the two pieces being joined. The following are the five basic types of commonly used joints :

1. Lap joint 2. Butt joint
3. Corner joint 4. Edge joint
5. T-joint.

1. *Lap Joint*. Refer to Fig. 9.74.
- The lap joint is obtained by overlapping the plates and then welding the edges of the plates.
- The lap joints may be *single traverse, double traverse and parallel* lap joints.
- These joints are employed on plates having thickness less than 3 mm.

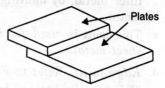

Fig. 9.74. Lap joint.

2. *Butt joint* :

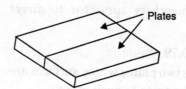

Fig. 9.75. Butt joint.

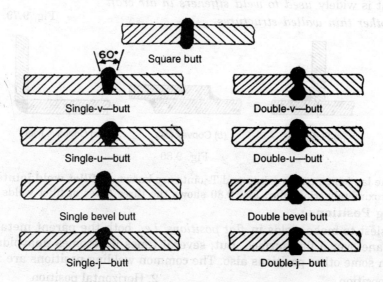

Fig. 9.76. Various types of butt joints.

- The butt joint is obtained by placing the plates edge to edge as shown in Fig. 9.75.
- In this type of joints, if the plate thickness is *less than 5 mm*, bevelling is *not* required. When the thickness of the plates ranges *between 5 mm to 12.5 mm*, the edge is required to be bevelled to V or U-groove, while the plates having thickness *above 12.5 mm* should have a V or U-groove on both sides.
- The various types of butt joints are shown in Fig. 9.76.

3. Corner joint. Refer to Fig. 9.77.

- A corner joint is obtained by joining the edges of two plates whose surfaces are at an angle of 90° to each other.
- In some cases corner joint can be welded, without any filler metal, by melting off the edges of the parent metal.
- This joint is *used for both light and heavy gauge sheet metal.*

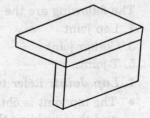

Fig. 9.77. Corner joint.

4. Edge joint. Refer to Fig. 9.78.

- This joint is obtained by joining two parallel plates.
- It is *economical for plates having thickness less than 6 mm.*
- It is *unsuitable for members subjected to direct tension or bending.*

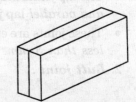

Fig. 9.78. Edge joint.

5. T-joint. Refer to Fig. 9.79.

- It is obtained by joining two plates whose surfaces are approximately at right angles to each other.
- These joints are suitable up to 3 mm thickness.
- T-joint is widely *used to weld siffeners in air craft and other thin walled structures.*

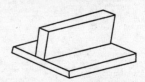

Fig. 9.79. T-joint.

(a) Flush fillet   (b) Convex fillet   (c) Concave fillet

Fig. 9.80

Note. The lap joints, corner joints and T-joints are known as **fillet weld joints.** The fillet cross-section is approximately triangular. Fig. 9.80 shows the three types of fillet welds.

Welding Positions :

It is easiest to make welds in *flat positions, i.e.,* both the parent metal pieces lying in horizontal plane over a flat surface. But, several times it becomes unavoidable to weld the workpieces in some other positions also. The common welding positions are :

1. Flat position   2. Horizontal position
3. Vertical position   4. Overhead position.

MANUFACTURING PROCESSES

1. **Flat position.** Refer to Fig. 9.81.
   - In this welding position, the welding is done from the upper side of the joint and the welding material is normally applied in the downward direction.
   - On account of the downward direction of application of welding material this position is also sometimes called as *downward position*.

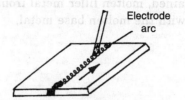

Fig. 9.81. Flat position.

2. **Horizontal position.** Refer to Fig. 9.82.

In this case, the weld is deposited upon the side of a horizontal and against a vertical surface.

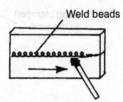

Fig. 9.82. Horizontal position.

3. **Vertical position.** Refer to Fig. 9.83.
   - In this position, the axis of the weld remains either vertical or at an inclination of less than 45° with the vertical plane.
   - The welding commences at the bottom and proceeds upwards.
   - The tip of the torch is kept pointing upwards so that the pressure of the outcoming gas mixture forces the molten metal towards the base metal and prevents it from falling down.

Fig. 9.83. Vertical position.

4. **Overhead position.** Refer to Fig. 9.84.
   - In this case, the welding is performed from the underside of the joint. The workpieces remain over the head of the welder.
   - The workpieces as well as axis of the weld all remain in approximately horizontal plane.
   - It is reverse of flat welding.

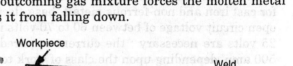

Fig. 9.84. Overhead position.

9.4.2.2. Electric arc welding

Arc welding is the system in which the metal is melted by the heat of an electric arc. It can be done with the following methods :

   (i) Metallic arc welding.
   (ii) Carbon arc welding.
   (iii) Atomic hydrogen welding.
   (iv) Shielded arc welding.

(i) Metallic arc welding. Refer to Fig. 9.85. In metallic arc welding an arc is established between work and the filler metal electrode. The intense heat of the arc forms a molten pool in the metal being welded, and at the same time melts the tip of the electrode. As the arc is maintained, molten filler metal from the electrode tip is transferred across the arc, where it fuses with the molten base metal.

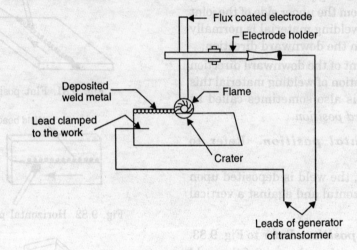

Fig. 9.85. Metallic arc welding.

Arc may be formed with direct or alternating current. Petrol or diesel driven generators are widely used for welding in open, where a normal electricity supply may not be available. D.C. may also be obtained from electricity mains through the instrumentality of a transformer and rectifier. A simple transformer is, however widely employed for A.C. arc welding. *The transformer sets are cheaper and simple having no maintenance cost as there are no moving parts.* With Arc system, the covered or coated electrodes are used, whereas with D.C. system for cast iron and non-ferrous metals, base electrodes can be used. In order to strike the arc an open circuit voltage of between 60 to 70 volts is required. For maintaining the short arc 17 to 25 volts are necessary ; the current required for welding, however, varies from 10 amp to 500 amp. depending upon the class of work to be welded.

The great *disadvantage* entailed by D.C. welding is the presence of *arc blow* (distortion of arc stream from the intended path owing to magnetic forces of a non-uniform magnetic field). With A.C. arc blow is considerably reduced and use of higher currents and large electrodes may be restored to enhance the rate of weld production.

The field of application of metallic arc welding includes mainly low carbon steel and the high-alloy austenitic stainless steel. Other steels like low and medium-alloy steels can however be welded by this system but many precautions need be taken to produce ductile joints.

(ii) Carbon arc welding. Refer to Fig. 9.86. Here the work is connected to negative and the carbon rod or electrode connected to the positive of the electric circuit. Arc is formed in the gap, filling metal is supplied by fusing a rod or wire into the arc by allowing the current to jump over it and it produces a porous and brittle weld because of inclusion of carbon particles in the molten metal. It is therefore *used for filling blow holes in the castings which are not subjected to any of the stresses.*

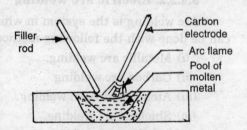

Fig. 9.86. Carbon arc welding.

The voltage required for striking an arc with carbon electrodes is about 30 volts (A.C.) and 40 volts (D.C.).

MANUFACTURING PROCESSES

A *disadvantage* of carbon arc welding is that approximately twice the current is required to raise the work to welding temperature as compared with a metal electrode, while a carbon electrode can only be used economically on D.C. supply.

(*iii*) **Atomic hydrogen welding.** Refer to Fig. 9.87. In this sytem heat is obtained from an alternating current arc drawn between *two tungsten electrodes in an atmosphere of hydrogen.* As the hydrogen gas passes through the arc, the hydrogen molecules are broken up into atoms and they recombine on contact with the cooler base metal generating intense heat sufficient to melt the surfaces to be welded, together with the filler rod, if used. The envelop of hydrogen gas also shields the molten metal from oxygen and nitrogen and thus prevents weld metal from deterioration.

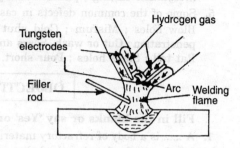

Fig. 9.87. Atomic hydrogen welding.

The welds obtained are homogeneous and smooth in appearance because the hydrogen keeps the molten pool. *Atomic hydrogen welding being expensive is used mainly for high grade work on stainless steel and most non-ferrous metals.*

(*iv*) **Shielded arc welding.** In this system molten weld metal is protected from the action of atmosphere by an envelope of chemically reducing or inert gas.

As molten steel has an affinity for oxygen and nitrogen, it will, if exposed to the atmosphere, enter into combination with these gases forming oxides and nitrides. Due to this injurious chemical combination metal becomes weak, brittle and corrosion resistance. Thus several methods of shielding have been developed. The simplest (Fig. 9.88) is the use of a flux coating on the electrode which in addition to producing a slag which floats on the top of the molten metal and protects it from atmosphere, has organic constituents which turn away and produce an envelope of inert gas around the arc and the weld. Welds made with a completely shielded arc are more superior to those deposited by an ordinary arc.

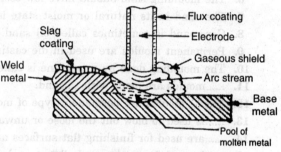

Fig. 9.88. Shielded arc welding.

1. MOULDING AND CASTING

HIGHLIGHTS

1. A *mould* may be defined as a negative print of the part to be cast and is obtained by the pattern in the moulding sand containers (boxes) into which molten metal is poured and allowed to solidify.
2. The various hand tools used in foundry are :
   Shovel ; Riddle ; Rammers ; Strike-off bar ; Ventwire ; Stick ; Lifter ; Swab ; Bellow ; Trowels ; Gate cutter ; Draw screws and rapping plate ; sprue pins ; Mallet ; Gaggers ; Clamps, Cotters and Wedges.
3. "Casting" means the pouring of molten metal into a mould, where solidification occurs.

4. The various casting processes in use are :
   Sand casting ; shell moulding ; permanent mould casting ; die casting ; centrifugal casting ; investment casting ; plaster casting ; slush casting.
5. Some of the common defects in casting are :
   Blow holes ; Misrum ; Cold shut ; Mismatch ; Drop ; Flashes or pins ; Fusion ; Metal penetration ; Cut or wash, Scars and blisters ; Hot tears, Sponginess ; Scab ; Swell ; Buckle ; Rat tail ; Slage holes ; Pour short.

OBJECTIVE TYPE QUESTIONS

Fill in the blanks or say 'Yes' or 'No' :

1. A ...... is a body of refractory material, which is set into the prepared mould before closing and pouring it, for forming the holes, recesses, projections, undercuts and internal cavities.
2. Refractoriness or thermal stability of core can be increased by giving a thin coating of graphite or similar material to the surface of the core.
3. Metal cores are usually made of aluminium.
4. Horizontal core is usually ...... in form.
5. Strickles are frequently used with core boxes to obtain shapes of large works.
6. The moulding sand should have low coefficient of expansion.
7. The sand in its natural or moist state is called ...... sand.
8. Core sand is sometimes called oil sand.
9. Permanent moulds are used in die casting.
10. The moulding done by a machine is called......moulding.
11. ......moulds are made with......sand.
12. The......moulds are permanent type of moulds.
13. ......is used to blow out the loose or unwanted sand from the surface and cavity of the mould.
14. ......are used for finishing flat surfaces and joints in a mould.
15. ......is used for cutting a shallow trough in the mould to act as a passage for the hot metal.
16. In foundry work a......is used for driving the draw strike into the pattern and then rapping it.
17. Indirect arc furnace is used for melting all types of metallic alloys.
18. Cupola furnace is most commonly used for melting and refining pig iron.
19. Investment casting is also known as lost-wax casting.
20. Blow holes appear as......in a casting.
21. ......is a shift of the individual parts of a casting with respect to each other.
22. ......are thin projections of metal not intended as a part of the casting.
23. Metal penetration defect occurs as a rough and uneven external surface on the casting.
24. A shear is a shallow blow.
25. A......is along, fairly, shallow, broad, vee depression that occurs in the surface of flat casting.

Answers

1. Core	2. Yes	3. No	4. cylindrical	5. Yes
6. Yes	7. green	8. Yes	9. Yes	10. machine
11. Loam	12. metal	13. Bellow	14. Trowels	15. Gate cutter
16. mallet	17. Yes	18. Yes	19. Yes	20. cavities
21. Mismatch	22. Fins	23. Yes	24. Yes	25. buckle.

THEORETICAL QUESTIONS

1. What are the various types of tools and equipment used in a foundry ?
2. Describe briefly the following :
   (i) Flour moulding
   (ii) Bench moulding
   (iii) Pit moulding
   (iv) Machine moulding.
3. What is a core ? What is its use ?
4. How many types of cores are there ? Explain them with the help of sketches.
5. What are core prints ?
6. What is a core box ?
7. Enumerate the characteristics of a good moulding sand.
8. Define a 'mould.'
9. Enumerate various types of moulds.
10. List the various types of hand tools commonly used in foundry.
11. Explain briefly, with neat sketches, any two of the following tools :
    (a) Riddle
    (b) Shovel
    (c) Bellow
    (d) Gate cutter.
12. What are crucible furnaces ? Where are they preferred and why ?
13. Describe with a neat sketch the construction and working of a cupola furnace.
14. Describe briefly the following melting furnaces :
    (i) Reverberating furnace.
    (ii) Open hearth furnace.
    (iii) Induction furnace.
15. Explain the defects in casting.
16. Explain briefly the following casting methods :
    (i) Shell casting
    (ii) Die casting
    (iii) Investment casting.
17. Explain briefly, with neat sketches, the following defects in casting :
    (i) Blow holes
    (ii) Mismatch
    (iii) Drop
    (iv) Sponginess.
18. Explain briefly various cleaning operations usually performed on a casting.
19. Describe briefly various methods used to inspect the castings.

2. FORGING

HIGHLIGHTS

1. *Forging* is the process by which heated metal is shaped by the application of sudden blows or steady pressure and characteristics of plasticity of material is made use of.
2. *Hand forging* is employed for small quantity production and for special work.
3. *Upsetting* is process of increasing cross-sectional dimensions when forging.
4. *Drawing down* is the process of increasing the length of a bar at the expense of its cross-sectional area.
5. *Cutting out* is the process of cutting large holes of various shapes by using a hot chisel over a hole in the sewage block.
6. *Forging machines* : Drop stamp ; steam hammer ; Pneumatic hammer ; Hydraulic hammer.

7. *Forging defects* : Cracks ; Hair cracks ; Incomplete component ; Mismatched forgings ; Scale pits ; Fibre flow lines discontinued ; Oversize components ; Burnt and overheated metal.

OBJECTIVE TYPE QUESTIONS

Fill in the blanks or say 'Yes' or 'No' :

1. ...... is the process by which heated metal is shaped by the application of sudden blows or steady pressure and characteristics of plasticity of material is made use of.
2. Hand forging is employed for large quantity production.
3. Accuracy obtained in hand forging is less than that of drop forging.
4. In hand forging a flattener is used to give smooth finish to flat surfaces.
5. ...... are used to make recesses of any shope in hot metal.
6. An hardie is used to nick the bar and to shape the cold work.
7. Hot set is used to cut hot metal.
8. A ...... is used to cut plates to curves.
9. ...... is the process of increasing cross-sectional dimensions when forging.
10. ...... is the process of increasing the length of a bar at the expense of its cross-sectional area.
11. ...... is the process of cutting large holes of various shapes by using a hot chisel over a hole in the sewage block.
12. A ...... machine is one which is used to shape a metal article while the material is in hot plastic state.
13. A steam hammer operates on the principle of steam engine.
14. The ...... hammer is used for large castings and in cases where a heavy pressure is required.
15. The main advantage of the hydraulic forging process is that it gives a definite squeeze and the time element permits the material to flow.

Answers

1. Forging	2. No	3. Yes	4. Yes	5. Punches
6. Yes	7. Yes	8. gouge	9. Upsetting	10. Drawing down
11. Cutting out	12. forging	13. Yes	14. hydraulic	15. Yes.

THEORETICAL QUESTIONS

1. Define the term 'forging'.
2. How is 'forging' classified ?
3. Draw a neat sketch of smith's forge and explain its construction and working.
4. How is 'hand forging' operation carried out ?
5. Explain briefly with heat sketches the various tools used in smithy.
6. State the uses of the following tools used in smithy :
   (i) Smith's ball peen hammer
   (ii) Chisel long cold set
   (iii) Swages, top and bottom
   (iv) Set hammer
   (v) Punches
   (vi) Gouge.
7. Explain briefly the follow smithy operations :
   (i) Upsetting
   (ii) Drawing down
   (iii) Cutting out.

MANUFACTURING PROCESSES

8. Describe briefly any two of the following forging machines :
   (i) Drop stamp
   (ii) Steam hammer
   (iii) Pneumatic hammer
   (iv) Hydraulic hammer.
9. Describe with sketches the following equipment of smithy shop :
   (i) Smith's forge ;
   (ii) Anvil ;
   (iii) Swage block.
10. What do you understand by drop forging ? Explain its process.
11. Sketch and describe steam and board drop hammers.
12. What are the common forging defects and why they occur ?
13. Why are forge parts generally heat treated ?

3. WELDING

HIGHLIGHTS

1. *Soldering* is an operation of joining two or more parts together by molten metal.
2. *Brazing* is a soldering operation using brass as the joining medium.
3. *Welding* is the method of joining metals by application of heat, without the use of solder or any other metal or alloy having a lower melting point than the metals being joined.
   Welding processes :
   1. *Pressure welding* :
   — Forge welding
   — Resistance arc welding.
   2. *Fusion welding* :
   — Gas welding
   — Electric arc welding
   — Thermit welding
      • Tungsten inert-gas (TIG) welding
      • Metal inert-gas (MIG) welding
      • Submerged arc welding
      • Electro-slag and electro-gas welding
      • Electron beam welding
      • Ultrasonic welding
      • Plasma arc welding
      • Laser beam welding.

OBJECTIVE TYPE QUESTIONS

Fill in the Blanks
1. Metals can be joined by ...., and .... processes.
2. .... is an operation of joining two or more parts together by molten metal.
3. .... is a soldering operation using brass as the joining medium.

4. When hard soldering the chief flux used is ....
5. The brazing equipment mainly comprises .... and ....
6. The operation of silver soldering is somewhat easier to carry out the ....
7. TIG is a .... electrode process.
8. Welds made by the sub-merged arc welding process have .... strength and ....
9. .... is a method of uniting iron or steel parts by surrounding the joint with steel at a sufficient high temperature to fuse the adjacent surfaces of the parts together.
10. .... is often considered the fourth state of matter.

Answers

1. Riverting, soldering and brazing   2. Soldering   3. Brazing
4. Borax   5. Blow pipe, Brazing hearth   6. Brazing
7. Non-consumable   8. High, ductility   9. Thermit welding   10. Plasma.

THEORETICAL QUESTIONS

1. Define the term *'welding'* and name the various welding techniques.
2. Explain briefly the following types of flames :
   Neutral flame, Carburising flame and Oxidising flame.
3. Name and briefly explain the various equipment used in gas welding.
4. List the advantages and disadvantages of gas welding.
5. Give the comparison between A.C. and D.C. arc welding.
6. Write short notes on any three of the following :
   (a) Gas shielded arc welding
   (b) Submerged arc welding
   (c) Thermit welding
   (d) Plasma arc welding
   (e) Laser beam welding.
7. Explain briefly any two of the following :
   (i) Metallic arc welding
   (ii) Carbon arc welding
   (iii) Atomic hydrogen welding
   (iv) Shielded arc welding.